THE PROVINCE OF ALL MANKIND

THE PROVINCE OF ALL MANKIND

HOW OUTER SPACE BECAME AMERICAN FOREIGN POLICY

STEPHEN BUONO

CORNELL UNIVERSITY PRESS
Ithaca and London

Copyright © 2025 by Stephen Buono

All rights reserved. Except for brief quotations in a review, this book, or parts thereof, must not be reproduced in any form without permission in writing from the publisher. For information, address Cornell University Press, Sage House, 512 East State Street, Ithaca, New York 14850. Visit our website at cornellpress.cornell.edu.

First published 2025 by Cornell University Press

Library of Congress Cataloging-in-Publication Data

Librarians: A CIP catalog record for this book is available from the Library of Congress.

ISBN 9781501781964 (hardcover)
ISBN 9781501782923 (paperback)
ISBN 9781501781971 (pdf)
ISBN 9781501781988 (epub)

for Sam

But the Rocket has to be many things, it must answer to a number of different shapes in the dreams of those who touch it—in combat, in tunnel, on paper—it must survive heresies shining, unconfoundable . . . and heretics there will be: Gnostics who have been taken in a rush of wind and fire to chambers of the Rocket-throne. . . . Kabbalists who study the Rocket as Torah, letter by letter—rivets, burner cup and brass rose, its text is theirs to permute and combine into new revelations, always unfolding. . . . Manichaeans who see two Rockets, good and evil . . . of a good Rocket to take us to the stars, an evil Rocket for the World's suicide, the two perpetually in struggle.

—Thomas Pynchon, *Gravity's Rainbow*

Contents

List of Illustrations viii
Preface ix
Acknowledgments xi
List of Abbreviations xiii

Introduction   1

Part One: Dreams

1. Imagination   13
2. Interplanetary Men   49
3. Star of Hope   83

Part Two: Nightmares

4. Lunartics!   121
5. The Cosmic Bomb   153

Part Three: Waking Up

6. A Celestial Magna Carta   191
7. Stairway to Heaven?   223

Conclusion   252

Notes 267
Index 317

Illustrations

1. James Mangan points to Celestia 2
2. Apocalyptic Martians, *The War of the Worlds* 23
3. The Grand Lunar, *The First Men in the Moon* 25
4. "The Change," *In the Days of the Comet* 28
5. Kurd Lasswitz 32
6. David Lasser 44
7. Arthur "Val" Cleaver 55
8. Arthur C. Clarke 59
9. Olaf Stapledon 69
10. Frank Malina 75
11. The Star of Hope 89
12. "Sorry, That Pass Is No Good Here . . .," *Washington Star* 112
13. Lunar construction vehicle, *Project Horizon* 122
14. "Soviet World Domination under Preparation" 143
15. "President Kennedy's New Frontiers" 180
16. Outer Space Treaty signing ceremony 192
17. "Isn't it time you were getting me in orbit?," *Christian Science Monitor* 199
18. "Peace in Space, Good Luck to Earth," *Washington Post* 213
19. Fractional Orbital Bombardment System, schematic drawing 217
20. *Earthrise* 228

Preface

"We are going." So goes our declarative refrain promising that human beings, sometime soon, will return to the moon after more than half a century. From Launch Pad 39B on Meritt Island in Florida, NASA's Artemis 3 mission will send four astronauts on an unforgettable ride toward Earth orbit, then on to lunar orbit, and finally the moon's south pole, an alien terrain unknown to previous visitors.

No doubt a nervous energy will temper their excitement. For unlike the equatorial—and therefore light-flooded—Apollo sites, Artemis 3 will descend on blackness. The sun will linger just above the horizon, casting long, dark veils over the land. With headlamps the astronauts will venture out among soaring mountains and deep craters that combine to create regions of permanent shadow.

A similar dusk hangs over the politics of space exploration. China is scheduled to make a crewed landing near the south pole before 2030 as well, an unwelcome prospect for NASA administrator Bill Nelson, who has warned that Chinese astronauts could occupy the area "like the Spratly Islands" if they get there first. By the time Artemis 3 takes flight, Russia will have pulled out of the International Space Station after two decades of close cooperation with the United States. Since 2007 all three countries, and India as well, have tested kinetic antisatellite missiles and thereby generated thousands of pieces of orbital debris. Initiatives to ban these and other space weapons continue to languish month after month at the United Nations.

It was not until I began researching this book that I noticed anything happening in space, or looked forward to anything that *would* happen. I was not, and am neither still, a Musk fanboy, a Trekkie, or even a buff. And certainly, upon entering graduate school, I had not considered writing about space exploration, space technology, or space policy. So then what?

I came of age during the war on terror. Yellow ribbons lined the trees and fences of my boyhood paper route. High school peers volunteered just after graduation to empower Bush's surge. Images from Abu Ghraib seared themselves onto adolescent brains. My early education pitched the conflicts in Iraq and Afghanistan as merely the latest episodes of US hubris and folly, evidence of the "tragedy," as it was often put, of US foreign policy. Marilyn B. Young, who spent most of her life teaching and writing about the subject, wrote at the time that across her long years, the sequence of US wars she studied seemed less like a progression than a continuation, "as if between one war and the next the country was on hold." It rang true.

I was determined to illuminate some lost chapter of this still-unfolding story. Why not a history of foreign policy schools in the United States, places like MIT's Center for International Studies or Princeton's Woodrow Wilson School that trained a generation to wield global power after World War II? Or what about a full-length biography of George W. Ball, the only senior official in the Lyndon Johnson administration to object to the US commitment in Vietnam? Compelling and deserving subjects, still pluckable by future dissertators.

But imagine my surprise upon first combing through the early corpus of US space diplomacy. For all the embattled rhetoric and hyperbole of the space race, policy and law for the cosmos seemed to have developed along admirable lines. NASA successfully guarded civilian spaceflight from a jealous air force. Humanism counterbalanced the ego of the US and Soviet space programs. The two rivals even cooperated on an international treaty governing the exploration and use of outer space for all nations. This seemed not tragedy but triumph.

Of course, things are never quite what they seem. I asked: Were peaceful US initiatives in space merely low-hanging fruit for cynical cold warriors and naïve arms controllers? Did US leaders from John Foster Dulles to Adlai Stevenson, who avidly pronounced their wish for peace in space, engage in bald hypocrisy given the development of military space technology? Was the lauded Outer Space Treaty, as Senator Strom Thurmond (R-SC) put it, "nothing more than a scrap of paper"?

Here were questions demanding answers, and a story worthy of the mid-twentieth century, witness to the profundity of the atomic bomb, the digital computer, the birth control pill, and, yes, the space rocket. The ensuing research consumed years and sent me along paths that branched far beyond policy into science-fiction, engineering, philosophy, and religion. This book is the result of those rewarding incursions. I have come to my conclusions. The reader will reach her own.

Acknowledgments

The road that led to this book stretched and snaked its way among many places, almost too many to count. At each stop I picked up intellectual debts. Those debts, more surely innumerable, deserve mention.

Two unforgettable years at Stanford University's Freeman Spogli Institute for International Studies (FSI) helped me fashion my work into a book of which I could be proud. Special thanks go out to my mentor and friend David Holloway, who diligently read every word and shaped key portions of the text, especially my evolving thought on the "Interplanetary School of IR." Members of FSI's Nuclear Reading Group—especially Jeffrey Ding, Ryan Musto, Sulgiye Park, and Scott Sagan—made invaluable suggestions for chapter 5, "The Cosmic Bomb." I also thank Rose Gottemoeller and Harold Trinkunas for their advice and support. Kathy and Justin Welsh made California feel like home.

As an Ernest May Fellow in 2022-2023, I spent a fantastic year at the Belfer Center for Science and International Affairs at Harvard, where my research benefited from judicious colleagues in the Applied History Working Group. I thank Fred Logevall in particular for poring over the entire manuscript and for his guidance.

At Indiana University, Bloomington, where this project began, Nick Cullather was a steady hand on what was often a veering ship. He provided a reliable stream of advice, redirection, and reassurance—all from his trusty typewriter. De Witt Kilgore, Ed Linenthal, and Stephen Macekura supplied numerous comments and suggestions. Michael McGerr saw the germ of something worthwhile in an earlier work and helped it grow. During that time, I had the opportunity to share early stages of my project with the International Policy Scholars Consortium (IPSCON), a network of historians, political scientists, and area studies scholars attempting to bridge the gap between academic research and contemporary issues in world affairs. For their comments and encouragement, I thank IPSCON's Hal Brands, Frank Gavin, Jim Steinberg, and Phil Zelikow.

Scholars from across the United States and around the world helped me through the years and time zones. I thank especially Aaron Bateman, Megan Black, Mark Bradley, Greg Eghigian, Michael Falcone, Anne Foster, Alexander Geppert, Daniel Immerwahr, Neil Maher, Patrick McCray, Teasel Muir-Harmony, Brian Odom, Stephen Walt, and Salim Yaqub.

This book would have floundered, too, without financing. A travel grant from the Eisenhower Foundation kickstarted my research in Abilene, Kansas. The Indiana University History Department supplied several small research awards that allowed me to attend conferences and visit archives. An Arthur M. Schlesinger Jr. Fellowship from the Kennedy Presidential Library allowed me to spend an extended period in Boston combing through national security files. In 2019, I received an Aerospace History Fellowship from the American Historical Association and the National Aeronautics and Space Administration. Without it, completing this book would have been impossible.

For breaking up the solitude of writing I thank my family and friends. Jacob Hagstrom was a sounding board for ideas and laughs. My siblings Jonathan, Kevin, and Sarah were a constant source of encouragement. My parents—Jeff and Cheryl Buono—cheered on a roving, penniless child. The Eilenberg Family—Heléna, Steven, Michael, and Jenna—celebrated every milestone. And my wife, Samantha, showed unerring faith in my work. I dedicate this book to her.

Last, I want to acknowledge the late Martin J. Sherwin, thanks to whom I became a historian. Avanti, Marty!

Abbreviations

AAAS	American Association for the Advancement of Science
AAF	Army Air Forces
ABM	antiballistic missile
AFSWP	Armed Forces Special Weapons Project
ABMA	Army Ballistic Missile Agency
ACDA	Arms Control and Disarmament Agency
AEC	Atomic Energy Commission
AIS	American Interplanetary Society
ALSD	Apollo Lunar Surface Drill
AMF	American Machine and Foundry, Inc.
ARPA	Advanced Research Projects Agency
ARS	American Rocket Society
ATS	Applications Technology Satellite
ASAT	antisatellite
ADSID	Air-Delivered Seismic Intrusion Detector 7.
BIS	British Interplanetary Society
BoB	Bureau of Budget
BBS	Ban on Bombs in Space
BMD	ballistic missile defense
BMEWS	Ballistic Missile Early Warning System
CIA	Central Intelligence Agency
CIC	Counter Intelligence Corps
COPUOS	Committee on the Peaceful Uses of Outer Space
DDR&E	Director of Defense Research and Engineering
DEW	distant early warning
DISMs	delayed impact space missiles
DNA	Defense Nuclear Agency
DoD	Department of Defense
EMP	electromagnetic pulse
ENCD	Eighteen Nation Committee on Disarmament
FBI	Federal Bureau of Investigation

ABBREVIATIONS

FOBS	Fractional Orbital Bombardment System
GALCIT	Guggenheim Aeronautical Laboratory (California Institute of Technology)
GCA	ground-controlled approach
GCD	general and complete disarmament
GOP	Grand Old (Republican) Party
GPALS	Global Protection Against Limited Strikes
GPS	Global Positioning System
HF	high frequency
ICBM	intercontinental ballistic missile
IGY	International Geophysical Year
IR	international relations
IRBM	intermediate-range ballistic missile
JATO	jet-assisted takeoff
JCS	Joint Chiefs of Staff
JPL	Jet Propulsion Laboratory
LM	Lunar Module
LTBT	Limited Test Ban Treaty
MAD	mutually assured destruction
MHV	miniature homing vehicle
MIRV	multiple independently targeted reentry vehicles
MIT	Massachusetts Institute of Technology
MOL	Manned Orbiting Laboratory
NABS	nuclear-armed bombardment satellites
NACA	National Advisory Committee for Aeronautics
NAS	National Academy of Sciences
NASA	National Aeronautics and Space Administration
NATO	North Atlantic Treaty Organization
NAVSTAR	Navigation Signal Timing and Ranging
NIE	National Intelligence Estimate
NPT	Treaty on the Non-Proliferation of Nuclear Weapons
NRL	Naval Research Laboratory
NSC	National Security Council
NSF	National Science Foundation
NRO	National Reconnaissance Office
NSAM	National Security Action Memorandum
NSDD	National Security Decision Directive
OIMS	Society for the Study of Interplanetary Communications
OKB	Experimental Design Bureau
OST	Outer Space Treaty

PSAC	President's Scientific Advisory Committee
R&D	research and development
RAND	Research and Development Corporation
RBS	Random Barrage System
RFNA	red-fuming nitric acid
SAC	Strategic Air Command
SAINT	satellite interceptor
SALT	Strategic Arms Limitation Talks/Treaty
SDI	Strategic Defense Initiative
SDIO	Strategic Defense Initiative Organization
SLBM	submarine-launched ballistic missile
SPAD	satellite protection for area defense
SR	Systems Requirements
UAR	United Arab Republic
UN	United Nations
UNCLOS	United Nations Convention on the Law of the Sea
UNESCO	United Nations Educational, Scientific and Cultural Organization
UNGA	United Nations General Assembly
USAF	United States Air Force
USBM	United States Bureau of Mines
USIA	United States Information Agency
USSF	United States Space Force
VfR	*Verein fur Raumschiffahrt* (Society for Space Travel)
VHF	very high frequency
WAC	without altitude control
WGER	Working Group on Extraterrestrial Resources
WMD	weapons of mass destruction
WPA	Works Progress Administration

THE PROVINCE OF ALL MANKIND

Introduction

Once upon a time, in a cold and faraway city, one man became lord and protector of the entire universe. The time was January 18, 1949, the city Chicago, Illinois. There, only weeks before, James T. Mangan had been sitting atop his apartment building gazing at the cosmos when he wondered aloud: who owns outer space? Unusual though the question may have been almost a decade before the launch of *Sputnik*, it came naturally to this local eccentric. When his business partner, sitting beside him, replied that there certainly was "plenty of stuff out there," Mangan resolved to stake claim to all of space. A bold proclamation to be sure, but that morning the fifty-two-year-old walked to the Recorder of Deeds in Cook County—followed by television crews and photographers from *Life* magazine—and submitted a charter for a new nation: Celestia, or more formally, the Nation of Celestial Space. After some understandable confusion on the part of the recorder and a brief legal scuffle, the state's attorney granted the peculiar request. With the flutter of a pen and a blotch of wax, Mangan became "First Representative" of the most expansive nation in history (figure 1).[1]

Acquiring recognition for Celestia was another matter entirely. Mangan sent letters to the secretaries of state in seventy-four countries, but none replied. A student at the University of Tennessee claimed for himself "the southern half of all outer space" to protest Mangan's

FIGURE 1. James Mangan points toward his new nation, Celestia, outside Chicago's Adler Planetarium in December 1948, shortly before he made his claim "official." Wallace Kirkland/ The LIFE Picture Collection/Shutterstock

Yankee pretention to the universe. And when the Soviet Union finally launched the world's first artificial satellite in October 1957, Mangan could only watch helplessly from the sidelines as the Reds "trespassed" on his sovereign-less domain. Despite creating a national flag, passports, postage stamps, and gold currency for his new nation, Mangan seemed unable to convince outsiders of Celestia's legitimacy.[2]

The public's tickled reaction to Celestia frustrated Mangan, for he viewed its founding as an eminently serious, even noble undertaking. This was no self-indulgent public relations stunt, he insisted, but rather an effort to create "a bulwark of international peace." The Nation of Celestial Space was to be a model of political harmony and enlightened thinking, free from war and bitter historic rivalries. The First Representative quickly banned weapons from Celestia, even if they might be necessary to defend space from external threats. In September 1949, he issued diplomatic notes to the United States, Britain, and the Soviet Union forbidding nuclear tests in his territory.[3] Celestia was "a moral nation dedicated to peace everywhere," as Mangan wrote in a press release.[4] By example, he would show that "in space, right is more powerful than might," and that cosmic politics would operate "more on the principle of moral persuasion than force."[5] When Mangan began selling Earth-size parcels of space for one dollar apiece, he urged his potential customers to view the matter philosophically: "If you owned something 8,000 miles in diameter and 25,000 miles in circumference, you might realize that war is something to be laughed at." The Nation of Celestial Space might just lend citizens "enough bigness of thinking . . . to make them feel international squabbles are petty."[6]

One could be forgiven for consigning Celestia to the realm of historical curiosity. That in 1958 Mangan flew the Celestian flag outside United Nations headquarters in Manhattan seems a charming conclusion to a story that, however amusing, nevertheless carried little significance for the postwar world. But in fact, Mangan's ambitions for a space-nation devoid of "the world's pressures and poisons" both drew on a deep well of political thought connecting outer space with peace and reflected its continued relevance.[7] Convictions that the cosmos somehow transcended human conflict only grew stronger in the 1950s and 1960s as the incipient space age evolved into a full-blown technology race between the United States and the Soviet Union. The utopian ideas driving Mangan's crusade assumed greater salience as contemporaries came increasingly to fear destruction from orbital weapons, pollution of the heavens with earthly hatreds, and the extension of the Cold War to yet another exciting realm of human ingenuity.

Indeed, the story of Celestia is illustrative not because Mangan's ideas were so unique or unusual, but because they were pervasive. The Nation of Celestial Space was only the most playful example of what later writers would call "sanctuary" politics, a collection of ideas that (1) predicted spaceflight would bring about a more harmonious era in international relations and (2) argued that if such an exploration-induced revolution were to truly occur, it was imperative that governments preserve outer space as a haven free from war, imperialism, and other adolescent habits to be shed on humanity's breach into space and, by extension, political adulthood.[8] It was no coincidence that Earth, for so many dreamers, was but a cradle.

Notions of a "sanctuary" in space tapped into the term's deep etymology. The word derives from fourteenth-century Europe where, in Late Latin, a *sanctuarium* was a sacred place set aside for worship. Initially connected to purely ecclesiastical ideas, by the middle of the sixteenth century the definition had broadened to include a place of refuge, since by that time the use of churches as asylums for fugitives and debtors had endured since the time of Constantine. A third and final definition arrived in 1879, when the American bison, then nearly extinct, needed "land set aside for wild plants and animals to breed and live."[9]

Seventy years later, in the throes of a totalizing cold war, support for a space sanctuary absorbed elements of all three translations. When the *Apollo 8* astronauts read the Book of Genesis to a worldwide audience from lunar orbit in 1968, they proved that space was a place to "touch the face of God" and thereby inch closer to His image.[10] At the same time, religious and secular rhetoric alike referred to the cosmos as a site where human beings might find safety from the social and political diseases of their primitive, earthbound existence: the nuclear arms race, chronic inequality, ecological degradation, racism, and war. The logical corollary, and the one with which these pages are most interested, was that far from engaging in a hostile technology race, the spacefaring juggernauts should "set aside" the cosmos as a pristine frontier, an environment where the *human* animal could strive for a future divorced from the violence of history.

Sanctuary politics held that peace in space would reap benefits for earthly statecraft. Communications satellites would increase contact and thus understanding between nations. Reconnaissance satellites would monitor arms control agreements and help build trust. Disarmament in outer space could open doors for earthly demilitarization, perhaps even the abolition of nuclear weapons. "Space may well be the sea in which the human race will someday find an island of peace," Senator Lyndon Johnson (D-TX) told CBS affiliates in *Sputnik*'s wake. By 1973, merely fifteen years later, the international lawyer Edward R. Finch anticipated that space exploration might usher in "the true internationalization of nations into a conglomerate of States of the World governed by law."[11]

Such convictions enjoyed special resonance in the United States, where the economic and technical prospects for exploration were substantial, to say nothing of the demand to answer the Soviet gauntlet.

From the 1920s through the post-*Sputnik* years, Americans drew on budding "astrocultures" in Europe to articulate a vision of the human future in space in which national animosities and tribal hatreds melted in the heat of discovery.[12] Breaching Earth's atmosphere would foster in man "an interplanetary mind" elevated above parochial concerns and dedicated instead to egalitarian care of the entire human family.[13] Consumers of the era's copious science-fiction stories and popular science magazines came to understand space as either a site of destruction and civilizational collapse, or alternatively, of cosmic wisdom and restraint.

Thus, when the services began advanced research on rocket boosters, satellites, and other spacefaring vehicles in the 1940s and 1950s, many Americans demonstrated a preference for restraint by condemning what they considered to be the perversion of space exploration—of science itself—by the Cold War. Why, they asked, should this new vista in humankind's mastery of nature fall under the purview of militarists? Why should space exploration be yet another avenue of national competition rather than an adventure that all nations could undertake together?

A generation of US officials wrestled with these questions. In the fifteen years after *Sputnik*, policymakers in the Eisenhower, Kennedy, and Johnson administrations fretted over the introduction of space-based weaponry, the bankrupting of the nation by lavish expenditures on military space programs, and the probability that technological developments would outpace the capacity of international law to regulate new hardware. To avert these problems, they built a diplomatic menu that subsumed many of the utopian ideas offered by stargazers like James Mangan: joint missions with the Soviets; technology transfers to Europe and the Global South; United Nations jurisdiction of space activities; and pursuit of a binding treaty to govern the peaceful use of space for all.

Beliefs about the pacifying effect of exploration inevitably touched domestic space policy as well. Debates over whether the fledgling National Aeronautics and Space Administration (NASA) should be a civilian- or military-controlled institution, for instance, hinged on ideas about the purity of space and the necessity of stopping exploration from becoming a surrogate—or a spark—for war. Later, amid the *Apollo 11* mission, public officials were keen to portray the first moon landing as the accomplishment of all humanity rather than heroic American astronauts. In this way, early US space policy strove not only to prevent the extension of the arms race to space but also to pitch spaceflight as the catalyst for a harmonious new era in international life. Here was sanctuary politics grown full flower.

The Province of All Mankind is a political, cultural, and intellectual account of this history. The book charts the long maturation of the sanctuary idea and its startling—in retrospect, miraculous—expressions in US policy from *Sputnik* through the first moon landing. In so doing, it narrates the emergence of the cosmos as a profound new arena of US foreign relations and international law, one many leaders deemed worthy of special protections from Cold War conflict and hence thick with future consequence.

Two observations ensue from such a narration. First, the most important political assumptions about spaceflight surfaced long before rockets, satellites, and lunar probes were matters of fact, that is to say long before the space race, or even the Cold War generally, were features of the international scene. From the late nineteenth century, Americans and their contemporaries in Europe drew a powerful and enduring association between space exploration and the transcendence of international rivalry and war. What follows is in large measure a story about how, in the construction of spaceflight as a political project, *imagination* was as crucial as space technology itself.

Second, and more surprising, is that the idealism of the spaceflight imagination often jibed with the hard-headed, realist assumptions guiding US national security policy at the height of the Cold War. "Space for peace," as Mangan called it, appealed to cold warriors because it enhanced the image of the United States as a responsible, moral, and peace-loving power; because, from a space sanctuary, the United States could safely spy on the Soviet Union with satellites; and because, in the final analysis, extending the arms race to space simply did not benefit national safety. For a time, doves and (some) hawks nurtured the same egg.

Chronicling the origins and gestation of the sanctuary idea in American space policy presents a stark historiographical puzzle. Was not space exploration, after all, mainly an avenue of Cold War *competition*?[14] Was not space technology a crucial arm of *propaganda*?[15] Did not promoters of peace in space struggle against its *militarization* by jealous superpowers?[16] There was, to be sure, a very real "space race."[17] Walter McDougall put it best in his landmark 1985 account, . . . *The Heavens and the Earth*. Humanity's ascent into the cosmos was "a time of fear and euphoria both," and yet: "in the formative years of the Space Age, *the euphoria faded, the fear remained*." There was no reorientation of global politics, no political reconciliation between the superpowers, no bludgeoning of poverty and disease by novel engineering derived from NASA. "For the present and foreseeable future," he wrote, it was instead the victory of technocracy—that is, state-driven research and development—that "defines the character of the Space Age in history."[18]

Forty years and dozens of new works later, the realism of this interpretation, if not the details about technocratic methods, still stands tall in our histories and our collective memory. We remain fixed in McDougall's foreseeable future. And why not? Despite a new burst of creative energy in space, contemporary international relations remains so riven by mistrust, hostility, and great-power competition that pundits have long since declared a Second Cold War, one in which space rivalry will undoubtedly play a central role.[19] In December 2019, as Donald Trump announced the creation of a US Space Force that would protect national interests in the medium, he dubbed space a "warfighting domain," a designation the North Atlantic Treaty Organization (NATO) affirmed when, that same month, it declared space a fifth operational realm alongside land, sea, air, and cyber.[20] In 2021 China twice tested a hypersonic, maneuverable glider that descends from orbit to deliver a warhead at speeds exceeding 3,800 miles per hour, an event many observers dubbed another "*Sputnik* moment."[21] Over recent years China, Russia, and India have honed antisatellite (ASAT) weapons capable of torpedoing the space-based architecture upon which the modern economy is entirely—and precariously—dependent. Little wonder, then, why political scientists are so pessimistic about the future of international relations in space. Their book titles betray the basic plot: *Dark Skies . . . Heavenly Ambitions . . . Crowded Orbits . . . Scramble for the Skies . . . Astropolitik*, even *Original Sin*.[22]

It is easy enough, too, to read today's foreboding developments back into what historians may soon refer to as the "First" Cold War. Despite high rhetoric about the potential for space exploration to induce US-Soviet cooperation, real collaboration proved a nonstarter. Timid scientific exchanges succumbed in the early 1960s to "the tyranny of realism," as political scientist Donald Kash wrote in an early account.[23] Instead, US spending on military space projects ballooned from $814 million in 1961 to more than $2 billion by the end of the decade.[24] The leap was equally pronounced in the Soviet Union, where the military space budget reached $1.2 billion by 1968.[25] From these seedling investments did the ancestors of our contemporary military-space regime burst forth: the US Air Force's (USAF) Dyna-Soar space plane and Manned Orbiting Laboratory (MOL), the Kremlin's Fractional Orbital Bombardment System (FOBS), as well as a host of direct-ascent and co-orbital ASAT weapons. These were years when Nikita Khrushchev brandished space rockets like giant clubs, when Barry Goldwater campaigned for president on American military preponderance in space, and when both Washington and Moscow drew up blueprints for military installations on the lunar surface. The two sides even tested nuclear weapons in space.

But this is only half the story. Consider what was left when the moondust cleared at the end of the 1960s. By that time, despite the fear induced by *Sputnik* and the political antagonism of the ensuing space race, the United States had spearheaded the development of a special United Nations committee to govern the peaceful use of the cosmos for all people. Upon NASA's founding in 1958, the federal government resolved that space activities "should be devoted to peaceful purposes for the benefit of all mankind," language that international lawyers and diplomats enshrined in the first United Nations General Assembly (UNGA) resolutions on space exploration.[26] Far from a new theater of conflict, outer space became a tranquil nest for passive satellites that ultimately did more to temper the Cold War than escalate it.[27] The USAF's ambition to leverage space as "a new high ground" had collapsed under budgetary, technical, and ethical (read: political) pressures. The Dyna-Soar program collapsed in 1963, the MOL in 1969. The Soviet Union later canceled FOBS having never orbited a single warhead. Instead, negotiators from the two space giants collaborated on an international agreement that banned nuclear weapons from space, forbade military installations and maneuvers on the moon and other celestial bodies, and outlawed claims of national sovereignty in space. This landmark Outer Space Treaty (OST) declared that the exploration and use of outer space was henceforth "the province of all mankind."[28]

What are we to make of these events? Two continental superpowers, two vast military-industrial complexes, waged a decades-long cold war having never staged weapons in orbit, on the moon, or anywhere else in space. Neither side triggered a war, or even so much as a Checkpoint Charlie-like military confrontation, outside Earth's atmosphere. Bucking the violent depictions of space that *Star Wars* and *Battlestar Galactica* later beamed into American living rooms, the military constellation in space consisted merely of ancillary systems for communications, navigation, reconnaissance, meteorology, and early warning. Compared with the foreign policy misadventures of the 1950s and 1960s—CIA-backed coups in Guatemala, Iran, and the Congo; a bungled invasion of Cuba; and a devastating proxy war in Vietnam, just to name a few—US space policy exhibited remarkable sobriety, even prescience.

The answer is that the irenic vision of space propounded by the likes of James Mangan, dismissed though it was by realist critics, achieved victory in a duel of ideas about the future. It was a contest between the sanctuary worldview and what political scientist Lincoln Bloomfield called "the dark side of space," wherein human capacities for violence and greed would first extend to and then escalate in the cosmos.[29] Did satellite television, for instance, mean education or propaganda? Would our artificial moons keep nuclear war at bay or inaugurate it from on high? Would astronauts "conquer" space as Edmund Hillary or Augustus Caesar? "Colonize" it as Roald Amundsen or Cecil Rhodes?[30] It was a conflict, in other words, between dreams and nightmares.

For all the ink devoted to space racing, saber rattling, and panic-inducing "*Sputnik*" moments," the historical record shows that, at least for a time, the dreams had gained the advantage. In the dozen years it took to complete the Cold War space race (1957–1969), the sanctuary paradigm, forwarded *in spite of* the Cold War, kept conflict out of space. Despite the rigorous competition that characterized space exploration in these years, sanctuary politics became a vehicle through which to critique great-power rivalry, to express fatigue with its political rules, and to suggest alternatives. It encouraged Americans to think differently about the future of international relations. Some, like Mangan, entertained pure utopianism. Others, like John F. Kennedy, merely opened themselves up to optimism even as they waged the "long twilight struggle" against communism.[31] In this way, the sanctuary idea did as much to diminish the influence of the Cold War as the space race did to reinforce it. To reverse McDougall's adage, it grounded incipient space politics more in hope than fear.

Half a century later, the dream of a space sanctuary continues to have staying power. But whatever cultural and political battles it may have won during the 1950s and 1960s, recent history shows that the war is still ongoing. Read the epitaph again: Pynchon teaches us that our creations, derived from our nature, are both noble and terrible. Space technology is the sum of an uncountable number of individual choices. Not all will be noble, not all terrible. May the balance be ever in our favor.

Part One

Dreams

Chapter 1

Imagination

Cliché pervades the story of the space age. "The *Sputnik* shock." "We choose to go to the moon." "One small step for man." Each retelling fuels the story's power and prosaism. Try as we might, we cannot divorce ourselves from the pageantry of *Apollo* and *Soyuz*. The unfolding drama is too meaningful, the images too inspiring.

Take *Earthrise*, the photograph seen round the world. On Christmas Eve in 1968, *Apollo 8* astronauts Frank Borman, James Lovell, and Bill Anders witnessed a spectacular sight during their orbit around the moon. In the porthole of their capsule, they saw Earth floating in an ocean of blackness. "Oh my God!" Anders shouted. "Look at that picture over there! There's the Earth coming up." Truth revealed itself to the voyagers. The planet, far from a grand political map bespotted by nation-states, was a tiny, solitary, and fragile blue ball.

And it was humankind's to share. The astronauts and commentators back on the ground agreed: the view of the Earth from space proved that, as Borman recalled, "it really is one world."[1] From such an exalted perch—captured in Anders's famous picture (see chapter 7)—national, religious, and racial boundaries were invisible. Hatred and war seemed trite. In a commemoration of the flight that appeared on the front page of the *New York Times* the following day, poet Archibald MacLeish wrote that "To see the Earth as it truly is, small and blue and beautiful in that eternal silence where it floats, is to see ourselves as riders on the Earth together, brothers on that bright loveliness in the eternal cold—brothers who know that they are truly brothers."[2]

That viewing Earth from space would dissolve national, ethnic, and racial animosities was central to the romantic politics that infused US space policy for two decades after *Sputnik*: if governments were to preserve the transcendent perspective offered by space and with it build a more peaceful world, they had better leave their demons behind. Indeed, the photograph did as much to mark the ascendance of sanctuary politics as any presidential speech, United Nations resolution, or front-page editorial. Beyond the realm of policy, too, writers both at the time and since credited *Earthrise*—along with NASA's 1972 photograph *Blue Marble*—with sparking modern environmentalism, harkening the march of globalization, and ushering in an era defined by interconnectivity, goodwill, and shared humanity.[3]

Yet for all the enthusiasm *Earthrise* stoked about the dawn of a brighter age, and for all the scholarly attention devoted to it in subsequent years, the event was a genesis neither for globalist culture nor the sanctuary strain in American space policy. Both the view of the whole Earth from space and the impact that view might have on human civilization had been a subject of fascination since Plato.[4] In 1931, nearly forty years before *Apollo 8*, the American rocket enthusiast and social reformer David Lasser imagined a group of astronauts who, peering back at Earth from space, achieve the cosmic mindset that Lovell, Anders, and Borman later attained. "A Great spirituality fills us," says one of Lasser's fictional spacefarers, "a humbleness and a yearning for the continuance of this immense peace. . . . Cities, empires, states; dreams and ambitions; conflict and confusion are infinitely remote, part of the dream-world of that slowly turning globe." Lasser included a drawing, remarkable for its resemblance to *Earthrise*, of the astronauts' unique vista.[5]

Still earlier depictions of a transformative, interplanetary perspective—what the philosopher Frank White has called the "Overview Effect"—appeared in the budding science-fiction literature of the late nineteenth and early twentieth centuries.[6] In space, one of the protagonists of H. G. Wells's classic *The First Men in the Moon* (1901) experiences a profound disembodiment from his previous beliefs, attitudes, and fears. "Effervescing with new ideas—new points of view," he gains an appreciation for the "earth's littleness and the infinite littleness of my life upon it."[7] The space-plying Marxist of Aleksandr Bogdanov's *Red Star* (1908) looks back on the plight of his earthly comrades with cool dispassion.

"Back there blood is being spilled," he observes, "yet here stands yesterday's revolutionary in the role of a calm observer."[8]

To be sure, *Earthrise* was a powerful and symbolic event. Yet the mere existence of these accounts, decades prior, suggests that perhaps NASA's historic images were not as mind-altering as scholars have previously supposed. For nearly eighty years before *Apollo 8* established a relationship between space exploration and the transcendence of human conflict, a host of writers across the United States and Europe had already envisioned that relationship in their minds. Understanding human beings as belonging to a single global community, in short, required acts of political imagination long before anyone set eyes on *Earthrise* or *Blue Marble*. The astronauts who first captured the whole Earth from space were but three participants in a vast intellectual, cultural, and political discourse about the cosmos that reached back decades before human spaceflight was practicable. This "interplanetary" discourse predated, and subsequently matured alongside, the physical beginnings of modern rocketry after World War I. It encompassed science-fiction authors, journalists, philosophers, engineers, and scientists of all stripes, including many of the thinkers and doers that historians have for years associated with the opening of the space age: novelists H. G. Wells and Arthur C. Clarke; rocket theorists Konstantin Tsiolkovsky and Hermann Oberth; and political activists like Lasser. Consequentially, it also included many individuals who would eventually become the architects of the American and Soviet space programs.

Today, these figures are immortal heroes in the story of human spaceflight. Yet our histories and our popular memories of the space age have rarely considered that beyond their contributions to technology, the founding generation of rocket enthusiasts also forwarded a set of moral arguments about what the final purpose of space exploration should be.[9] These influential thinkers were not merely scientific and technical creatures but political and philosophical ones, too. Although space travel could be a goal unto itself for some early dreamers, many harbored lofty assumptions about the benefits that would accrue from escaping the bounds of Earth: social cohesion, economic prosperity, an end to war.

What united the realms of engineering, science fiction, and boosterism together was a conviction that outer space was a blank slate onto which human beings could etch new, more egalitarian structures of governance. Leaving behind the hoary animosities that had characterized earthly politics, they agreed, human societies could start anew in space by implementing enlightened rules that transcended national or imperial interests. At the same time, the cooperative processes and cosmopolitan attitudes propelling humans into space could be transplanted back to Earth, the cooperation of astronauts in space reflected in the deeds of the political leaders who had sent them there. Here was interplanetary thought laid bare.

Whether they realized it or not, adherents of this utopian vision were engaged in sound political theorization. As all good political scientists do, they attempted to establish an abstract relationship, a hypothesis, between two phenomena. Those in the business refer mainly to *independent* or *x* variables—things that cause variation in the observable world—and *dependent* or *y* variables, which rely on the quality of independent variables. Many scholars now agree, for example, that the presence of democratic institutions among states (an independent variable) makes those states less likely to go to war with one another (the dependent variable). Others have argued that if a country has unstable borders (an independent variable), then it is more likely to have an autocratic and intolerant political climate (the dependent variable). In the interplanetarians' own elegantly simple equation, world peace was the dependent variable, the phenomenon to be explained. The *i*ndependent variable, the force that would produce this outcome, was human spaceflight.

To bolster their theory, interplanetary writers also supplied what social scientists refer to as a "casual mechanism." This is an explanation of *how* an independent variable produces a given outcome, how *x* causes *y*. As practitioners know well, this is no easy task; real-world phenomena are not subject to laboratory experimentation, and it is difficult to establish whether a given variable is working independently from others in a system as complex as interstate relations.

Yet interplanetary thinkers did not shrink from this task. They forwarded a series of ideas and concepts that directly linked space exploration to the achievement of global order and international harmony. First, they portrayed space exploration as a vessel to channel humanity's nervous energy, a "pressure valve" through which international tensions could find constructive release.[10] "By providing an outlet for man's exuberant and adolescent energies," predicted Clarke in 1946, "astronautics may make a truly vital contribution to the problems of the present world." Over time, exploration would "begin to color Man's psychological outlook," particularly the "active—even aggressive—minds of those peoples, such as the Americans, British and Russians, for whom the problem is most acute." Amid mounting mistrust between the United

States and the Soviet Union, such a solution had seemingly arrived just in time. "In many ways," Clarke observed, "the very dynamic qualities of astronautics are in tune with the restless, expansive mood of our age."[11]

A second presupposition was that because exploring outer space was such a difficult and dangerous enterprise, human beings would have no choice but to cooperate with one another, and that this cooperation would manifest itself in every area of political life. The massive infusions of capital, technology, and labor required for spaceflight would force even the wealthiest nations to pool resources with others. Budding international organizations would provide both a political framework for this collaboration and a visible signal of the impending postnational age. Once in space, humans from opposite ends of the world would fight the harsh elements rather than each other. The urgency of surviving in weightless, airless, and waterless environments would push once rival pioneers into scientific, economic, and social partnerships.

Finally, interplanetary thinkers proposed that the process of exploring and colonizing outer space would contribute to the social, psychological, and perhaps even the physical evolution of human beings. From the late nineteenth-century writings of Russian cosmists to the science-fiction novels that pervaded the early Cold War, they wrote about the edifying, pacifying, and mind-broadening effects that exploration—and, in some cases, the space environment itself—would have on human beings. This process would begin, of course, with the Overview Effect. Borderless and vulnerable, the sight of the planet from space would stimulate feelings of oneness with humanity and a tranquil disassociation from worldly affairs. Over time, collaborative efforts to tame the wilderness of space would make *Homo sapiens* a more magnanimous, beneficent, and empathetic species. Humans would become, in Tsiolkovsky's language, nothing less than "perfect."[12]

The promoters of this political vision constituted a distinct interpretation of international relations, call it the "Interplanetary School" of IR. Culled from varied professional worlds and separated by time and geography, interplanetary theorists nevertheless huddled around a discrete collection of ideas that linked humanity's first steps into the cosmos and the next, more harmonious stage in global history. Often referring to their predecessors, the most prominent figures in this intellectual movement displayed a self-awareness about their interpretation of the future and its place alongside the far gloomier theories of international affairs that proliferated from the end of the nineteenth century through the Second World War.

Indeed, it is startling in historical perspective to compare the optimism of the Interplanetary School with the prognoses of geopolitics, political realism, and race war that had gained global readerships by the middle of the 1930s. Taking seriously the school's major figures, their most important texts, and the ways they departed from prevailing political forecasts opens a window onto why "sanctuary" rhetoric came to possess such emotive power in the United States at the dawn of the space age, and ultimately how that rhetoric, quixotic though it may have seemed to cold warriors after *Sputnik*, manifested itself in early US space policy.

The Interplanetary Imagination

In 1905 the dream of socialism hemorrhaged in Russia. On January 22, "Bloody Sunday," Imperial Guardsmen in Saint Petersburg opened fire on unarmed protestors as they marched toward the Winter Palace demanding reforms of Nicholas II. The shootings sparked massive protests and strikes in cities across the empire. At the same time, Tsarist forces suffered humiliating defeats to the Japanese at Port Arthur, Mukden, and Tsushima. Within weeks, chaos engulfed the entire country. The regime shot, hanged, or drove into exile thousands of revolting peasants and workers. Sailors mutinied in Sevastopol, Vladivostok, and Kronstadt. Armenians and Caucasian Tatars massacred each other in a protracted ethnic conflict. Jews perished in renewed pogroms. Nine million square miles had fallen into darkness.[13]

Searching for an escape, one of the most talented and fervent revolutionaries, Aleksandr Bogdanov—who with Lenin had helped establish the Bolshevik faction of the Social Democratic Labor Party just two years before—penned a short, accessible novel with the 1905 revolution as its backdrop. *Red Star* begins in Saint Petersburg, where the novel's narrator and protagonist, a revolutionary and mathematician named Leonid, befriends a fellow partisan who reveals himself to be a visitor from Mars. Menni, as the alien is known, convinces Leonid to embark on a quest to the Red Planet to study it. Upon his arrival, the Earthling is overcome with awe and bewilderment at the contrast between the chaos of Russia and the transcendent harmony of Menni's society.

The forces driving Bogdanov's fictional revolution resemble those unfolding in reality: peasants earn little more than scraps and are unable to sell or mortgage the land they work ("the agrarian problem"); ethnic and national minorities violently resist Russification ("the national problem"); and the government has banned strikes and labor unions ("the labor problem").[14] Not so on Mars. There is no agrarian problem because there are no agrarians; farming is completely industrialized and food is plentiful. There is no national problem because there are no nationalities; Martians are ethnically homogenous and speak a single language. And there is no labor problem because labor is voluntary, carried out merely in the interests of self-fulfillment and edification. There is no state, no politics, and no private property. A centralized bureau of statistics, equipped with computer-like machines, determines the material needs of all and ensures fair distribution. Education, science, decision making, technology, and art are all collectivized. And while Martians possess wildly varying capacities and talents, there is nothing resembling class or rank. Even gender is neutralized, as there are no discernable physiological or behavioral differences between males and females. Socialism has been perfected.[15]

The story is not without conflict, of course. The Martians struggle constantly against their natural environment. Leonid experiences profound disorientation, even mental illness, in his new locale, and he learns of a Martian council debate over whether to colonize Earth. There is just enough murder, romantic complication, and political intrigue to throw Martian utopia into doubt. But for everyday Russians in the throes of a violent revolution, *Red Star* was a catharsis. Bogdanov's narrative permitted readers to imagine a world in which the social and political maladies then afflicting them simply ceased to exist. Leonid's excursions into Martian society opened a porthole into the disorder of human civilization and allowed one to envision alternatives.[16] As historian Richard Stites observes in his English translation of the novel, the real drama of the story lies in "the juxtaposition of a unified, harmonious, serene, and rational life on Mars with the chaotic, barbarous, and self-destructive struggles of the people and social classes of twentieth-century Earth."[17]

Red Star was but one of a score of turn-of-the-century works anticipating the dawn of space travel and the evolutionary changes it would render on human beings and their societies. The entire genre of interplanetary fiction from 1890 to 1930 can be read as an escape from, or alternatively a recommendation for how to solve, two sets of problems. The first was domestic: urban congestion, disease, chronic wealth inequality, and political corruption. The second afflicted international relations: imperialism, nationalism, racism, and war.[18] To imagine worlds in which these ailments disappeared, authors reached out in time (perfect civilizations in the distant future), in geographic space (harmonious communities on lonely, remote islands), and to new strata of technology (machines that make war obsolete or provide unlimited food and energy for all). But as a narrative device and as a method of visualizing better worlds, outer space was unmatched. The cosmos provided a reliable tabula rasa onto which writers could project their societies' most transcendent ambitions for social and political change. Years before the rocket revolution made spaceflight a matter of practicality, visionary writers pitched outer space as a viable setting for, or a *catalyst* of, political experimentation. Removed from the physical environment of Earth, or merely exposed to alien forces, human beings could, quite literally, start over.

Hebert George Wells was many things. For British workers, he was a Fabian champion of socialism. For Europe's war-torn masses, he was tireless advocate for world peace, human rights, and especially global governance. Anticipating the arrival of tanks, aircraft, space travel, satellite television, and the atomic bomb, he pioneered the genre of intellectual gameplay now known as futurism. He was a first-rate philosopher and political thinker. He was a talented sketch artist and cartoonist. And, for a sequence of remarkable women that included Margaret Sanger, the American birth control activist, he was a lover and companion.[19] But for millions of ordinary people the world over, H. G. Wells was, above all, a writer of fantastic stories that took readers to times and places they could as yet scarcely imagine. Indeed, he birthed modern science fiction.[20] Though his writing career would ultimately span more than a half-century, Wells exploded onto the scene in the waning hours of the nineteenth century with four "scientific romances." The first three were *The Time Machine* (1895), *The Island of Dr. Moreau* (1896), and *The Invisible Man* (1897). In 1897, reeling from poor health and still uncertain about his future as an author, he sat down in the cramped dining room of his home in Woking to finish the fourth, his masterpiece.[21] It was called *The War of the Worlds*.

First serialized in *Pearson's Magazine* in 1897 and published by William Heinemann and Harper the following year, *The War of the Worlds* is Wells's grisly tale of a Martian invasion of Earth. The aliens first arrive in Woking, no less, where they begin laying waste to England in tripodic "fighting machines" equipped with chemical and thermal weapons. Caught completely unawares and possessing inferior technology, humankind is helpless to stop them. Martian craft eviscerate British defenses. They pluck fleeing humans from the ground and harvest their blood for sustenance. The onslaught ends only when the Martians succumb to bacterium for which they have no immunity, "slain, after all man's devices had failed, by the humblest things that God, in his wisdom, has put on this earth."[22]

Wells had conjured a vision of outer space as a hostile world where supreme beings, equipped with supreme weapons, plotted humanity's destruction and the absorption of Earth's resources. Far from the cooperative adventure promised in Jules Verne's earlier works—*From the Earth to the Moon* and its sequel *Around the Moon*—space was now a source of violence, death, and cold indifference.

This was not Wells's intent. He embraced the novel's immediate success with ambivalence, for he sought not only to entertain readers but educate and reform them: *The War of the Worlds* was a warning not of predatory dangers looming in space but rather the dangers that lurked within humanity. More than an invasion drama, the novel was a penitence for human failing and its malevolent expressions in world politics, imperialism especially.[23] "We men," says Wells's narrator in chapter 1, "must be to [the Martians] at least as alien and lowly as are the monkeys and lemurs to us. . . . Before we judge of them too harshly we must remember what ruthless and utter destruction our own species has wrought, not only upon animals . . . but upon its own inferior races."[24] Britain alone sufficed as an example. By the time of the novel's release, the Raj had turned India, once teeming with megafauna, into an ecological wasteland.[25] During the rubber boom of the late nineteenth century, the Anglo-Peruvian Company enslaved, tortured, and mutilated a generation of Putumayo River people. And the genocidal subjugation of Tasmania at the hands of British settlers had all but wiped out the island's aboriginal population. "Are we such apostles of mercy," the narrator asks, "to complain if the Martians warred in the same spirit?"[26]

Not only did the Martians demonstrate the hypocrisy of imperialism; on a much grander timescale, they offered a remedy to humanity's evolutionary downfall. As numerous Wells scholars have attested, the writer was captivated—often haunted—about the implications of human evolution.[27] Born in 1866, just seven years after the release of *On the Origins of Species*, he had been a pupil of the biologist Thomas Henry Huxley, who in his seminal lecture "Ethics and Evolution" had demolished Christianity's interpretation of Darwinism, which held that humanity was the product of Nature's special design, and therefore God's. Huxley, and soon Wells, were convinced that evolution would as likely produce evil as good in man, that humanity "might be damned as surely by the laws of evolution as by original sin."[28] As early as 1891,

Wells wrote that survival of the fittest inevitably accompanied "zoological retrogression," that each furtive step toward biological perfection was joined by one toward "degradation." Species constantly changed, of course, but "not necessarily in an upward direction."[29] Wells complained that humankind was "still mentally, morally, and physically what he was during the later Paleolithic period," that for an eon to come it was "likely to remain . . . at the level of the Stone Age."[30] Taking the longest view, he reasoned that inevitably humans would go the way of the dodo.[31]

Wells's earliest fiction had reflected this pessimism. In *The Time Machine*, the working class of 1890 have evolved 800,000 years later into a separate species from the upper classes. The Morlocks, as the race of proletariats is known, live a dark, brutal life underground. Though richer, the evolved capitalists, the Eloi, are no better off. Natural selection has made them decadent, childlike, weak, anxious, and apathetic. The bestial humanoids of *The Island of Dr. Moreau* are even more pitiful creatures. "Remorseless as nature," Dr. Moreau, the novel's psychotic antagonist, performs cruel vivisections on animals to make them "human." He is evolution incarnate, embodying the "plasticity of living forms" Nature imposes over time.[32]

The War of the Worlds, however, marked a shift in Well's thinking (figure 2). Frustrated with the glacial pace of social and political change both in Britain and the wider world, Wells abandoned evolution in favor of an abrupt change—a "cosmic happening"—that might shake humanity off its Huxleyan march to the grave.[33] Mars's invasion of Earth was that divine intervention, an apocalypse that brought at once judgement *and* deliverance, angels of death *and* salvation. As Jeanne and Norman Mackenzie note in their masterful biography of Wells, the arrival of the Martians was "a visitation, a warning of Old Testament severity that there is no hope for mankind unless it sees the error of its ways and repents."[34] And so it goes in the novel. In the epilogue, as Britons begin to rebuild their shattered society, the narrator posits that for all its destruction the alien assault, "in the larger design of the universe," had imbued humankind with a sense of humility and promoted "the conception of the commonweal of mankind." He speculates, too, that the Martians had paved a technological path by which *homo sapiens* themselves might become a spacefaring species. "The broadening of men's views that has resulted can scarcely be exaggerated," he intones on the final page. If Martians could reach Earth or Venus, perhaps humanity could escape to other planets when mother Gaia finally succumbed

FIGURE 2. By the late 1890s, H. G. Wells had decided that evolution would not deliver homo sapiens to the level of consciousness required for lasting peace. With *The War of the Worlds* (1898), he reached for an apocalyptic force whose destruction would bring about the profound social, political, and technological change needed for human development. The Brazilian artist Henrique Alvim Corréa rendered the Revelation-like catastrophe of the Martian invasion in a 1906 French translation. Wikimedia Commons.

to the death of the sun. Spaceflight and hence survival, gifts from the alien gods.[35]

"The Heavens," as Wells admitted in his autobiography, held a unique power for the writer. As a child his mother had discovered him "in the small hours . . . inspecting the craters of the moon" with a telescope the boy had found and assembled himself.[36] In an 1894 essay, he rued the disappearance of cosmically oriented religions from which ancient people had derived meaning and a sense of their place in the world. He was intimately familiar with the imaginative tradition of space travel writing going all the way back to Lucian's *Icaromenippus* and *True History* in the second century.[37] For Wells, the cosmos served as a durable reference point by which human beings could aspire to higher stages of consciousness and reestablish a sense of modesty, of their "infinitesimal littleness" as he phrased it in another short story. After experiencing the monstrosity of Dr. Moreau's island, for instance, Wells's protagonist turns to astronomy: he looks to "the glittering hosts of heaven," where "whatever is more than animal within us must find its solace and its hope."[38] It was a theme to which he returned again and again in his fiction. After the success of *The War of the Worlds*, Wells published two additional novels that helped build an anticipatory politics around space exploration and the rejuvenation it might one day deliver to men and nations. Both warrant close attention.

The First Men in the Moon was serialized in *The Strand Magazine* from December 1900 to August 1901 and was published in hardcover that year. The plot takes off when its businessman protagonist, Mr. Bedford, meets a reclusive physicist named Cavor, who is in the process of developing a new metal capable of negating gravity. After this material—dubbed "Cavorite"—proves a success, the two men embark on a trip to the moon in a spaceship lined with the revolutionary substance. When they arrive, Bedford and Cavor quickly lose themselves in the local vegetation, where they encounter a race of insectoid beings they call Selenites, who have created a seamlessly organized civilization in the bowels of the lunar soil. There are no tribes or political factions. They speak a single language. Violence is absent. They are "in intelligence, morality, and social wisdom ... colossally greater than men." What distinguishes Selenites is their physiological specialization and diversity: each individual is molded for a single purpose, "a perfect unit in a world machine." Farmers are equipped with scoop-like limbs and sinewy muscles. Scholars and administrators have massive brains. Workers who traverse dark tunnels ferrying food from the surface to the inner cities are naturally luminous. Cavor marvels at the perfect synchronization of social purpose with physical form. "With knowledge the Selenites grew and changed," he observes, while humans "stored their knowledge about them and remained brutes—equipped."[39]

Though the Selenites are peaceful beings, drama and conflict ensue. Bedford and Cavor separate to look for their spaceship after two

FIGURE 3. Mr. Cavor, the scientist-protagonist in H. G. Wells's 1901 novel *The First Men in the Moon*, wins an audience with the Grand Lunar, the embodied accumulation of his civilization's knowledge. Having never committed violence in the course of their history, the moon's denizens, the Selenites, are horrified by the earthling's descriptions of war.

contentious encounters with the ant-like beings, some of whom they kill. Bedford stumbles upon the ship and returns to Earth, miraculously crash-landing just off the coast of his native Britain. Cavor, meanwhile, injures himself and falls into captivity, where he beams a series of fragmented radio communications to Earth detailing his decent into Selenite society. Back home, Bedford is able with the help of a Dutch electrician to cobble together the essence of his partner's messages, which form the climax of the novel.

The First Men in the Moon culminates when Cavor recounts his audience with the Grand Lunar, leader of the Selenites (figure 3). His supreme knowledge is evidenced by his exaggerated head size, which stretches many times the length of his own body. Servants shower this "swollen globe" with a cooling spray. Through translators who have learned Cavor's tongue, the Grand Lunar proceeds with a battery of questions about Earth. Why did humans build their cities on the surface, instead of underground where it was safer? What made up Earth's atmosphere? Why did Earthlings speak so many languages? And what, pray tell, was "the democratic method?"[40]

What holds the monarch most spellbound is the phenomenon of war, which he and the rest of the Selenites find utterly incomprehensible. The Grand Lunar is "astonished" to learn that Earthlings "were still not united in one brotherhood, but under many different forms of government" that battle ceaselessly for resources and supremacy. "You mean to say," he asks in disbelief, "that you run about over the surface of your world—this world, whose riches you have scarcely begun to scrape—killing one another for beasts to eat?" Cavor sheepishly affirms. "But do not ships and your poor little cities get injured?" the king continues. "Make me pictures," he demands of the scientist. "I cannot conceive of these things." When Cavor further details the "story of earthly War," the audience of enthralled Selenties are visibly shaken.[41]

It is not until after Cavor has spilt the beans about human nature—"the strength and irrational violence of men, of their insatiable aggressions, their tireless futility of conflict"—that he realizes his mistake. In a final, desperate message to Earth, Cavor cries that he had been a fool to explain war to the Grand Lunar and feverishly attempts to relay instructions to manufacture his gravity-defying Cavorite. But it is too late: the communication is abruptly cut off, presumably by the Selenites. Cavor is never heard from again, and the novel closes on ghostly allusions about his fate.[42]

As with *The War of the Worlds* and indeed all Wellsian romances, *The First Men in the Moon* was a bald manifestation of the author's political worldview. By 1900 Wells had already synthesized a good deal of his intellectual influences—Enlightenment liberalism, the late nineteenth-century peace movement, international socialism, and Huxley's theories of "ethical evolution"—into a coherent set of ideas about what the future of international relations should and should not entail.[43] Over subsequent decades these convictions would burst forth in a torrent of writing. In *Anticipations* (1901) and *New Worlds for Old* (1908), Wells proposed that political globalism was inevitable and laid out his proposals for world government. His 1908 novel *The War in the Air* attacked humanity's funneling of technological innovation toward armaments. In 1917 Wells vowed himself "an extreme Pacifist" in *War and the Future*, but believing that the Great War would crush "the German will-to-power" and thereby create the necessary conditions for world peace, he supported Britain's entry.⁴⁴ That a total conflagration among great powers would pave the way for nonviolence and world federalism was a theme he had already explored at length in *The War That Will End War* (1914), *The World Set Free* (1914), and *In the Fourth Year* (1918), a key text that laid the case for a League of Nations. In other writings he would rail against imperialism and nationalism, support state control of weapons manufacturing, advocate the spread of English as a language of political unification, and lay out plans to create an educated, cosmopolitan citizenry that would embrace one-worldism.⁴⁵

The First Men in the Moon was an early, if fictional, expression of these ideas, one that helped solidify space adventure as a political genre. All of Wells's futurist ambitions are personified in the Selenites, his fears in the failings of human beings. His insectile antagonists have applied their ample technology purely to organization and industry. Bedford, in contrast, pursues the techno-magic of space travel only to access the moon's bountiful resources. Though they differ widely based on their social function, the Selenites are bound by a common language and by their collective knowledge, stored in the undulating brain of the Grand Lunar. Humans, meanwhile, have fallen from the Tower of Babel; they struggle endlessly against one another, tribe against tribe, nation against nation, tongue against tongue. As the two Earthlings struggle to find the way back to their spaceship, Cavor questions the wisdom of returning home in the first place, for if his secret to antigravity were to escape, "Governments and powers will struggle to get hither, they will fight against one another, and against these moon people; it will only spread warfare and multiply the occasions of war." "In a little while," he muses, "if I tell my secret, this planet to its deepest galleries will be strewn with human dead." In a memorable passage, he resolves to hide Cavorite from the world. "Science," he decides, "has toiled too long forging weapons for fools to use. It is time she held her hand." Other scientists would have to figure it out again on their own "in a thousand years' time."⁴⁶

The moralism of the Wellsian cosmos peaked in 1906 with the publication of *In the Days of the Comet*, a second, more sanguine attempt at a "cosmic happening." The novel follows the story of William Leadford, who suffers at the beginning of the novel from joblessness, class-based alienation, and especially unrequited love. When Nettie Stuart, the object of his affection, jilts him for the son and heir of a wealthy family, the disillusioned young man buys a revolver, planning to murder both Nettie and his rival.

FIGURE 4. The protagonist of *In the Days of the Comet* (1906) inhales the fog created by the disintegration of a comet in the Earth's atmosphere. Instantaneously, he loses all impulses toward violence, lust, obsession, and envy. "The air was changed and the Spirit of Man that had drowsed and slumbered and dreamt dull and evil things, awakened, and stood with wonder-clean eyes, refreshed, looking again on life."

As this drama unfolds, an enormous comet approaches the Earth with greater and greater velocity. Earthlings look on helplessly as the celestial block of ice and dust grows larger in the night sky. Just as Leadford is about to commit his crime of passion, the comet enters the atmosphere and disintegrates into a thick, soporific green fog (figure 4). The nitrogen in the air changes form and becomes a respirable gas that completely upends the psychological condition of all who breathe it in.[47]

Indeed, the fog single-handedly ushers in "The Change," a metamorphosis within every human being—"a vast, substantial exaltation"—in which violence, selfishness, and passion give way instantaneously to reason, altruism, and peace. Leadford drops the revolver. After imbibing the fog, Wells's hero can no longer understand why he had murderous feelings toward Nettie or her lover in the first place.[48] He instantly experiences a "clear-headedness" that divorces him from "the tumid passions and entanglements" of his personal life.

Nor is this comet-induced revolution in human psychology and temperament confined to intimate relationships. Awakened individuals working for their respective governments bring newly enlightened minds and widened perspectives to bear on international life. One of the novel's characters, a cabinet member in the British government who before The Change had been planning a war with Germany, now pledges that he will end war once and for all. "We have chattered and pecked one another and fouled the world—like daws in the temple, like unclean birds in the holy place of God," he soliloquizes. "No more of this!" [49]

Across the globe the evaporation of personal insecurities and hatreds trickle up to institutions, communities, and nations. The "narrowness, the intensity, the confusion, muddle, and dusty heat of the old world" was "over and done":

> It was as if at the very moment of the awakening those barriers and defences had vanished, as if the green vapors had washed through their minds and dissolved and swept away a hundred once rigid boundaries and obstacles. They had admitted and assimilated at once all that was good in the ill-dressed propagandas that had clamored so vehemently and vainly at the doors of their minds in the former days. It was exactly like the awakening from an absurd and limiting dream.[50]

At the end of the novel, diplomats from around the globe convene to frame a constitution that will govern a World State. Leadford explains that after inhaling the fog, national leaders could no longer truly comprehend, or bring themselves to care about, political jurisdiction or authority. The "web of conflicts, jealousies, heated patchings up and jobbings apart, of the old order," he reports, "they flung it all on one side." Inaugurating the global confederation, the British cabinet member leads a universal refrain: "Let us begin afresh!"[51]

In the Days of the Comet was not a reversal of Well's Martian invasion but rather a softening of it. Compared with the rapacious aliens who force human beings into cooperation through violent conflict, the comet's tail was a benign yet equally purgative force. As Wells acknowledged years later, without bloodshed the passing bolide "does the work of centuries of moral education in the twinkling of an eye, and makes mankind sane, understanding and infinitely tolerant."[52] In a letter to his editor, Wells wrote that he had intended the novel "to be a beautiful dream" in which humanity might find its exaltation, where it might exist "above the law."[53] Of all the narrative devices that he would deploy in his early fiction to achieve this dream—time travel, evolution, and technological saltation—contact with cosmic power was the most reliable. Any why not? Looking back, he considered it the mission of his body of work to give "concrete working expression to a world-wide 'Open Conspiracy' to rescue human society" from harmful tradition and "reconstruct it upon planetary lines."[54]

H. G. Wells was the most consequential "interplanetary" thinker of the twentieth century. He was influential not only because he was among the first to articulate the coming politics of space exploration, nor because he stated his ideas so plainly in his novels. Rather, Wells's significance owed to his popularity. His scientific romances reached millions of people across Europe and North America. By 1903 *The War of the Worlds* had already been translated into a half dozen languages.[55] Russian publishers both serialized the novel in a Saint Petersburg journal and issued a book version the same year of the English-language release. The German literary community, believing Wells a serious philosopher, invited him to deliver a speech on "The Common Sense of World Peace" at the Reichstag in 1929. Even in France, home to Jules Verne, Wells was widely considered the most important voice in science fiction. Despite the relatively tepid reception to *In the Days of the Comet*—attributable, as it turned out, to the narrative's embrace of free love—Wells scrambled to arrange for the simultaneous translation of the novel into French, German, Dutch, and Italian.[56] It was serialized to popular appeal in the United States, where, after a half dozen visits over his lifetime, he was regarded a modern prophet.

George Orwell later wrote of Wells: "I doubt whether anyone who was writing books between 1900 and 1920, at any rate in the English language, influenced the young so much. The minds of all of us, and therefore the physical world, would be perceptibly different if Wells had never existed."[57] This was a keener and more prescient observation than Orwell realized, for among these youths were the energetic minds that would eventually propel the spaceflight revolution in the 1920s and 1930s. The Russians Konstantin Tsiolkovsky and Sergei Korolev (who was Ukrainian); the Germans Herman Oberth, Wernher von Braun, and Walter Dornberger; and the Americans Robert Goddard and John "Jack" Parsons had each adorned their bedside tables with Well's romances.

More immediately, Wells's stories had a profound impact on other science-fiction writers in Europe and the United States who reached out to space in the formative days of the genre. After the success *Cosmopolitan* magazine enjoyed in serializing *The War of the Worlds* in 1897, the *Boston Post* and New York's *Evening Journal* teamed up, without Wells's permission, for a sequel to his blockbuster entitled *Edison's Conquest of Mars*.⁵⁸ Its author, the noted astronomer Garrett Serviss, conjured a reversal of the original novel in which Earthlings mount an invasion of Mars, one that nevertheless fulfils the promise to achieve the "commonweal of mankind." The story's main protagonist, as the title suggests, is Thomas Edison, who engineers a "disintegrator" capable of reducing any object to atoms, as well as an antigravity device powered by electric repulsion. These technologies promise to be Earth's salvation from another invasion, but the cost of building them surpasses the treasury of any single government. To meet the challenge, every nation convenes in Washington, D.C., to pledge its contributions to "a gigantic war fund" that will finance spaceships and disintegrators for an assault on the Red Planet. Queen Victoria, Emperor Wilhelm, Czar Nicholas, and Emperor Mutsuhito (Meiji) of Japan, among other royalty, each vouch their allegiance to the common cause.

On their way to this world congress, Edison and the story's narrator, a humble professor read to be Serviss, run across the massive armadas transporting the various delegations to Washington. Far from connoting a military arms race or an impending war, the glistening navies are a symbol of unity and peace. "Side by side, or following one another's lead, these war fleets were on a peaceful voyage that belied their threatening appearance," the professor remarks. "There had been no thought of danger to or from the forts and ports of rival nations which they had passed. There was no enmity, and no fear between them when the throats of their ponderous guns yawned at one another across the waves." The disparate warriors were now, "in spirit, all one fleet, having one object, bearing against one enemy, ready to defend but one country... the entire earth."⁵⁹

Equally reflective of the Wellsian strain in space adventure was *Auf zwei Planeten* (Two Planets), the most prominent work of philosopher Kurd Lasswitz, widely considered to be the progenitor of science fiction in Germany (figure 5).⁶⁰ In October 1897, only months after the first serializations of *The War of the Worlds* appeared in Britain, Emil Felber Verlag published *Two Planets* to warm receptions in Berlin and Weimar. Like Wells's classic "first contact" story, Lasswitz's account relates a historic meeting between human beings and Martians. The novel begins in the Arctic Circle where three German explorers on a balloon expedition are marooned on an island close to the North Pole. The island serves as a forwarding base for another expedition, a Martian resource survey of Earth. The Martians, beneficent by nature, rescue two of the explorers (the third, Hugo Torm, goes missing) and encourage them to return to the Red Planet as guests.⁶¹

FIGURE 5. Kurd Lasswitz, in 1903, six years after the publication of *Auf Zwei Planenten*. Wikimedia Commons.

Frictions emerge in the search for Torm when belligerence and miscommunication result in a confrontation between the Martians and a British warship. Despite their superior weapons, the Martians refuse to act violently, and some are taken hostage. The alien visitors issue an ultimatum: return our comrades or become subject to a stifling quarantine. The English, proud and confident after the Martians had yielded, refuse, and prepare an offensive against the intruders. Yet when the British attempt to break the blockade near Portsmouth, the aliens' superior technology destroys nearly the entire Royal Navy, ending British power for good. The world's largest empire collapses overnight, and a set of new wars break out over the globe as warring factions compete for the former colonies.

From this display of obstinacy and violence, the Martians conclude, in very colonial fashion, that Earthlings are too undeveloped—morally, philosophically, and psychologically—to govern themselves. "The Martians," writes literary historian Ingo Cornils, "must reluctantly take on 'the white man's burden.'"[62] The planet's diplomatic representative declares the Earth a Martian protectorate. With the backing of massive military force, the Martians outlaw war and enforce global disarmament. Still defiant, Lasswitz's native Germany suffers the same fate as Britain; when the emperor resists the order to demilitarize, Martian airships appear over Berlin, where their enormous magnets pull every conceivable weapon into the sky. Initially the Martians pacify only the warring states of Europe, but eventually their rule extends over the globe as the race becomes ever more convinced of its mission to civilize the Earthlings. A Martian newspaper concludes that human beings are "wild animals, and we have to tame them."[63]

Martian rule, though bestowing certain benefits—hunger is eradicated, for example—becomes increasingly despotic. A human resistance forms (of course) in the United States, where a group of scientists have made certain improvements to Martian technology. The rebels strike the Arctic base as well as the Martian space station hovering above it. The Martians, despite the knowledge that they can easily quell this uprising, consent to negotiate an armistice. They agree to restrict relations between the planets to light-beam communications until a long-term policy could be hashed out at home. When technical difficulties threaten to derail the negotiations, representatives of Earth and Mars stand out on the planks of spaceships in the stratosphere and shout the final terms of peace, face to face, to the other.[64]

Reaction to *Auf zwei Planeten* was mixed. When first published in 1897, Wilhelmine Germans criticized the Martians that Lasswitz wrote so highly of as *Typen der internationalen Friedensapostel* (types of international peace advocates). After World War I, however, the novel enjoyed high circulation in Germany and the book was translated into several European languages. When a new edition appeared in 1930, Verlag sold at least 70,000 copies, considerable compared with other popular works of the day. As a young man Wernher von Braun "devoured" the novel.[65] When the National Socialists, who considered the story "too democratic," came to power, the fate of the book was sealed until after the Second World War. By 1934 it was out of print.[66]

Much like Wells's Mars story, *Two Planets* is above all an allegory for and critique of European imperialism and "Nordic" supremacy. The Martians turn the table on the colonial powers, subjecting them to rule that is suppressive but purportedly civilizing. But the novel also imparted explicit political ideas about the cosmos. Note that in *Two Planets* the Martians are a beneficent, magnanimous, and peaceful race until they encounter Earth and its environs. On the Red Planet, the Martians adhere to regulations and customs that must have appeared enviably utopian to a late nineteenth-century German audience. They are incapable by nature and culture from compelling any other sentient being—initially, even the lowly Earthlings—from engaging in any undesired action. By law they are required to inform themselves of current events by reading a variety of newspapers with opposing views. A narrative tour of Mars reveals that the aliens have engineered a telescope-like apparatus called the "Retrospective," which allows them to look back on the past and issue judgment based on the most accurate information.

When the enlightened race is exposed to the Earth's heavier gravity, its thicker atmosphere, and the hostility of many of its inhabitants, however, its innate magnanimity falls away. Slowly, the aliens become avaricious and come to disregard the rights and freedoms they had so cherished back home. As one of the principal Martian characters, a captain in the spacefaring fleet, explains amid Britain's violent resistance: "The humans are insane.... I hear that the human beings have named our planet after the god of War. We wanted to bring peace, but it appears that our contact with this wild species in throwing us back into barbarism."[67]

Mars is home to a mature, politically enlightened race; Earth, to a pugnacious, tribal one. The contrast was intentional. By conjuring a civilization more morally and ethically advanced than those on Earth, Lasswitz sought to show the steps humanity could take to reach higher levels of political consciousness, to a higher state of "maturity."[68] He also aimed to show that by exposing itself to other worlds and the infinity of space, human societies might develop traditions, customs, and bodies of thought capable of creating a united people like the one on Mars. Perhaps merely escaping the corrupting influence of Earth's atmosphere and its gravitational pull might open men's minds to new, more virtuous possibilities—it had worked for the Martians.

The emphasis of these early space novels on peace, reconciliation, and cooperation is particularly remarkable when considered alongside the sudden proliferation of future war stories in the United States and Europe in the last quarter of the nineteenth century and in the first quarter of the twentieth. Beginning in the 1870s, in the wake of managerial and mechanical innovations that modern nation-states had displayed in both the US Civil War and the Franco-Prussian War, writers sought to capture reading audiences with prognostications about cataclysmic, even apocalyptic, conflicts they determined to be in wait just over the horizon. Unlike most space-themed science fiction that emerged over the period, these forecasts were unequivocally dark and violent. George Tomkyns Chesney's 1871 novel *The Battle of Dorking*, which describes the invasion of Britain by a massively armed Germany, launched a wave of invasion literature that pitched national homelands against ravenous barbarians or advanced, but dispassionately covetous, civilizations. In nearly all cases, authors reflected contemporary political and racial hatreds. Pierton Dooner's *Last Days of the Republic* (1880), King Wallace's *The Next War: A Prediction* (1892), J. H. Palmer's *The Invasion of New York; Or, How Hawaii was Annexed* (1897), M. P. Shiel's *The Yellow Danger* (1898), and Jack London's "The Unparalleled Invasion" (1910) were only the most widely read examples of an avalanche of Yellow and Black Peril literature that Americans and Europeans consumed wholesale in the years before World War I. In all, publishers released more than four hundred such stories between 1871 and 1900.[69]

As *War of the Worlds, Edison's Conquest of Mars,* and *Auf Zwei Planeten* attest, military themes and anxieties about civilizational conflict suffused early space novels, too. The presence of futuristic weapons, cruel but superior aliens, and widespread violence represented an acknowledgment of both contemporary fears and the popularity of war adventures at the end of the nineteenth century.

But these and other fictional accounts of interplanetary contact said something quite different about war. They suggested, first, that war would be a unifying experience for human beings, that conflict with distant aliens would serve to make intraspecies conflicts obsolete. To garner the necessary arms to defeat an advanced Martian enemy, human societies would finally act as one. As Lasswitz understood it, this war-induced cooperation would produce a postwar order based on universal sacrifice in battle. Wartime alliances would persist. For other interplanetary writers, moreover, outer space was the site, or the very cause, of the transcendence of war. Taking the view of aliens not only permitted an investigation of what made human beings so pugnacious in the first place but also offered a means to imagine psychological substitutes in which war was inconceivable or simply unnecessary. While possessing superior weaponry, alien visitors often had societies that outlawed war altogether. Interplanetary writers used alien civilizations to proffer new ethical values, laws, and institutions that diverged radically from the ones with which they knew well and sought to criticize. Small wonder that cosmic fiction grew only more popular after 1914.

Twins in the Cradle: Rocketry and Politics

Wells, Bogdanov, Lasswitz, and other writers provided the loose, imaginative material from which the succeeding generation of spaceflight "pioneers" drew inspiration and in some cases hard theoretical speculation. Over the quarter-century between the first appearance of *The War of the Worlds* and the coalescing of the spaceflight movement in the mid-1920s, the most important minds of that movement—Tsiolkovsky, Goddard, and Oberth—sought to complete the bridge between plausible technology and fantastical romance that the great science-fiction novelists had begun to construct. Tsiolkovsky offered an equation for the minimal horizontal speed required to orbit the Earth in a 1903 essay. Goddard, without knowledge of this work, presented a detailed analysis of weight-to-thrust ratios and the results of his experiments with various fuels in a seminal 1919 article for the Smithsonian Institution. Three years later, Oberth submitted a study of rocketry to the University of Heidelberg for his doctorate, which the school rejected as impractical and far-fetched. Nevertheless, with money borrowed from his wife, he published his research as a book, *Die Rakete zu den Planetanraumen* (The Rocket into Interplanetary Space), which garnered intense interest among a growing cadre of spaceflight enthusiasts in both Europe and the United States. Largely independent from one another, all three men came to the same set of basic conclusions: that human space travel was possible; that rockets, powered by liquid oxygen and liquid hydrogen, would be the most effective means of escaping Earth's gravity; that reaching escape velocity would require rockets with multiple stages; and that the basic formula for engine performance and vehicle trajectory, among other problems, were already within reach.[70]

These arguments, particularly those contained in Oberth's and Goddard's work, helped propel human spaceflight from the outermost margins of scientific theorization to the frenzied discussions of public fascination in the years after World War I. Yet the mathematics and physics of rocketry did not grow up alone. Alongside the flowering of new technical speculations about space travel grew several philosophical, metaphysical, and political claims about the purpose of human activity in the cosmos. Often these claims embodied the cultural assumptions that had undergirded the turn-of-the-century scientific romances: that exploration would inaugurate a period of human evolution capable of pacifying and enlightening the species; that cooperation in space would mitigate and then erase international conflict; and that human beings would leave behind on Earth all the historic "diseases" of society. Just as the rocket pioneers had drawn inspiration from Verne's crude mathematical calculations in *From the Earth to the Moon*, so too did spaceflight enthusiasts from a variety of specializations draw on the obvious moral lessons contained in stories like *In the Days of the Comet* and *Two Planets*.

Notions that outer space was a font of greater wisdom, morality, and egalitarianism found early expression, it should come as no surprise, in Russia. From the late nineteenth century through the Soviet period, ideas about space predominated in a native philosophy known as Cosmism. This body of thought, as literary historian George M. Young has pointed out, can be difficult to define. It is an "oxymoronic blend" of religion, science, esoterism, paganism, and utopianism, "higher magic partnered to higher mathematics." Its ranks included Bolshevik radicals, both Eastern and Western scientists, philosophers, poets, physicists, and novelists. It also incorporated Pan-Slavism and the theologies of the Russian Orthodox Church. Pitching their ideas as an alternative to the destructive intellectual traditions in the West, Cosmists sought to elevate to scientific discourse topics typically reserved for science fiction, esoteric literature, and the occult. At heart their writings centered on the perfection of human beings as a species; omniscience, omnipotence, and transcendence to higher forms of consciousness were common themes.[71]

More than any other goal, however, Cosmists sought to defeat death itself. They pursued not only immortality for the living but also resurrection for the dead. Nikolai Federovich Fedorov, Cosmism's leading light, referred to immortality as the "Common Task" of every person. All problems were at their base a struggle between life and death, Fedorov preached: "All philosophies, while disagreeing about all else, agree on one thing—they all recognize the reality of death, its inevitability, even when recognizing, as some do, nothing real in the world. The most skeptical systems, doubting even doubt itself, bow before the fact of the reality of death."[72]

The solution? Mobilize every nook and cranny of science to the cause. In his writings, collectively titled *Filosofiia obshchego dela* (The Philosophy of the Common Task), Fedorov explained that humans, after they die, disintegrate into infinitesimal particles, which could be collected and reconnected using some future technology. By reassembling those already passed, and by achieving immunity from old age and disease for the living, humanity could finally achieve "brotherhood" among the countless generations that had occupied the planet. This was, in effect, resurrection, a fact that attracted the interest of many Christians, including Fyodor Dostoevsky and Leo Tolstoy, and one that attracted the interest of Russians still reeling from the physical ravages of the Crimean War.[73]

Space exploration was a key ingredient to the Cosmists' ambitions for human immortality. For starters, Fedorov and his disciples believed that the corporeal particles that emanated from disintegrating bodies would disperse and eventually escape Earth's atmosphere. Reassembling human beings from these disparate particles meant collecting them from the moon, the planets, and the furthest reaches of space. Second, once these bodies were again made whole, the resurrected would need room to live. Space again provided the answer. As Fedorov explained, conquering space was "an absolute imperative, imposed on us as a duty in preparation for the Resurrection. We must take possession of new regions of Space because there is not enough space on Earth to allow the co-existence of all the resurrected generations."[74]

Cosmism, and more precisely the Common Task, may seem peculiar if not outright bizarre, even to a twenty-first-century audience increasingly familiar with the basic ambitions of transhumanists, who seek to enhance human intellect and physiology through integration with technology. Yet cosmism enjoyed a wide readership in prerevolutionary Russia, and counted among its ranks scores of prominent scientists, philosophers, intellectuals, and officials, including Alexander Bogdanov. One of Fedorov's followers, perhaps his most consequential, was Konstantin Tsiolkovsky, the eccentric schoolteacher from Kaluga widely credited as a "grandfather" of modern astronautics and rocketry.[75]

Born to a middle-class family in Izheyskoye in 1857, Tsiolkovsky's early life was beset by misfortune. At age ten, while sledding amid an early frost, he contracted a cold that developed into a bout of scarlet fever. Tsiolkovsky lost most of his hearing. Because of his new impairment, he was unanimously turned away from elementary school. Undeterred, his mother Maria Ivanovna spent countless hours teaching him how to read and write at home. Crucially, she also bestowed some basic arithmetic. Maria died in 1870, leaving the boy of thirteen mostly alone with nothing but books to keep him company.[76]

Despite these setbacks, Tsiolkovsky achieved theoretical successes that historians now agree were foundational to human spaceflight. With no formal education, he traveled to Moscow to learn and work at the city's leading public library, where he became a pupil of Fedorov. Although Fedorov and the young Tsiolkovsky never spoke of spaceflight, the latter developed an intense interest in the subject, largely inspired by the stories of Jules Verne. Beginning in the early 1880s, Tsiolkovsky began to write a series of scientific papers that soon blossomed into a full-bodied collection of works on flight. After 1884, his research, including experiments done in a self-made laboratory in his apartment, focused on designs for an all-metal balloon (an airship), streamlined airplanes and trains, hovercraft, and rockets for interplanetary travel. His designs were prescient, especially his multistage rocket fueled by liquid oxygen and hydrogen. He argued, correctly, that liquid fuels could help propel spacecraft at the minimal horizontal speed to achieve a low orbit of the Earth. Other contributions included designs for airlocks, steering thrusters, space stations, coolants, and closed-cycle biological systems to provide food and oxygen to space colonies. All were rough approximations to technologies that the Soviet and US space programs would develop a half-century later. In 1895, inspired by the recently constructed Eiffel Tower in Paris, Tsiolkovsky even sketched the first-ever design for a space elevator.[77]

Though these experimental and theoretical accomplishments have made Tsiolkovsky a household name in rocketry, the "Kaluga eccentric" regarded them as secondary to his metaphysical work, what he collectively called "cosmic philosophy." Cosmist in its attitude and content, Tsiolkovsky's writings connected the colonization of outer space with the betterment of humankind—intellectually, spiritually, and particularly morally. It is unlikely that he ever read Lasswitz's *Two Planets* (a Russian translation was never published), but Tsiolkovsky came to the same basic conclusion: Earth was but the seedbed of human civilization, and the planet had a suppressive and deleterious impact on the behavior and social organization of human beings. For Tsiolkovsky, human evil was the product of a parochial viewpoint that encompassed only Earth, the sun, and perhaps brightly lit stars in the night sky. Holding such narrow worldviews—"moved by a rough egoism of [a] short earthly life"—humans naturally resorted to selfishness and violence. Avoiding war therefore required people to cultivate a "cosmic mindset," to consider themselves microscopic entities in a much larger scheme. Universalism must replace localism; holism must replace provincialism; and "perfection" must replace "the bad." Tsiolkovsky was convinced that intelligent life existed in the universe and that if humans were to meet them as "celestial neighbors," they first must achieve a measure of emotional and intellectual maturity toward which space exploration was but the first step. Through the transformative power of space travel, human beings would develop the emotional and social capacities necessary to create perfect societies on other worlds, far from the narrow worldviews of their species' childhood on Earth. Distant planets would provide an ever-widening home where benevolent civilizations could achieve the total elimination of "suffering."[78]

During the interwar period, Tsiolkovsky helped usher in the world's first true space fad. Popular science writers such as Yakov Isidorovich Perel'man and Nikolai Rynin helped articulate the mathematical formulae of Oberth, Goddard, and Tsiolkovsky for a laymen audience. Young avant-garde painters in the Amaravella collective produced mythical and spiritual images of outer space as a "place" for humankind to fulfill the quest for immortality. Particularly popular was Aleksey Tolstoi's novel *Aelita*—and filmmaker Yakov Protazanov's movie adaption of the same name—in which a lonely engineer travels to Mars aboard a self-made rocket and discovers an advanced civilization. In the wake of a May 1924 news article in *Izvestiia* detailing Oberth's claims about the possibility for spaceflight, a group of engineers and science educators based in Moscow formed what would be the first of dozens of enthusiastic "rocket societies" aimed at raising public support—and money—for spaceflight experiments, the Society for the Study of Interplanetary Communications (*Obshchestvo Izucheniia Mezhplanetnykh Soobshchenii*, or OIMS). Later that year, at a Moscow event advertised as "The Truth about . . . Professor Goddard's Moon Projectile," so many people pushed to gain admittance to the conference hall that the police had to be called in to restore order. Demand for the talk was such that organizers staged it again a few days later.[79]

The Soviet Union's early obsession with space travel, though initially based on the discoveries of Oberth and Goddard, came to include both Tsiolkovsky and his cosmic philosophy as integral components of the spaceflight dream. When in 1927 the Association of Inventors, an enthusiastic group of students and workers, staged the world's first exhibition on rocketry, Tsiolkovsky was the centerpiece. The organizers, who referred to themselves as "cosmopolitans" and "citizens of the universe," declared Tsiolkovsky the prophet of human spaceflight and the Soviet Union the true birthplace of the space age. The exhibition, attended by nearly 12,000 people in the two months that it was open to the public, pitched outer space as the future site of human civilization, one in which it would achieve not only higher states of technology but perhaps also a slice of socialist utopia.[80]

Spreading out from Berlin, Oberth's home base, the popular spaceflight phenomenon proceeded apace in the Weimar Republic. As historian Michael Neufeld has observed, widespread beliefs about the linear trajectory of technology and human progress, intense nationalism, and the growth of a distinct consumer culture all combined to make Germany fertile soil for popular rocketry. In mid-1927, Max Valier, an Austrian science writer, along with Johannes Winkler, a World War I veteran who had gained engineering experience at the Junkers Plane and Motor Works in Dessau, established the *Verein fur Raumschiffahrt*

(Society for Space Travel, or *VfR*) to pool research, raise money for rocket experiments, and publicly promote the dream of spaceflight in their journal *Die Rakete* (The Rocket). Before long they were joined by Oberth as well as another young science writer, Willy Ley, whose popular English-language books would soon bring space travel literature to a broad audience in the United States.[81]

The German space craze peaked in 1928–1929, when Valier teamed up with Fritz von Opel, heir to his family's car-manufacturing fortune, to stage thrilling rocket-car experiments. Two thousand invited guests packed the Avus racetrack in Berlin in May 1928 to witness Opel's RAK2, powered by two dozen black-powder rockets, reach a record speed of 147 miles per hour. Just a year later, another crowd rushed into the Ufa-Palast am Zoo cinema to view the premier of Fritz's Lang's *Frau im Mond*, a space romance based on a novel written by his then-wife Thea von Harbou. The German film company UFA hired both Ley and Oberth as scientific consultants and commissioned them to construct a working liquid-propelled rocket to publicize the film at the premiere (Oberth's design was too ambitious, and the rocket was not completed in time). The effect seemed immediate. Suddenly rockets appeared on advertisements, in parades and fireworks displays, and as toys and models for children. Still reeling from the hyperinflation brought on by the debt payments instituted at Versailles, Germans hungrily consumed the escapist fantasies of interplanetary colonization, new resources, and political utopia that this incipient "astroculture" provided.[82]

Issues of war and peace insinuated themselves in the German space fad. This included Oberth himself, whose monumental contributions to technology occluded his political views. As early as 1931 he expressed hope that the rocket would prove so terrible a weapon that it would "force the world, in self-protection, to outlaw all war."[83] Even as he volunteered his services to the Nazis (who found him too irascible and insufficiently trained), the Transylvanian understood that exploration was "bound up with the future of human culture." Later, after the Soviets orbited *Sputnik*, he felt it necessary to ponder, in his otherwise specialized treatise *Man into Space*, the political conditions that should precipitate space travel. "Research and progress are only possible," he'd conclude, "so long as human society values them and if there is common effort instead of the frittering away of energy in wrangling over language, religion, parties, systems of government, and battles for trade and world markets, with one side fearful of disclosing to the other side the knowledge and experience it has gained." Concerned that the Cold War would influence the direction spaceflight would take, Oberth urged that it was "not necessary that it should be Mars, the god of war, who fathers all the projects that await us." He believed, on the contrary, "that the evolution of space people can be hastened only by men and women whose minds are not solely concerned with arms production."[84]

Even more disillusioned with the military domination of rocketry was Willy Ley, Oberth's young colleague at the *VfR* and arguably the most significant proselytizer of spaceflight in Europe. Though he was for a time a member of the National Socialist Party, Ley was a true believer in the universality of science, and he set about trying to integrate the disparate "interplanetary" societies that had sprouted across the industrial world. In 1928 he stopped contributing articles for the Nazi press and opted instead for *Vorwarts*, the official newspaper of the Social Democratic Party. The following year, he published *Die Starfield Company*, an idealistic celebration of international cooperation set in the 1980s. The novel describes a future in which rockets and planes make global travel seamless and politically integrative. To combat alien "air pirates," the West's largest airline company, Transcontinental, teams up with the Starfield Company, India's largest airline. An interracial love story emerges between the German- and Indian-born chief executives of the two companies, a friendship that trickles up to nations.[85]

In time, Ley "wanted nothing to do" with military rockets or the Nazis, who considered him a xenophile.[86] In the first half of 1932 he spent hours in Berlin libraires digging up materials on ballistics, aerodynamics, and weapons in a vain attempt to debunk the military utility of rockets, what appeared as *Grundriß einer Geschichte der Rakete* (Outline of the History of the Rocket) later that year. In March, after reading Carl Spohr's story "The Final War" in publicist Hugo Gernsback's popular *Wonder Stories*, he wrote to the magazine with a lesson from his own novel: instant travel, and with it global consciousness, would purge humanity of its violent and selfish instincts. "If men of one nation learn and see enough of other nations they will lose the idea of war against a nation in whom they have friends," he gushed, concluding: "there is a hope."[87] But within a matter of months Hitler became chancellor, the Reichstag went up in flames, and the newly empowered state began burning books, including Ley's favorite, Lasswitz's *Two Planets*. Using company stationary to write a letter authorizing a vacation to London, in 1935 he escaped first to the United Kingdom, then the United States.

By then space fever had already spread to America. On April 4, 1930, twelve enthusiasts crammed into the midtown Manhattan apartment of G. Edward Pendray, a newspaper editor, science-fiction author, and rocketry promoter, where they founded the American Interplanetary Society (AIS). The group's name was telling: the vast majority of the AIS's founding members were authors of scientific romance; only three had been trained in science or technology. Nearly all were contributing writers for Gernsback's pulp magazine, initially called *Science Wonder Stories*, launched the previous year. For most of the 1930s, the group's rolls and its bank account remained small, but its participants were passionate. They met for bimonthly meetings at the Museum of Natural History, exchanged copious notes, and published their papers in the *Bulletin*, a mimeographed publication that kept members abreast of rocketry developments in Western Europe, Russia, and in their own backyard.[88]

The Society's first president was the socialist science-fiction author and publisher David Lasser. Born in 1902 to Jewish immigrants from Russia, Lasser came of age amid the political turmoil of the Progressive movement. Lying about his age, he left high school at age sixteen to enlist in the Army during World War I. He was gassed on the front lines in France and honorably discharged in 1919. Remarkably, after the war MIT admitted him for a bachelor's in engineering administration despite his never having finished high school. Though he never used this training in his professional life, it was his degree that attracted Gernsback, who hired Lasser to help authors authenticate their submissions to *Science Wonder Stories* with sound engineering principles. Throughout his time at the magazine, Lasser took up various social causes, above all fighting for the rights of workers and the unemployed (figure 6). In 1933 the Socialist Party named him head of the Unemployed Leagues in New York, a conglomeration of socialist and communist advocacy groups that sought to relieve workers employed by the Works Progress Administration (WPA).[89]

As with many of the European revolutionaries of spaceflight, Lasser's enthusiasm mixed with his politics. He considered the advent of rocketry a hinge upon which the entire fate of humanity rested. "Its power of good and evil are so equal and so opposite," he observed in the AIS's *Bulletin*, that the ends to which states used rockets would serve "as a test of our right to inherit the Earth."[90] In 1931 he delivered his first AIS presidential address, "The Rocket and the Next War," in which he, like Oberth, predicted that future conflicts would include the pulverization

FIGURE 6. David Lasser goes door to door in 1937 for the Worker Alliance of America. Like H. G. Wells, his socialist politics infused his dreams for spaceflight. Library of Congress.

of city populations and industrial centers by rockets. Also like Oberth, he entertained the "rather naïve belief" that when "wars become as terrible as the rocket can make it . . . war will cease, because then, I think, the organized opposition of the earth's people will prohibit them."[91] The rocket presented a moral choice: if used merely for scientific discovery it offered "a means to a newer, higher and better civilization"; but if "ever used to the extent that I have pictured it," Lasser argued, "I would rather that the knowledge of it be erased from our minds, and that we lose also its wonderful possibilities in peace."[92]

Lasser continued this line of thinking in his best-known work, *The Conquest of Space*, released that same year. Providing a legible distillation of earlier work by Goddard and Oberth, the book was at once technically sophisticated and readable, inspiring and serious. In an interlude from his technical explications, Lasser speculated about the impact space exploration would have on international relations. Solutions to the mechanical challenges of rocketry, he suggested, were "too large to be localized in any group or nation." He could foresee the building of the first spaceship "only as a joint effort of [a] united earth." At a time when discriminatory policies racked American civil society, he predicted that space travel would break down "racial jealousies" and unite nations in a "communion of joy."[93] In the appendix, he included his annual report to the AIS from 1931, in which he proposed founding an International Interplanetary Commission as a forum for scientists around the globe to exchange data on rocket development and cooperate on experiments.[94]

Interplanetary Thought in Context

Just as the novels of Wells, Lasswitz, and others offered relief from the dreary outlook of future war literature, the coalescing of interplanetary thought in Europe and the United States was a welcome reprieve from an array of theories regarding the global order that proliferated on both sides of the Atlantic from the end of the nineteenth century through the 1930s and 1940s. These theories were far more pessimistic about the unfolding of international relationships and the possibility for conflict. Geography, race, and nationalism, argued many intellectuals, were immutable forces closing in on modern international life, continuously prodding human beings toward civilizational warfare and internal collapse. In 1904 British geographer Halford Mackinder suggested that competition over the territory of Eastern Europe, "the Heartland" of the world, would define global conflict. Like the historian Frederick Jackson Turner, who published his famous frontier thesis just a decade before, Mackinder conceived of the world as a now-closed political system. "Every explosion of social forces, instead of being dissipated in a surrounding circuit of unknown space and barbaric chaos," he wrote, "will be sharply re-echoed from the far-side of the globe, and weak elements in the political and economic organism of the world will be shattered."[95]

The totality and barbarity of World War I only magnified intellectual skepticism about the prospects for peace. Political scientists such as G. Lowes Dickinson and James Bryce, though they fought for the League of Nations, wrote deterministically about the rationality—and hence inescapability—of political competition and war. From the fifteenth century onward, Dickinson observed in *The European Anarchy* (1916), the calcification of sovereignty as the reigning policy among European powers meant that international policy was dictated by "Machiavellianism."

The imperial pursuit of territory, wealth, and prestige would drive nations into war so long as a legal vacuum existed on the international level. Without common law and a supranational entity to enforce it, "moral sentiments ... will be defeated by lack of confidence and security." Peace would be nothing but war-in-wait. Bryce similarly remarked that international politics existed in "a State of Nature." Quoting Plato, he raised the possibility that war was "the natural relation of every community to every other."[96] Verdun and the Somme were extrapolations, to a global stage, of humanity's innate proclivities toward selfishness, paranoia, and violence.

Interwar thinkers were equally fatalistic about race relations. As Robert Vitalis has shown, the "race problem" was at the very heart of the birth of international relations as an academic discipline. The most popular IR textbook of the decade, Raymond Leslie Buell's *International Relations*, observed that the most fundamental problem of the contemporary world was the "restless energy of Caucasian people" in their "search for new markets" and "demand for cheap labor." Basically, conflict could be broken down into two dilemmas: "1) the extension of the white man into the colored world; and 2) the entrance of colored people into the white man's world." The zoologist Madison Grant became a political scientist almost overnight with the publication, in 1916, of *The Passing of the Great Race*, which purported to detail the history and migration of distinct racial groups across Europe. Drawing on then-prevalent theories of genetics and Darwinian evolution, as well as eugenicist writing, Grant set out to convince the general reader that the "maudlin sentimentalism" behind open immigration policies was "sweeping the nation toward racial abyss." If the American "Melting Pot is allowed to boil without control," he foreboded, "the type of native American of Colonial descent will become ... extinct." Grant's protégée, the political scientist and historian T. Lothrop Stoddard (also a member of the Ku Klux Klan), followed suit with his 1921 study *The Rising Tide of Color*, which narrated the disintegration of white supremacy and empire from population growth among non-whites, rising nationalism in the Global South, and industrialization in China and Japan. Above all, *The Rising Tide of Color*, *The New World of Islam* (1921), *Revolt against Civilization* (1922), and Stoddard's other early works were calls to strengthen what he termed "bi-racialism," essentially the strengthening of Jim Crow segregation, particularly anti-miscegenation laws. Keeping the races separate, he argued, might "exorcise the dread spectre of race war."[97]

Prognostications about the future of global race relations fared no better on the other side of the black/white divide. For W. E. B. du Bois,

World War I had epitomized white imperialism, "the real soul of white culture." It foreshadowed, he thought, larger wars to come between the oppressors and the oppressed. "The Dark World is going to submit to its present treatment just as long as they must and not one moment longer." The "War of the Color Line," du Bois predicted, "will outdo in savage inhumanity any war this world has seen."[98]

More than anything else, these gloomy outlooks—and the roots of IR as a scientific project—were responses to World War I. Scholars wanted to know how such a calamity had come about and how the survivors could avert the next one. The same can be said about the maturation of "interplanetary thinking" in Europe and the United States between 1919 and 1939. Recognizing that technology had been crucial to both the lethality of the Great War and the prospect for the political, economic, and social integration of the world, they set about casting the rocket as a history-defining object, an instrument at once of civilizational destruction and salvation—a technological embodiment of Wells's Martians.

But interplanetary thinkers also pitched space exploration in opposition to incipient IR scholarship. For interwar political scientists, territory and resources were finite, racial hatred innate and immutable, and war and anarchy natural processes akin to evolution. All were logical outgrowths of human behavior and history. The Interplanetary School rejected these hypotheses. Squabbles over territory and resources would be unnecessary, for both were infinite in space. Diverse populations would transcend racism through cooperative efforts to probe the cosmos; far from a race war, the human family would come together in space. Political experimentation outside of Earth's atmosphere would ensure that progress, not anarchy, was natural law.

E. H. Carr, the political scientist and historian to whom most IR scholars attribute the genesis of political realism, would have regarded this collection of ideas as quintessentially "utopian." Interplanetary thought mapped neatly onto the definition he laid out in his famous book *The Twenty Years' Crisis*. Work like Tsiolkovsky's or Lasser's were characterized by "wishing ... over thinking, generalization over observation." Based almost entirely on aspiration, their theories were "purposive," putting ends before means and ambition over analysis, priorities that limited the possibility of turning ideas into action.[99]

This appraisal, while sharp, is fair. Interplanetary thinkers conjured an appealing vision after World War I (critics would call it a "fantasy") but did not spend much time interpreting the facts. They extracted rocketry from its political contexts, or invented new political contexts in which exploration was supposed to unfold. Futurists and romantics, they failed to perceive the full weight of the technology's value to modern nation-states competing in a system that had caused the Great War in the first place. Later paragons of realism would make the contrast even sharper. Whereas Hans Morgenthau wrote that "lust for power" was an indestructible feature of human nature and the root cause of conflict, interplanetarians held that traveling and living in space could *alter* human nature by reaching a new and higher consciousness.[100] Whereas Carl Schmitt hailed the Westphalian system of sovereign states and wars conducted—and contained by—those states, the Interplanetary School cheered for a postnational synthesis.[101] And while Raymond Aron eschewed predictions and thought that political theory should never go beyond the teachings of history, the interplanetarians hunkered down in the future and relished prophecy.[102]

What made their dream endure through yet another world war was, as Carr conceded of all utopian writing, a "simplicity and perfection" that lent it "an easy and universal appeal." The global crave for peace would prove as intense in 1945 as it had been in 1919, a fact that helped sustain the legitimacy of the interplanetary idea through the beginning of the space age. Theories about the pacifying and globalizing effect that exploration would have on international relations provided all the things that Carr admitted realism lacked: "a finite goal, an emotional appeal, a right of moral judgement, and a ground for action."[103] One can fairly judge that instead of ignoring the facts completely, interplanetary thinkers, appreciating the facts with clarity, presented their vision as a plausible alternative to realism. As Carr himself noted, utopianism and realism were necessary in balance; the Interplanetary School supplied that balance at a crucial moment when the rocket, no longer a fairytale, emerged as an all-too-real instrument of violence.

CHAPTER 2

Interplanetary Men

Southeast London was aflutter on Saturday, November 25, 1944. Shoppers queued at the Woolworth's near New Cross Station to purchase tin saucepans that had just arrived. Down the street, workers at Deptford Town Hall collected their wages and panned out to local vendors. Newsboys sold headlines about the Allied march toward Berlin. It was half-past noon.

Suddenly, a bolt from the blue. Without warning a massive "ghost bomb" smashed through the roof at Woolworths and exploded.[1] The store and surrounding buildings heaved and collapsed, instantly killing dozens and scattering an ankle-deep pile of debris that stretched to the train station, almost half a mile away. Nearby civilians rushed to the dust-covered scene to search for survivors. They found only one.

Having played victim to similarly merciless attacks since September, Londoners knew it had been a V-2 rocket, Hitler's *Vergeltungswaffe* (vengeance weapon). Unlike its sluggish predecessor, the V-1, the four-story contrivance that struck Woolworth's was invincible. Traveling at nearly five times the speed of sound, the rocket had left its mobile launch pad in the Netherlands only minutes before and flew noiselessly. Antiaircraft guns, fighter planes, and radar were helpless. More than five hundred of these rockets would pound the British capital before war's end. The yawning moat that had been the English Channel had become a rivulet for which future conquerors would need neither bridge nor boat nor battalion. Germany had birthed the first ballistic missile, and the world would never be the same.[2]

Few appreciated the tragedy of this new reality better than the weapon's artificer, Wernher von Braun. Conscripted into the German military apparatus from the penniless-but-passionate *VfR* (Society for Space Travel) in 1933, von Braun had been from his childhood a "technological utopian" who dreamt of taking humanity (himself included) into space with rockets.[3] Aware that the needs of the state were beginning to contort his dream of a space rocket, and long skeptical of its military utility, he had engineered the V-2 with some ambivalence.[4] In a later interview for the *New Yorker*, von Braun recalled that he and his comrades had "felt a genuine regret that our missile, born of idealism . . . had joined the business of killing."[5] In his novel *Mars Project* (1952), von Braun personified his lingering doubt in a character named General Brader, commander of the US Space Forces, who in a speech to a world Congress reveals that his engineers had been "animated by secret visions of reaching into the heavens" and haunted by the military's cooptation of their technology.[6] Von Braun, in fact, witnessed the weapon's destruction firsthand as the US Army transported him across the Atlantic via London and Paris in September 1945.[7] "We had designed it to blaze the trail to other planets," he recalled, ruefully, "not to destroy our own."[8]

Von Braun embodied the ethical dilemma of the rocket. Its power, both to reach space and deliver mass destruction, had been proven simultaneously. When the German team at Peenemünde successfully launched its A-4 prototype to an altitude of fifty-six miles on October 3, 1942, von Braun's boss, program director Walter Dornberger, could scarcely contain his excitement. "Today," he exclaimed, "the spaceship is born!"[9] Yet it was with troubled minds and heavy hearts that many observers embraced the dawn of the rocket, for it was not "spaceships" that rained down on Britain, France, and Belgium. Had humanity, arriving at a crucial fork on the road of science, chosen the wrong path? Could space technology be divorced from war making? What influence would the rocket have on politics, society, war, diplomacy, and governance? What, indeed, were its implications for human civilization writ large?

A new generation of interplanetarians took up these questions as states began to hatch the V-2's terrible offspring. Ballistic missiles brought to center stage the intellectual heirs of H. G. Wells, Konstantin Tsiolkovsky, and David Lasser. The trauma of the Second World

War—in particular, the fact that a fascistic state had perverted the space rocket into a device of terror and violence—ushered "cosmic philosophy" into maturity. But contrary to expectations the V-2, rather than popping the interplanetary balloon, continued to inflate it. Earlier ideas about the relationship between space exploration and political transcendence found greater purchase after the mechanics of rocket propulsion, for both good and evil, had been demonstrated. It was at this watershed that James Mangan's notion of a "bulwark of international peace" in space suddenly became relevant: if the hopeful predictions of the Interplanetary School were to materialize, human beings and their governments needed to make the right choices—about space technology, about exploration, and about their political contexts—at the outset. Arthur Clarke captured the mood in *The Exploration of Space*, first issued in 1951:

> We stand now at the turning point between two eras. Behind us is a past to which we can never return, [even] if we wish. Dividing us now from all the ages that have ever been is that moment when the heat of many suns burst from the night sky above the New Mexico desert—the same desert over which, a few years later, was to echo the thunder of the first rockets climbing toward space. The power that was released on that day can take us to the stars, or it can send us to join the great reptiles and Nature's other unsuccessful experiments.[10]

The defeat of fascism, the founding of the United Nations, and the arrival of the Universal Declaration of Human Rights in quick succession gave postwar interplanetarians reason to be optimistic.[11] There was just one problem, very new but one to which George Orwell had already bestowed a name: he called it the Cold War.

Of Rockets and Ethics

Considering the growth that occurred in state-led rocket and missile projects during the 1940s and 1950s, it is easy to forget just how implausible spaceflight seemed before World War II. The gusto with which the trans-Atlantic rocket societies expounded their vision of the future and the energy they brought to proselytizing spaceflight obscured their limited influence before the revelations of Peenemünde. None of the interwar societies had membership greater than seven hundred. In the fall of 1929 the *VfR*, lacking funds, suspended publication of *Die Rakete*.

A few years later, to project a more serious, practical orientation, AIS leaders changed the name of their organization to the American *Rocket* Society (ARS). For many, the technologies were too farfetched, the distances too great, and the environment of space too toxic for human travel, let alone habitation. Lasser, Ley, and other popular science writers chaffed at the label of "crackpot" or "daydreamer" that occasionally cropped up in the press. Goddard, for one, kept the AIS at arm's length to distance himself from the perceived eccentricity of organized spaceflight enthusiasts. "People must . . . realize that real progress is a succession of logical steps and not a leap in the dark," he complained in a letter to the Smithsonian Institution.[12]

Just as the spaceflight movement seemed to be peaking, depression, nationalism, and political crisis forced public experimentation and dialogue mostly underground. In the new Soviet Union, where the space fad had flourished under the New Economic Policy of the 1920s, popular rocketry found itself buried amid the throes of the Cultural Revolution, the Great Terror, collectivization, and the First Five-Year Plan, which expunged cosmists from view and instead embraced the utilitarian, technology-centered rocket promoters. In Germany the public experiments so widely attended in 1928 gave way to secret government research for the army. Early in 1932 high-ranking officers visited the *VfR*'s testing grounds, the *Raketenflugplatz*, to view a (failed) test of the group's Repulsor rocket and subsequently offered the society a contract for a launch demonstration. Von Braun, working on a doctoral dissertation in physics at the University of Berlin, had much of his project classified after Hitler came to power. By the mid-1930s, the interwar rocket societies found themselves cut off from one another by national loyalties, travel restrictions, economic hardship, and, in particular, government censorship in rocket technology.[13]

The war changed everything. So long a subject of rumination, speculation, and experimentation, the rocket had in a matter of months become an inescapable, tangible fact. Glittering images of spaceships and moon rockets that had recently graced the covers of pulp magazines now seemed less like cheap advertisements than prescient blueprints of machines bursting off the page into physical reality. But the rocket's arrival was a rude awakening. The state's absorption of rocket engineers legitimated their theoretical work but, at the same time, contorted it. Little time passed, indeed, between the first successful test of the V-2 to full-scale production. Once Dornberger's team (and its slave labor) proved the missile's efficacy, they began rolling out like sausages:

from September 1944 to February 1945, prisoners at the Mittelwerk tunnel complex outside Nordhausen assembled between six and seven hundred V-2s per month, more than twenty a day.[14] Over the same period, the German army launched more than 1,000 rockets against targets in the United Kingdom, France, Belgium, the Netherlands, and on advancing Allied forces in the fatherland. They killed 2,754 people and injured 6,523 more.[15] Violence and validity went hand in hand: once people witnessed the terrible power of the V-2, the once romantic and quixotic speculations of the interwar societies appeared not only feasible but imminent. The Nazis' *Wunderwaffe* ushered rocketry from the margins of scientific inquiry to the very center of international politics and military policy.[16]

Imagine, then, the awe—nay, the stupefaction—of the average citizen as she read the daily headlines in August 1945: not only had the Nazis built a weapon capable of reaching across continents and, presumably someday, oceans, but the Americans had built a bomb that harnessed the most elemental power of the known universe. Observers immediately drew connections between the destruction wrought on Hiroshima and Nagasaki with the terror released a year earlier by Hitler. What if would-be aggressors could combine the awesome power of atomic weapons with the speed of the ballistic missile? Americans, though they could enjoy temporary comfort considering their monopoly on the bomb, shuddered to think of their two oceanic buffers evaporating. Only months after the end of the war Manhattan Project veteran Niels Bohr proclaimed that defense against bomb-tipped missiles was "impossible." A *fait accompli* seemed to have emerged when in the summer of 1946 the US Army initiated tests, with reassembled V-2s taken from Germany, of missile-tracking radar over the New Mexico desert—the United States was rushing to perfect the combination of these weapons and craft a workable defense against them. By 1952 the Army would begin tests of the MGM-5 Corporal, the first tactical missile authorized to carry a nuclear warhead—it was also the first man-made object to reach outer space. "The lay mind has been too stunned by release of nuclear energy in the atomic bomb to question or doubt what recently would have been visionary forays into the ionosphere," observed one journalist. "Hope and fear mingle in a prayer that a new world conscience and morality can overtake the leaping stride of scientific marvels."[17]

Indeed, for many the coming missile age reflected a moral and ethical crisis. Idealistic notions about cooperative interplanetary travel and the unification of the species through colonization of other planets, though critics marked them romantic and implausible before the war, appeared not only viable after August 1945 but preferable to the perilous strategic environment promised by the marriage of the bomb to the rocket. In Sumerian times, the *Baltimore Sun* editorialized, a criminal with a weapon could do little harm outside his neighborhood. Now, roughly five thousand years later, "when the bad man has an atomic bomb as the warhead of a stratosphere-plying, radar-directed, round-the-world rocket," new moral principles were needed. The ballistic missile called for "ethical standards as much stronger than those of Sumeria as the bomb is more terrible than the Sumerian knife."[18] The exploration of outer space deserved "a better justification" than the development of new weapons, implored another column. "It is one of the tragic paradoxes of our time that man should win such stupendous victories in the realm of science, only to see them all perverted to the uses of people intent on destroying other peoples."[19] The intercontinental ballistic missile (ICBM) demanded fresh philosophical insights and hopeful narratives.

Despite the sense of dread that pervaded postwar society, these were not hard to come by. As with popular discourse about atomic energy, space technology offered a weighty choice between peaceful, civilian uses and destructive, martial ones. After the war, a new generation of interplanetary writers continued to offer space exploration as a panacea for contemporary social and political challenges. With a new sense of urgency, these writers adapted the ideas of progressive-era and interwar thinkers to emphasize the ethical responsibility of governments to funnel space technology research through constructive channels. Though spaceflight enthusiasts had had difficulty persuading the public about the viability and importance of their ideas before the 1930s, in the wake of Hiroshima they finally found a receptive audience.

The regenerated "cosmic philosophies" of earlier periods found earnest expression in Britain, where the British Interplanetary Society (BIS) proved a hotbed for both technical and political debates over the coming age of space.[20] The BIS had been an enthusiastic sibling to the rocket groups that had cropped up in the interwar period: the AIS in the United States, the *VfR* in Germany, and OIMS in the Soviet Union. Although it benefited from only minor patronage from its establishment in 1933, the BIS quickly recovered from a wartime lull, its membership growing from 250 in 1946 to 2,500 only seven years later. Departing from the more experimental rocket groups in the United States and Europe, the BIS developed a reputation as a platform for discussion and speculation rather than the development of technology. From its inception,

FIGURE 7. Arthur "Val" Cleaver envisioned an "interplanetary project" to remake international relations through human spaceflight. © National Portrait Gallery, London

the Society's experimental efforts were hampered by the Explosives Act of 1875, which prohibited private testing of liquid-fuel rockets in the United Kingdom.[21] The privations of war made things even more difficult. As BIS chair Arthur "Val" Cleaver recalled, the Society "organized into groups and did design studies and wrote papers. . . . But when it came to actually building hardware and doing the job, it was too expensive for England to undertake."[22] In the absence of resources, and without pressures to succeed experimentally, BIS members opened their doors freely to conjecture.

Cleaver led the charge. In October 1947, spaceflight advocates, technologists, and amateur rocketeers gathered in London to hear his annual BIS address. Having served as the chief project engineer for the de Havilland Engine Company and considering his wartime research on rocket propulsion, Cleaver was one of the most respected minds in aeronautics (figure 7). The audience was keen to hear his thoughts on the technical and fiscal prospects for human spaceflight, especially in war-ravaged

Britain. Cleaver supplied these in spades, but he dwelled far longer on the social and political impact of spaceflight. Indeed, delivered just a year before the founding of Celestia, he waxed on ideas and aspirations that closely resembled James Mangan's campaign for a pacific, utopian space nation (see introduction). Cleaver called it "the interplanetary project."[23]

In brief, this "project" entailed nothing short of a revolution in international relations through the exploration of space. Cleaver conceived of human spaceflight as a catalyst for world order and world peace. If motivated by curiosity and adventure rather than national aggrandizement, space exploration could supply an "outlet" for humanity's "basic instincts of aggressive and dynamic energy." Stripped of political rivalry, it would be a "substitute for the stimulus of war." To be sure, a restless and competitive species could find suitable replacements for conflict in many areas of science, but space exploration, for its capacity to capture the imagination, offered a particularly promising avenue for international reconciliation. For Cleaver, only a federation of states—perhaps even a World Government—could muster the resources necessary for interplanetary travel. He visualized the establishment of an Interplanetary Authority that would govern the development of space technology and ensure that activities in space were carried out in the interests of all. Grounded in the "lofty motives" of scientific brotherhood and modeled on the United Nations, this authority would represent the inauguration of a new age of global amity based on scientific and technological cooperation.[24]

Cleaver painted a vivid picture of earthly politics once national governments had united to escape the planet's gravity. "In a world free from the preoccupations, fears and compulsions arising from the threat of war, with all the wasteful expenditure of time and money on armaments which inevitably follows," he argued, "a truer standard of values would arise." Wars would be waged on poverty, disease, and malnourishment rather than on rival states. Money, manpower, and physical resources once devoted to arms racing could finally be diverted to education, public health, and social welfare. International scientific cooperation in space would manifest itself back on the ground—states would transplant cosmic partnerships to disarmament, international legal arbitration, and democratic governance. "Projects having as their aim the gaining of new knowledge and experience, inspired by love of adventure and beauty and the spirit of curiosity," Cleaver predicted, "would . . . absorb the attention of mankind."[25]

Acknowledging the violent application of space technology evident in the V-2, Cleaver called on the assembled spaceflight enthusiasts to resist militarization at every turn. It was incumbent on every person who aspired to "the interplanetary idea" to ensure that the world reject the development of space technologies as weapons of war. "We are all in this thing together, each with some microscopic grain of responsibility," Cleaver urged. Even if co-opted by warlords like Adolf Hitler, postwar advocates must remain committed to the dream of spaceflight and strive "to save all that we can for the cause of enlightenment at any given time."[26]

A Philosophy of Astronautics

It was no coincidence that as with the AIS in the 1930s, science-fiction authors proved central to many of the Society's discussions about space exploration in the early postwar years. Two such authors in particular, Arthur C. Clarke and Olaf Stapledon, forwarded the gospel of spaceflight that Lasser and others had carried in the interwar period. Using much the same language as their predecessors had, the two futurists propounded a vision of spaceflight as a civilization-defining experience that would heal international relationships and unite humanity in technological struggle. Adapting this astrofuturist argument to the new realities of the weapons revolution and the Cold War, Clarke and Stapledon set the table of expectations upon which Americans would draw in the years after *Sputnik*.

Arthur Clarke was a true disciple of the Interplanetary Project. By the time Cleaver delivered his 1948 address to the BIS, Clarke had already written a handful of articles urging the peaceful application of space technology and had served as chair of the Society. In fact, Clarke had been in the audience that evening. Inspired and emboldened by Cleaver's remarks, Clarke stood to comment. He considered the immediate postwar moment "a tremendous opportunity." If promoters of human spaceflight represented in the ranks of the BIS could achieve recognition as thought leaders in the field of rocketry, then perhaps they would "play a major—perhaps the major—part in the conquest of space." Considering the devastation that the V-2 had wrought on London only three years before, Cleaver's vision, he thought, had the potential to coax this revolutionary technology away from violent application. It was not too much to hope that civilian spaceflight enthusiasts in the Society could "play our part in seeing that astronautics becomes an instrument of good rather than of evil for the human race."²⁷ It is no exaggeration to say that in the 1940s and 1950s, Clarke was the most important figure in the attempt to see this vision through.

Clarke had grown up with the spaceflight movement. Born in 1917 on England's Bristol Channel, Clarke took an early interest in science after the early death, from lung cancer, of his father, a farmer who—aside from smoking cigarettes for most of his life—had been gassed on the front lines during the Great War. Left to help his mother care for his three younger siblings, he often retreated to what was in many circles still called "scientificition"; he hungrily read Verne, Wells, and the stories of space exploration featured in US pulp magazines (see figure 8). Apart from collecting fossils and assembling objects with his Meccano construction set, Clarke built rudimentary telescopes and experimented with small rockets. He even installed an intercom in his home. It was thus no surprise that when Philip Ellaby Cleator, a spaceflight enthusiast from Liverpool, founded the BIS in 1934, Clarke rushed to become one of its first members. Many of the Society's prewar meetings would take place in his apartment.²⁸

Clarke's interest in science proved fortuitous when wartime circumstances drew British youth into service. Whereas his father had served in the infantry, Clarke joined the Royal Air Force, which charged him with leading a team of field specialists working with ground-controlled approach (GCA) radars that MIT's Radiation Laboratory had developed in the United States. His experience with radar in World War II triggered his thinking about satellite communications, what in late 1945 emerged as a seminal paper in the popular science magazine *Wireless World*. In "Extra-Terrestrial Relays," Clarke proposed that radio communications could be bounced off satellites in geostationary orbit. More than a decade before *Sputnik*, this was a radical idea: only through telegraph and ionospheric communication could data be transmitted "over the horizon."²⁹ With merely three satellites, Clarke suggested, radio operators could achieve worldwide coverage. The paper helped legitimate Clarke as a serious scientist and helped lend his fiction—already established by the middle of the 1950s through *Prelude to Space* (1951), *The Sands of Mars* (1951), *Islands in the Sky* (1952), and *Childhood's End* (1953)—an air of authenticity and a sense of realism.

From the first, Clarke understood the severity of the dilemma posed by the joining of the rocket to the bomb. In February 1946 he appealed to authorities to transform the V-2 from a deadly weapon into an instrument of scientific investigation. He even showed the way: a rocket capable

FIGURE 8. A bookish Arthur C. Clarke in his study, 1936. His interplanetary worldview was grounded in boyhood readings of Verne and Wells. Courtesy of the National Air and Space Museum, Washington, DC (ID: NASM 9A 12591).

of traveling eight kilometers per second parallel to the Earth's surface would continue to circle the planet "forever" in a closed orbit; though the "vengeance weapon" could only reach a third of this speed carrying a cumbersome nuclear warhead, with only one hundred pounds of harmless electrical equipment the missile was capable of making the mark.[30] A month later he published an incisive essay in the *RAF Quarterly* that called for nuclear weapons to fall under the authority of the UN Security Council. By law, warring nations would be able to use solely conventional weapons. Only in emergency situations could use of the bomb be authorized. The weapons would be stored in remote facilities where teams of scientists (only those with a "supranational outlook" should staff the sites) would oversee the deployment of the warheads. Rather than a military-industrial complex, this system would be a first-class international web of scientific research. It might even serve as "a nucleus around which the scientific service of the world state would form, perhaps many years in advance of its political realization." The only defense against the bomb, he implored, was to prevent it from ever being used. Echoing others writing at the dawn of the ballistic missile, he concluded that "upon us, the heirs to all the past and the trustees of a future that our folly can slay before its birth, lies a responsibility no other age has ever known."[31]

Convictions about the threat of nuclear-tipped rockets went hand in hand with a certainty that the peaceful use of space technology would usher in an era of progress and international amity. On October 5, 1946, in his capacity as the new chair of the BIS, Clarke addressed a regular meeting of the Society at the White House Inn in London. His talk, "The Challenge of the Spaceship," was a futurist jeremiad on "Astronautics and Its Impact on Human Society." Clarke argued that the coming space revolution was one of only four great turning points in human history, all driven by the harnessing of technology. The taming of fire, the advent of agriculture, and the development of nuclear energy had been hitherto the most consequential watersheds, but none would be as significant as space exploration. Even the most rudimentary steps into the cosmos would be salutary. Research conducted in outer space and on other planetary bodies would inaugurate a revolution in astronomy, physics, chemistry, biology, and countless other disciplines; satellite photography would turn the guesswork of weather forecasting (and with it the growth of food) into a hard science; space stations and communications satellites would allow radio and television broadcasting to reach global audiences.[32]

These immediate technological consequences would not by themselves represent radical departures from previous milestones of progress. It was the indirect *political* and *psychological* forces resulting from exploration that would reflect true change. Copernican astronomy, Darwinian evolution, and Freudian psychology each pointed to a future in which exploration would initially have little practical use but would eventually have an incalculable effect on human consciousness. Through the "expansion of the world's mental horizons," human societies would unleash an unprecedented wave of creativity that would unmoor political, social, and economic relationships from previous restraints. It would be nothing short of "a new Renaissance."[33] More precisely, the mind-broadening effect of positioning Earth within a much larger cosmological scheme would make international problems seem smaller and thus less significant. It was reasonable to hope, Clarke told a receptive audience tired from war, that exploration would "have a considerable effect in reducing the psychological pressures and tensions of our present world." Assembling the necessary resources to probe the cosmos might "turn men's minds outward and away from their present tribal squabbles." "In this sense," Clarke concluded, "the rocket, far from being one of the destroyers of civilization, may provide the safety-valve that is needed to preserve it."[34]

Clarke's utopian vision of a world united through the exploration of outer space stood out in a postwar moment notable for its pessimism regarding technology. As global war had done a generation earlier, the fight against the Axis powers had undermined faith in the inextricability of science and progress. Not six weeks before Clarke's address, journalist John Hersey published his harrowing account of postbombing Hiroshima in the *New Yorker*, setting off a storm of global criticism. The social critic Lewis Mumford, once sanguine about the potential benefits of technological revolution, considered the bombing of Japan and Germany "a moral reversal," a selling of the American soul to technological power. He reminded readers of the *Saturday Review* that technology could be regressive; unbridled confidence in progress would lead to moral relativism and nihilism.[35] Meanwhile trials against Nazi doctors began in Nuremberg. The world learned of profane experiments designed to maximize the survivability of German soldiers in the field: prisoners at Dachau were involved in high-altitude tests to determine how crews from damaged aircraft could safely parachute to the ground; others were frozen to test the limits of hypothermia or forced to swallow seawater to ascertain whether it could be made drinkable. In Buchenwald, Natzweiler, and other camps, inmates were used to test immunizations for malaria, typhoid, tuberculosis, yellow fever, and hepatitis, or were exposed to mustard gas and phosphene to develop antidotes. Barbarism and genocide had accompanied scientific discovery.

That the world was standing on the threshold of "a new Renaissance" was even harder to believe in the context of widespread paranoia about the prospect of another cataclysmic world conflict. Throughout the 1940s, the editors of *Time* peppered the magazine's pages with references to and prognostications about "World War III."[36] In January 1945, well before the end of the war, intelligence analysts offered predictions about when the Soviet Union might be ready for a full-scale conflict with the United States. So began the tradition of "National Intelligence Estimates" that represented what was supposed to be a synthesis of Army, Navy, OSS, and State Department analyses of the future.[37] From the very beginning of the postwar period, stories of an apocalyptic third world war inundated popular culture, culminating in an October 1951 issue of *Collier's* in which twenty authors—including Edward R. Murrow, Hanson Baldwin, Hal Boyle, and Philip Wylie—painted a detailed and vivid "Preview of the War We Do Not Want."[38] Although a UN coalition headed by the United States ultimately defeats the Soviet Union in a nuclear war that lasts nearly four years, cities around the world are devastated by sustained bombing on both sides.

Fully aware of these shocking predictions, Clarke did not want to seem overly sanguine about opening space to humanity. There was always the possibility, he admitted, that exploration would open the door to "interplanetary imperialism" between spacefaring powers. The futurist offered that "the Solar System is rather a large place," but considered it an open question "whether it will be large enough for so quarrelsome an animal as *Homo sapiens*." Clarke observed that although formal imperialism was falling out of favor in Europe, there was evidence that talk of "conquering" outer space was stoking a neocolonial revival in the United States. There, only a few months earlier, a major in the army had advocated the use of Mars as a military base from which to intimidate the Soviet Union—or, in the case of a war, to destroy it. In World War II, he reasoned, it had been necessary to establish bases in the remotest parts of the world; in World War III, the thinking went, "we cannot limit such occupation to the earth alone."[39] At the same time, R. L. Farnsworth, the president of the US Rocket Society (not to be confused with the *American* Rocket Society), reissued a lengthy pamphlet on space missiles, which he thought would blaze a "New Trail to Empire." The American predicted, as Clarke did, that the process of exploring other planets would bring about another Enlightenment, only, he urged his countrymen—particularly through big business—to grab a greater measure of the pie before rivals could stake their claim. The moon, a possible "Lunar Empire," was the greatest prize. "We can no longer mislead ourselves that commissions and conventions can secure peace," Farnsworth had concluded. "Let us strive for the day when the American flag is planted firmly upon the volcanic ash of the Moon, and that thenceforth, in its entirety, it will become a possession of the United States!"[40]

Clarke inveighed against such thinking; in his address he referred to Farnsworth's ambitions as "sixteenth-century buccaneering." The pamphlet was playing directly into the hands of critics like fantasy author J. R. R. Tolkien and Christian theologian C. S. Lewis, who considered humanity too aggressive and covetous to extend its existence to space, "God's quarantine regulations." Farnsworth and other space colonists, despite their intentions, threatened to "reduce astronautics to a laughingstock." In a separate fulmination that Clarke never published, he expressed disbelief that those "mentally rooted so firmly in the past" could so clearly see the possibilities of spaceflight in the future.[41] So strongly did Clarke oppose the notion of national claims in space that he went so far as to recommend postponing the age of space should imperial designs gain ground in the halls of power: "If we intend to inflict on other worlds the worst excesses of a materialistic and spiritually barren civilization," he warned, "our case is lost before we begin to plead it. The 'quarantine' will have to remain in force for a few more centuries yet if many advocates of interplanetary travel think as Mr. Farnsworth appears to do." If imperialism survived the dawn of the space age, it was the "duty" of every spaceflight enthusiast to "hamper and delay" interplanetary travel "whatever our personal hopes and aspirations may be."[42]

"The Challenge of the Spaceship" was only the first and most robust example of a "philosophy of aeronautics" that Clarke propounded from the late 1940s, one that equated space exploration with the transcendence of international politics and war. As his predecessors had argued in the 1920s and 1930s, and as so many others would continue to argue after *Sputnik*, Clarke considered the spaceflight adventure "an outlet for dangerously stifled energies," a funnel through which to permanently channel humanity's "aggressive and pioneering instincts."[43] Though he employed the same language of "conquering" space that many of his contemporaries used, Clarke considered the colonization of space "the only form of 'conquest and empire' compatible with civilization."[44] He believed that the frontier of space, as Turner had argued of the American West at the end of the nineteenth century, would nurture within each would-be colonizer the traits necessary to sustain human society: curiosity, hardiness, and resourcefulness were common themes to both. But, unlike the violent and rapacious policies that had characterized the expansion of the United States to California and Oregon, the colonization of space would awaken cooperative rather than combative instincts. Traversing the hostile environments of foreign planets and space would require human beings to fight local conditions rather than each other and present challenges surmountable only through collaboration. "And once one has grown used to the idea of cooperation," he suggested, "it is very hard to get out of it."[45] Whereas racism and even genocide had prevailed in expansion to the West, a "variety and diversity of cultures" would become a celebrated by-product of interplanetary society.[46] Colonies in space would not resemble the earthly, tribal outposts of centuries past; these provinces would be as multifarious as the planets. The varied climates of the Earth's neighboring planets would inevitably shape the worldviews of their inhabitants, but all would abide by a set of coherent and universal moral principles grounded in egalitarianism and scientific rationality.

Clarke was particularly taken with David Lasser's work, and seconded his predecessor's forecasts about what space travel would mean for human development. Echoing Lasser, Clarke predicted that the view of the planet from space would fundamentally shepherd human consciousness away from the immediate needs of one's family or nation to those of the human family and the well-being of the Earth. "We all know the narrow, limited type of mind which is interested in nothing beyond its town or village, and bases its judgments only on parochial standards," he wrote in his seminal 1951 book, *The Exploration of Space*. "We are slowly—perhaps too slowly—evolving from that mentality toward a world outlook. Few things will do more to accelerate that evolution than the conquest of space." It was difficult, Clarke reasoned, to see "how the more extreme forms of nationalism can long survive when men have seen the Earth in its true perspective as a single small globe against the stars."[47] When C. S. Lewis suggested that humans were not yet mature enough to explore space, Clarke wrote to him that although the humanity was still in its "infancy," aeronautics would accelerate the development of *Homo sapiens*. National rivalries, he insisted, would finally appear "in their proper perspective" when seen against the infinity of the cosmos.[48]

Although space exploration offered new vistas that might unite humanity, it was vital that the moral and ethical development of human beings precede the first serious forays into space, a conviction Clarke held more tightly after *Sputnik*. Among the torrent of new scientific discoveries and innovations making front-page headlines in the 1950s, weapons systems seemed to stand out. Few failed to appreciate that the development of the birth control pill, the videotape recorder, the passenger jet, the microchip, and a vaccine for polio had been coterminous with that of hydrogen bombs, the B-52 Stratofortress, and, of course,

ICBMs. Recent history, particularly the world wars, demonstrated that if human beings permitted science to outpace the maturation of international relations, the "social system will breed poisons which will cause its certain destruction." Clarke repeated in a number of philosophical reflections that "superhuman knowledge" should be accompanied by "equally great compassion and tolerance." If "wisdom" failed to match science, humanity would "have no second chance."[49]

Part of the problem was that although the world had become "space conscious," collective awareness about exploration was still tied too closely to military strength and international prestige. Clarke railed against the "technoporn" of military innovation in the 1950s, the "gleaming weaponry and beautiful explosions" featured on television and in popular science magazines. One need not have looked further than the organization of viewing parties atop Las Vegas apartment buildings to witness nuclear tests in the Nevada desert. Clarke especially regretted the fact that so many people had interpreted the world's first satellite not as an opening salvo for the space age but rather a reflection of military capability, a threatening omen of Soviet power. The fear that Soviet satellites stoked in the West helped make the dawn of space exploration a "neurotic" moment: the "decadence" of modern art, the "sick jokes" circulating in nightclubs, the proliferation of self-help books, and the pale, emaciated bodies posing at fashion shows were all symptoms of a "malaise" that had taken hold of Western culture.[50]

"The opening of the space frontier," Clarke predicted, "will change all that." Rather than settle for space consciousness, humanity should aspire to become "space *minded*." This meant abandoning "spacemanship" to embrace exploration as a scientific and philosophic enterprise undertaken by the entire human race, not a political one pursued by rival technocracies.[51] If defined narrowly as the technological probing of other planets for resources, for propaganda, and for territorial aggrandizement, exploration offered no relief. If, however, humans were to embrace a more holistic definition of exploration that encompassed the betterment of human nature, the transcendence of humanity's material needs, and the strengthening of its internal bonds, the possibilities were endless.[52] He offered satellite communications as only the most conspicuous example of how space mindedness would change the world. Even a rudimentary system would quickly become the "nervous system of mankind."[53]

Science fiction permitted Clarke to manifest his political ambitions for space exploration unfettered. In the tradition of Lasswitz's *Two Planets*, in 1953 Clarke published *Childhood's End*, a utopian account of an alien invasion of the Earth that, far from destroying humanity, civilizes and enlightens it. Prophetically, the novel opens at the end of the twentieth century, when the United States and the Soviet Union are competing to launch the world's first spacecraft into orbit, one that will have obvious military implications. Visitors from outer space, whom the Earthlings come to know as "The Overlords" for their superior culture and technology, arrive in futuristic spacecraft over the planet's major cities to put an end to the "space race" and to take control of international affairs. The aliens do not reveal themselves; Overlord Karellen, the "Supervisor of Earth," speaks only with the UN secretary general. Before long, alien management of the foreign policy produces a "golden age" in human societies. Ignorance, poverty, disease, and crime virtually disappear. Education is protracted, with students attending university well into their thirties. Workers labor for their enjoyment and edification, not for sustenance, much of which the Overlords provide. Production of food and material goods are more or less automated. "By the standards of all earlier ages," the narrator reports, "it was Utopia."[54]

It was also finally "One World." A truly global culture flows from Overlordship. All communicate using English, literacy is universal, and television is accessible to even the remotest parts of the planet. Nationalism suffers a swift demise. The reader learns that "the old names of the old countries were still used, but they were no more than convenient postal divisions." As memories of war fade into the past, international life comes to be characterized by amicable relations, generosity, and cultural enrichment. It was, as Clarke described it, a "long, cloudless summer afternoon of peace and prosperity."[55] As with earlier fictional accounts, Clarke imagined space-based civilizations to be not only more advanced but also more magnanimous, more enlightened, and more moral as well. Contact with space contributes to the maturation of human societies on Earth.

Similar messages pervade Clarke's next novel, *Prelude to Space*, published the following year. The story's narrator is historian Dirk Alexson, who recounts events leading up to the launch of *Prometheus*, the world's first spacecraft capable of reaching the moon. At the end of the novel, when the massive ship is set to take flight, the global foundation responsible for the mission, "Interplanetary," issues a manifesto to ensure humanity's first forays into the cosmos are not carried out with selfish intentions. In a clear reference to Farnsworth and others who had advocated a formal claim to the moon, the manifesto rails against those who believe "the political thinking of our ancestors can be applied when we reach other worlds." Presaging the language that international lawyers would use to debate space treaties in the 1960s and 1970s, the declaration proclaims that any world humans would reach would be "the common heritage of all men." It even declares that if life exists on these worlds, the representatives of Earth would stake no claim for it; as a closing promise, the departing astronauts pledge they will "take no frontiers in space."[56]

As *Prelude to Space*, *Childhood's End*, and Clarke's other novels showed, fiction was a suitable vessel for interplanetary politics. Grounding their stories in a graspable technological future, science-fiction authors could conjure political worlds—governed by egalitarianism and enjoying permanent peace—that seemed equally graspable. It was to Clarke's benefit, and to the benefit of his ideas, that scientific romance had come to enjoy an even wider readership after World War II than it had in the era of Lasswitz and Wells, a fact attributable in no small measure to the development of rocketry itself.

A New Species of Man

It was through imagination that another British science-fiction author, William Olaf Stapledon, emerged in the mid-1940s as another oracle of humanity's social evolution in space. Born a month prematurely in a suburb of Liverpool, on the Wirral Peninsula in Cheshire, England, Stapledon was the only son of a merchant-mariner working on behalf of the shipping magnate Alfred Holt. He spent the first six years of his life in Port Said, Egypt, where he developed a fascination with the ships that passed through the Suez Canal. At night his father occasionally set up a telescope on the iron balcony overlooking the desert, a gaze that provided "Olaf" with an appreciation for the "appalling contrast between the cosmos and our minute home-lives."[57]

For the rest of his life, Stapledon would prove unable to shake his curiosities about either minute earthly events or the mysterious power of the cosmos. Indeed, in both his fiction and nonfiction works, he sought to integrate them. In 1905 he entered Balliol College, Oxford, where he earned a bachelor's degree in modern history. After graduation he spent six years teaching a variety of courses at the Manchester Grammar School and in the University of Liverpool's extramural program. By the outbreak of World War I, he had published his first book, a collection of twenty-four poems entitled *Latter-Day Psalms*, an exploration of "values beyond those affirmed by everyday experience."[58]

Stapledon was a committed pacifist in 1914. Modern warfare, he thought, was "a glib surrender of one's moral responsibility to an authority that was not really fit to bear it."[59] Nevertheless, as a conscientious objector he served in the Friends' Ambulance Unit in France and Belgium from July 1915 through the end of the war. In a published reflection on his wartime experiences, he wrote that as long as there was "a chance of serving those who nobly suffer through humanity's error[,] we cannot stay at home." "Because of our oneness with humanity," he convinced himself, "we dare not hold ourselves apart from the calamity." Despite his personal opposition to violence, the French government awarded him the Croix de Guerre for bravery, the 1914–1915 Star Riband, and the Victory Riband.[60] After the war, he returned home to pursue a doctorate in philosophy at the University of Liverpool. His thesis was the basis for his first book, *A Modern Theory of Ethics* (1929), an exhaustive study attempting to integrate biology, psychology, and modern ethics in light of the growing "disillusionment" he saw in postwar society. Following the breadth of his research, Stapledon held a variety of teaching posts in history, English literature, philosophy, and psychology, but he quickly abandoned the academy once his fiction writing could support him financially.

This did not take long. Though Stapledon never achieved the celebrity that Clarke would eventually acquire, by the mid-1930s he was one of the most prominent figures in science fiction and a central influence on literary contemporaries including H. P. Lovecraft, C. S. Lewis, and especially Clarke himself (figure 9). Stapledon's first four books were smash hits. *Last and First Men* (1930), *Last Men in London* (1932), *Odd John* (1935), and *Star Maker* (1937) inspired a generation of "future history" literature that anticipated the ascent and decline of human and alien civilizations on a cosmic scale.

It was Stapledon's status as a science-fiction author that brought members of the BIS into a packed room at the Charing Cross Road Art School in downtown London to hear him deliver a lengthy talk on the evening of October 9, 1948. His lecture asked simply: "Interplanetary Man?" In a fitting succession to Clarke's address two years before, the writer launched into a detailed analysis of the reasons human beings might wish to colonize other planetary bodies.

Stapledon proposed three possibilities. First, humanity would cull the vast mineral resources of other planets to advance its material wealth. Stapledon anticipated this would be a salutary development but warned of technology-induced hedonism and overabundance. Second,

FIGURE 9. Olaf Stapledon, philosopher of man and space. © National Portrait Gallery, London.

human beings might colonize space to achieve greater power, both over the environment and over each other. He predicted that the "rival imperialisms and ideologies" of Earth would extend to the cosmos, a process he envisioned destroying life on other planets. The third possibility—and the one about which Stapledon was most sanguine—was the gradual evolution of humankind, which implied making "the 'most' of man ... the 'best' of him."[61] He hoped that in their quest to master nature through science, human beings might shed the "diseases of infancy"—parochialism, imperialism, selfishness, violence—to become mature, adult members of first, a global community, and eventually, an interplanetary one.

Stapledon was something of an expert on human evolution. Though he could not speak with Darwinian authority—he was, after all, a philosopher, not a scientist—the BIS asked him to shed light on humanity's future in space from his position in the world of imaginative fiction. *First and Last Men* (1930), his first novel and the one to which he could most attribute his success in the genre, traced the development of no less than eighteen species of human beings across two billion years, beginning with the "First Men" living in Stapledon's time. The book describes a series of colossal earthbound wars among the First Men, characterized by aggressive nationalism, that ultimately lead to self-destruction. This species, although repeatedly reaching out toward some kind of egalitarianism, fails to develop the necessary outlook to sustain itself in the hostile conditions created by its history. "Socrates woke to the ideal of dispassionate intelligence, Jesus to the ideal of passionate yet self-oblivious worship," the narrator intimates in chapter 1. "Each, of course, though starting with a different emphasis, involved the other. Unfortunately, both these ideals demanded of the human brain a degree of vitality and coherence of which the nervous system of the First Men was never really capable."[62]

As Stapledon wrote in the preface to the novel, the goal of transplanting humanity into the distant future was to see it "in its cosmic setting," where such vitality and coherence would be possible. To escape the bewildering political and social realities of the present, he argued, human beings must be brave enough to contemplate alternatives outside the bounds of the present. He refused to believe that human morality was an end in itself, that moral behavior was simply what human beings regarded as moral. Rather, Stapledon proposed that humanity was a means to an end, an "instrument" to achieve a higher state of ethics completely dislocated from human nature in the present.[63] These were philosophical enterprises to which spaceflight was to prove central. By inaugurating the transformative project of space travel, human beings could escape the maladies besetting the First Men and achieve a unity between "thinking" and "feeling," a balance between rationality and spirituality that Stapledon, ever the transcendentalist, considered the apotheosis of human development.[64] In the unforgiving conditions of space, on distant planets, and in their relationships with other intelligent species in the universe, human beings could become something "over and above" what they had been on Earth.[65] Hence fiction's metaphysical value: to those who criticized the philosopher's forays into spectacular fiction, Stapledon replied that by writing the history of the future, by using "controlled imagination," people might entertain new values.[66]

Stapledon's address before the Society received worldwide coverage, but most reporters seized on the more fantastical elements of the speech to satirize its lofty speculations. Cartoonists conjured droopy, freakish figures that had been reengineered for life on other planets. Journalists in London admired the philosopher's "high-flying thought" and his "whoosh of imagination," but were generally disturbed by his vision of the future. In the United States, *Time* magazine correctly identified Stapledon as "something of a moralist," but cast a doubtful eye on his mission to create what it called a United Solar System.[67]

Yet these commentators, as one of Stapledon's biographers tells us, neglected to read the political subtext of the address. In his appeal for "genuine community" among the races of the solar system, Stapledon was commenting on the growing animosity between the United States and the Soviet Union. Humans, in their dealings with each other and with the other species in the universe, must "enter into their point of view" and cooperate with them for material prosperity and spiritual vitality. Each end of the ideological pole held some fundamental truths: a commitment to liberty in the West, for example, justice and fraternity in the East. "If war is avoided, and if in due season each side can learn from the other," Stapledon intoned, "the result may be a far more adult and spiritually enriched humanity than could ever have occurred without this cultural clash of mighty opposites."[68] Yet in assessing the likelihood that nations competing for influence on Earth might put aside their differences to cooperatively explore other planets, Stapledon tempered his characteristic optimism. If history were any guide, spacefaring powers would rush to annex any virgin celestial territories. The Cold War, he thought, would probably spread to Mars and beyond. "Must the first flag to be planted beyond earth's confines be the Stars and Stripes," he asked ruefully, toward the end of the lecture, "and not the banner of a united Humanity?"[69]

Stapledon's reference to the American flag was an explicit choice, for his gloomy outlook derived from strong feelings about the United States. The nation's technological sophistication, its pretention to world leadership, and its economic power suggested it would be a lynchpin of humanity's breach into space. If the pace and vigor of US rocket research were any indication, it would be the most important player in the opening of the space age. Human development orbits around American progress in *First and Last Men*. The United States is "universally feared and envied" for its industrial production and cultural vibrancy. Closely emulating real-world events at the time Stapledon delivered his BIS address, chapter 1 relates that in every corner of the globe, people consumed American products, local businesses unfurled from American capital, and American radios, television, and film "drenched the planet with American thought." The fictional Americans of Stapledon's novel, as they would become in reality after 1945, believed themselves to be the guardians of liberty, democracy, truth, justice, safety, and prosperity—"the custodians of the whole planet."[70]

These pretentions would not have mattered, Stapledon wrote of the First Men, had the United States "been able to give of her very rare best." In the end the "floods of poison" emanating from US culture corrupt the entire world. For all their accomplishments in literature, philosophy, astronomy, architecture, and organization, Americans are most responsible for the decent of the First Men into a new dark age. The most intelligent and productive members of society, those who had helped to create a modern economy and had forwarded important scientific research, were merely "a minority in a huge wilderness of opinionated self-deceivers, in whom, surprisingly, an outworn religious dogma was championed with the intolerant optimism of youth." The United States teemed with "bright, but arrested, adolescents." Surveying the ages, the narrator concludes of Americans: "One . . . can see their fate already woven of their circumstance and their disposition, and can appreciate the grim jest that these, who seemed to themselves gifted to rejuvenate the planet, should have plunged it, inevitably, through spiritual desolation into senility and age-long night."[71]

A damning judgement! But not one Stapledon was prepared to revise considering the atomic bomb and the cursed totems being built to transport them. Indeed, by the time "Interplanetary Man?" appeared in the *Journal of the British Interplanetary Society* in November 1948, the United States had fully integrated the German war booty—both the V-2 rockets and the engineers—into its own fledging arsenal, the perversion of the space rocket complete. In the initial postwar scramble for rocket programs none were left out: the Army Ordnance Department, Army Air Force, Navy Bureau of Ordnance, and Navy Bureau of Aeronautics each pursued incipient R&D agendas that covered the entire, dizzying spectrum of kinetic missilery: surface-to-surface, surface-to-air, air-to-surface, air-to-air. They contracted with McDonnell Aircraft, Martin Company, Northrop, AMC, Douglas, Bell, Goodyear, Boeing, General Electric, and Hughes. Convair won a $2 million contract to develop the MX-774B, a massive rocket capable of transporting a 5,000-pound atomic warhead across 5,000 miles: a precursor to the mighty Atlas.[72]

What did interplanetary theories have to say about *that*? By the time the US military began its first serious explorations of rocket technology in the mid-1940s, the Interplanetary School of IR had, despite its

European ancestry, migrated to American soil, primarily from flows of people and books. H. G. Wells had made a half-dozen visits to the United States in the years leading up to World War II, each with great fanfare. Arthur Clarke's novels reached thousands of Americans; *Childhood's End* sold out of its first print run. Stapledon came, too. As the only British delegate to the Cultural and Scientific Congress for World Peace in 1949, he warned Americans that if US-Soviet tensions rose any further, "there may be a war at any moment."[73] American reviewers lauded the ingenuity of *Star Maker* and *Last and First Men*, and leading newspapers reported on his seminal BIS address; Cleaver's too.[74] The German dreamers, as has been exhaustively documented, made it to the United States as well. After Ley emigrated in 1935, he contributed countless articles to American science-fiction magazines and published numerous best-selling books on spaceflight.[75] Hermann Oberth would go on to author two English-language space books of his own: *The Moon Car* (1959) and *The Electric Spaceship* (1960). Von Braun's influence on the US spaceflight imaginary hardly needs explication.

But the sordid events of a middle-age twentieth century proved that interplanetary dreams, however popular or pervasive, remained just that—dreams. Clarke's technical proposals had much support among British engineers, but little beyond. The cosmic philosophies articulated at the BIS made up what were essentially a series of manifestoes, soaring yet unfulfilled in a world consumed with the immediate requirements of the Cold War. Von Braun had made weapons. Oberth had offered his services to Hitler. And Ley, though he had achieved some success in tying together the disparate European rockets groups, found himself swimming upstream amid national postwar competition in rocketry. Space technology seemed destined for the same.

What was needed, then, was theory in motion, an interplanetary *praxis*. James Mangan's zany crusade for the Nation of Celestial Space (see introduction) in 1949 was one fitful step in this direction. Seeing that Cold War governments were militarizing an ever-expanding realm of physical and biological space—the ocean floor, the Sahara Desert, the human brain—Mangan determined to create a government of his own so that he might stop the infection at the edge of space. "Had some existing State, in particular a high-ranking world power, laid claim to celestial space ... consternation would now rule the people of all other nations," Mangan boasted years later. "Kind destiny allowed this *coup d'etat* to fall to mild hands."[76] Beyond a mere proclamation, he intended Celestia's constitution to be a legal instrument preserving a future for humanity "solely for Peace and Service to Man."[77] Hence his issuance of Celestian passports *only* to astronauts with strong moral character.

But in many ways, of course, Celestia did little more than science-fiction novels had done to advance material reality. Mangan succeeded in attracting curious attention, but had "space for peace" really earned any serious political capital? Did it convince powerful states to see the error in continued conflict and instead embrace its call for "magnanimity"?[78] Certainly not.

Yet at the dawn of the missile age, one important interplanetary practitioner in the United States sought to actualize, through science and engineering—through *things*—many of the utopian ideas set out by "Interplanetary Men" like Clarke, Stapledon, and indeed the Celestian leader himself.[79] This practitioner may yet have fallen through the cracks, as Mangan did, were it not for the fact that in the mid-1940s, he was America's foremost expert on rocketry.

Interplanetary Praxis

Born to Czech immigrants in 1912, Frank Joseph Malina grew up just northwest of Houston, Texas, what would eventually become the nerve center of the US manned spaceflight program. Like all his interplanetary predecessors, he had come to an early obsession with space through the fantastic tales of Jules Verne. It was a passion that led him to pursue a degree in mechanical engineering from Texas A&M, which he obtained in 1934. After graduating, Malina received a fellowship to continue his studies at the California Institute of Technology, where he quickly became swept up with its Guggenheim Aeronautical Laboratory (GALCIT), led by Hungarian émigré Theodore von Kármán, already an internationally renowned physicist and aeronautical engineer. Von Kármán supervised his doctoral dissertation on the "Characteristics of the Rocket Motor and Flight Analysis of the Sounding Rocket" (completed in 1940) and became a lifelong friend and colleague.[80]

At Caltech Malina cofounded, with his mentor's blessing, GALCIT's Rocket Research Project, through which the young engineer experimented with various fuels to develop a rocket powerful enough for high-altitude research. In the 1920s Robert Goddard had pioneered the use of liquid oxygen as a propellant, but the substance was difficult to work with and store. Together Malina and his colleagues—the machinist Ed Forman and his friend, the chemist and Thelemite occultist Jack Parsons—tried more utilizable fuel combinations. They developed and patented a hydrazine-nitric acid fuel that would later propel the Apollo

FIGURE 10. Frank Malina with the WAC Corporal rocket at White Sands, 1945. "My enthusiasm vanishes when I am forced to develop better munitions," he wrote before eventually abandoning work on rockets. NASA/JPL.

Service and Lunar Excursion Modules. More crucially at the outset, they also combined red-fuming nitric acid (RFNA) with aniline, a blend that, though it ignited spontaneously—a property that gave Malina's cohort the nickname "the Suicide Squad"—was easily handled and stockpiled.[81]

Malina had imagined applying his work on propulsion to sounding rockets, which would collect measurements and perform scientific experiments during their suborbital flights (figure 10). But by 1938, with war in Asia already raging and crisis building in Europe, he found himself briefing the National Academy of Science Committee on Air

Corp Research—founded by the future-minded Army General Henry "Hap" Arnold to keep abreast of cutting-edge technology for military use—on rockets for jet-assisted takeoff (JATO). In August 1941, four months before the United States entered the war against the Axis, the first solid-fuel JATOs assisted a small Ercoupe plane into the skies above Riverside, California; within the year a liquid-fuel JATO had lifted a light bomber, Douglas's A-20, off the Army Air Corps Bombing and Gunnery Range in Muroc, one hundred miles north of Los Angeles. With growing military interest in rockets and a core of talented engineers, in 1942 Malina, now a professor at Caltech, founded a new company with colleagues from the original Rocket Group to build JATOs: the Aerojet Engineering Corporation.[82]

Malina enjoyed the work but, as he explained to his parents, discovered that "my enthusiasm vanishes when I am forced to develop better munitions." When, as a graduate student, he worked part time for the Department of Agriculture studying soil movement in Dust Bowl-wrecked farmland, Malina found that he preferred "to keep on with SCS [Soil Conservation Service] as it is further from warmongering." His colleague Hsue-Shen Tsien, a brilliant engineer who would one day lead the Chinese missile program, had tried consoling him: better weapons were needed in the fight against fascism. But Malina felt "one cannot be certain that such a path will not boomerang."[83] Unlike Goddard, who accepted military patronage to support his rocket research without moral scruples, Malina struggled with the bargain he had struck with GALCIT to pursue the dream of spaceflight.[84]

Despite his enthusiasm for the rocket as a tool for science and a vehicle for cooperation, by the end of the war Malina found himself waist-deep in the military-industrial complex. In 1944 and again in 1946, he traveled abroad on behalf of the War Department for European Missions, collecting data on rocket developments in Britain and France and examining captured German weapons. At the same time, he traveled frequently to Washington as a consultant for the US National Defense Research Committee to obtain funding for the Jet Propulsion Laboratory (JPL), the semi-independent Caltech research facility that had absorbed GALCIT's various rocket programs.[85] Malina's travels were not only exhausting but deeply discomfiting. "I found that I was getting caught up more and more in trips to Washington in meetings with the army, navy, and air force, planning the next war," Malina recalled years later. "I found in these meetings that I was getting more and more disturbed[,] and I would break into cold sweats. I just hated the idea of, say, planning to use all this for bombarding people."[86]

On the eve of his first excursion—September 6, 1944, two days before the first V-2 attack struck Staveley Road in London, killing three—he was gripped by fears of witnessing the rocket's human toll. "As you know," he confessed to his wife, Liljan, "I have fought against ... an unmanageable fear of physical pain[,] not so much for myself, but when it happens to others."[87] And indeed, while in England Malina was privy to numerous reports about the physical damage wrought by the new weapon. On one occasion, a conference Malina attended at the British Projectile Development Establishment in Kent was "shook-up" by an incoming rocket. The incident left him "rather disturbed, but my colleagues continued ... as though nothing had happened."[88]

Back home, Malina tried to punctuate his military work with scientific endeavors. In the wake of British intelligence reports about the Germans' strides at Peenemünde, US Army Ordnance pressed the JPL to develop long-range missiles of its own. Von Kármán's team proposed a series of rockets that would represent a progression—captured in their names, which would correspond to ascending army ranks—of power and accuracy. First, there would come the Private, a small, unguided, solid-fuel rocket with a limited flight range; then, a larger, guided, liquid-fuel Corporal with a range of one hundred miles, up through Sargent and the officer ranks. Within this progression, Malina recommended the development of an intermediary rocket, between Private and Corporal, that would allow JPL engineers to test complex launch systems, fix guidance defects, and provide more experience before the expected leaps in size, fuel, and range. This transitional rocket would be a miniature WAC (without altitude control) Corporal that would carry twenty-five pounds of scientific instrumentation to an altitude of more than 1,000 feet, the world's first sounding rocket.[89]

Using the RFNA and aniline fuel that the Suicide Squad had developed earlier for JATOs, Malina achieved stunning success. On October 11, 1945, the WAC Corporal breached the stratosphere above the white sands of the Chihuahuan Desert in New Mexico at speeds approaching Mach 3. When it finally stopped ascending, it was nearly forty-five miles above the Earth, in the frigid and hostile mesosphere. It had flown more than twice the altitude expected. Four years later, when JPL launched a Bumper-WAC Corporal atop a German-made V-2, it reached an altitude of 244 miles, the first human-made object to reach outer space as it is understood today. In what was a cruel irony for Malina, the missile's final product, the MGM-Corporal, was the first guided missile authorized to carry a nuclear warhead.[90]

But by then, he was long gone. A week after the bombing of Hiroshima, Malina had filed for a leave of absence from JPL, and he never returned. Having glimpsed the Trinity site from the air, he felt increasingly anxious about the role he would be asked to play in engineering missiles powerful enough to ferry the atomic bomb to its victims. "I have never been convinced that what I was doing the past 10 years was right," he wrote his parents. "Technical developments are now so far ahead of other human arrangements that it appears nonsensical for one who sees the gap to make it even wider."[91] When on his second trip to Europe Malina sat in on a preparatory meeting of a UN committee to be called UNESCO (the United Nations Educational, Scientific, and Cultural Organization), he found a promising alternative to his military work. This new body offered a vision that Malina, ever the idealist, could embrace with confidence. "Since wars begin in the minds of men," read UNESCO'S constitution, "it is in the minds of men that the defenses of peace must be constructed."[92]

By early 1947, Malina was in Paris working as a counselor for UNESCO's Natural Sciences Department, where his first project was to study methods of decreasing national barriers to the free movement of scientists and equipment, a natural fit. Following on his research for the SCS, Malina also worked on the Arid Zone Program, which aimed to boost agricultural productivity in the parched regions still covering significant parts of the globe. "I think anyone that works in astronautics can't help but be somewhat world-minded," Malina later said of his motivations to work for the United Nations. "To us the world is one planet relative to the Moon or Mars and so forth." From his earliest days as an engineer, Malina had been a true believer in Albert Einstein's call for "some kind of world cooperation." He felt that the traditional notions of state sovereignty that had developed during the eighteenth and nineteenth centuries were "now reaching [their] limits."[93]

Malina's escape from weapons development was both a conscious choice and an abrupt change thrust on him by his own government. In September 1945 agents of the US Army's Counterintelligence Corps (CIC) had ransacked Malina's home looking for evidence of espionage ("blueprints" for the WAC Corporal had been found with an intercepted Russian courier in Paris). Malina had as a graduate student joined a local chapter of the Communist Party, and he expressed his belief in the importance of international scientific exchange to his parents and colleagues, but scant evidence supported the accusation that the engineer was engaging in any subversion. The Los Angeles Police Department began surveilling Malina in 1938; the Federal Bureau of Investigation (FBI) opened a file on him in 1942 and continued to compile reports for another thirty-one years, a span that covered the lion's share of his adult life. Though it lacked any link between Malina and the captured Paris documents, the government acted on its suspicion. In 1952, the Bureau pressured the Justice Department to issue a sealed indictment against Malina for making false statements to the government (allegedly, he had not listed his membership in the Communist Party USA and had said "no" when asked if he had ever belonged to any groups that wanted to overthrow the government). A secret warrant for his arrest, should he step foot in the United States, followed swiftly behind. After initially pressuring UNESCO to transfer Malina to American soil, the State Department did not renew his passport, stranding him in France.[94]

Emotionally and physically exhausted, and now independently wealthy from the sale of his valuable Aerojet stock, Malina quit UNESCO in 1953 to pursue a lifelong passion for art. As W. Patrick McCray has illuminated, Malina became one of a set of Cold War "thingkers" who combined traditional artistic practices with competence in science and engineering. From the mid-1950s, he experimented with painted string and wire to create depth and stimulating optical effects. He further leveraged his engineering background when he created a series of works, known as "electro-kinetic" art, that integrated painting with moving electrical components to create dynamic pieces. Indeed, having made seminal theoretical and technical contributions to American rocketry, Malina became, in his new creative profession, an international pioneer of so-called kinetic art. Many of his works (e.g., *Rocket Motor*, *Shock Waves*, and *Jet Plane*) suggest he never abandoned his original love for engineering. In 1965 he would finish *Cosmos*, a commission by Peramagon Press to represent the union of the arts and sciences. Eight feet wide, ten feet tall, and more than eight hundred pounds, the piece is a dazzling assemblage of lights, electric motors, and plastic parts painted in muted, translucent colors. Viewed altogether, it is a moving collage of astronautical images, a colorful swath of what he imagined astronauts and cosmonauts had gazed on in their first voyages into space. It was a culmination of Malina's strange arc from military rocketry to a quieter cosmos.[95]

Frank Malina personified, as Wernher von Braun did, the abiding tension that existed between the humanist, utopian impulses driving interplanetary thought on the one hand, and the real-world influence space technology promised to render upon war-making and politics on the other. Even as he looked on in pure elation as the WAC Corporal disappeared into the skies above White Sands, he knew to what ends rocket science was now inexorably moving. Ever the engineer, he was passionate about his work, but unlike von Braun his conscience (and, no doubt, the FBI) drove him away from the bargaining table between rocketry and state power.

These tensions characterized the interplanetary vision in toto. Increasingly through the early twentieth century, the utopian motifs and moralism of cosmic fiction competed with a rising tide of alien invasion literature that, removed from the socialism of Wells or the scientific internationalism of Ley, did more to entertain than educate. The new, more popular generation of science fiction tilted toward the externalization of contemporary fears rather than the fabrication of imaginative social and moral orders. The narrative result was violence without rebirth, conflict without change. In the 1930s, Lasser's optimism about the cultivation of an "interplanetary mind" sat uncomfortably alongside his warnings about the imminent temptation to bombard civilian populations with rockets.[96] Stapledon's hopes for "a commonwealth of worlds" based on cooperation and spiritual understanding competed in his lectures with predictions about the extension of "the coming struggle between America and Asia" to other planets.[97]

The uneasy duality between the transcendence promised by human spaceflight and the harsh political realities of the technology needed to achieve it grew out from the conviction, so widely held among the interplanetarians, that humanity's existence on Earth was but a wretched puberty to be left behind for a more ripened life in the cosmos. That Stapledon felt Americans were puerile "adolescents" in need of spiritual and ethical growth echoed a decades-long certainty that the Earthling represented an unfinished, and hence imperfect, project.[98] The notion that *Homo sapiens* persisted in a predeveloped state on its home planet traversed the entire span of interplanetary discourse. For Tsiolkovsky, humans were "infantile" compared with other beings in the universe that had enjoyed access to other planetary cultures or had simply inhabited longer-lived celestial bodies.[99] Writing amid an eager European literature regarding extraterrestrial intelligence, he observed that most of his fellow Russians could not conceive of aliens visiting Earth because they clung to the idea that humankind was "tied down" to the planet. Clarke and other writers often wrote of Earth as the "nursery" or "cradle" of human life. The species' precosmic experience represented merely its "childhood" or "infancy."[100]

Of course, the childhood thesis's natural corollary was that in space humanity would grow up. There human beings would build civilizations capable of outlawing war, abiding by the rule of law, abolishing material want and disease, and achieving universal respect and dignity for their diverse populations. As individuals, humans would undergo profound psychological transformations in space, which, as the characters in *In the Days of the Comet* had experienced, would expel the forces of racism, bigotry, selfishness, nationalism, and aggression. In their place, altruism, liberalism, worldliness, and pacifism would flourish. In time, humans would cease to be humans at all; they'd become something "over and above" humanity.[101]

What *was* new about interplanetary thought in the early postwar years—after the omen of the first V-2 attacks and the rocket's looming merger with the atomic bomb—was a shift in emphasis: from constructing the dreamworlds of the space future to notions of ethical responsibility in the present. Early advocates of spaceflight had conjured utopian dreams in part to stimulate enthusiasm and public expenditure for the rocket power needed to reach the heavens; but after 1944, when that power had been achieved and then distorted, voices suddenly called out to pump the breaks. "Morals and ethics must not lag behind science, otherwise (as our own recent history has shown) the social system will breed poisons which will cause its certain destruction," Clarke wrote in 1951. "With superhuman knowledge there must go equally great compassion and tolerance. When we meet our peers among the stars, we need have nothing to fear save our own shortcomings."[102] Philosophy, not technics, was the order of the day. Like so many others of his time, Stapledon thought atomic energy and space technology the djinns of Aladdin's lamp: its power for good or ill was equal. He hoped, therefore, that "science will be used wisely, instead of being abandoned to that blend of short-sighted stupidity and downright power-lust that has played so tragic a part in the application of science thus far."[103]

Whereas early interplanetarians had shaped popular understandings about outer space as a transformative place, the arrival of "vengeance weapons" demanded a crucial caveat: to profit from the mind-broadening benefits of cosmic conquest, human beings and their governments must sufficiently transform *before* the quest into space could begin. Much as he would have liked to have seen human beings touch the cosmos (he died of a heart attack in 1950), Stapledon held out hope that space colonization could wait "till mankind has attained a rather higher level of wisdom, and has a clear knowledge of the kind of world that would really favour human development."[104]

And the greater the rockets grew, the greater the plea. By 1956 the leader of Christendom felt it necessary to weigh in on the profundity of the coming age of space. "The common effort of all mankind toward a peaceful conquest of the universe should assist in impressing more deeply upon the consciences of men a sense of community and solidarity," intoned Pius XII at the Seventh International Congress of Astronautics in Rome. "The boldest explorations of space will serve only to introduce among men a new area of dissension if they are not undertaken with deep moral reflection and conscientious devotion to the higher interests of humanity."[105]

A year later, *Sputnik* screeched skyward atop the Soviet R-7 missile, the world's first ICBM.

CHAPTER 3

Star of Hope

Like so many Americans at the time, John McConnell was distressed. The date was October 31, 1957, and only four weeks prior he had read that the Soviet Union successfully orbited the world's first satellite, *Sputnik I*. By itself the artificial moon appeared benign. It was large and heavy—twenty-two inches in diameter and fully 184 pounds—but it contained only rudimentary electronic equipment, with its transmitters signaling an "eerie beep . . . beep . . . beep" to commercial radio stations across the United States. Yet *Sputnik* seemed to mark an end to the purported scientific and technological superiority of the United States over its rival; weakness in the nation's military posture, education system, and political will; and, crucially, a new vulnerability to surprise attack from Soviet ICBMs, identical to the rocket that had shot the historic "fellow traveler" into space.[1]

McConnell's anxiety, though, stemmed not from the gloomy reports of *Sputnik*'s implications for national safety but rather uncertainty about how his government would respond. To express his concern, McConnell penned a short editorial in the *Toe Valley Review*, his weekly newspaper in Bakersville, North Carolina. In "Make Our Satellite a Symbol of Hope!," he urged US leaders to launch a satellite in the spirit of friendship and unity rather than as an agent of competition with Soviet technocracy. A Pentecostal peace activist, McConnell considered the satellite not a threat to peace but instead an unprecedented opportunity to cultivate it.[2] He reasoned that if human beings had a singular goal toward which they could strive together, "the forces that make for peace and understanding [would] have the best chance to operate."[3]

Space exploration was such a goal. The technical obstacles to spaceflight, to say nothing of its expense, called for the evaporation of superpower competition and the collaboration of all nations. McConnell proposed the United States launch a satellite on behalf of the international community, a "Star of Hope." Engineers could equip the orbiter with a series of luminescent panels that would flash brightly in the night sky to remind stargazers of humankind's common interests and togetherness. McConnell suggested microfilming the signatures of everyone on Earth and attaching them to the satellite. "To create such a symbol would require no new discoveries," he pleaded. The brainpower and the technology were already available. All that was needed was the will to reject the Cold War's mandate to political conflict and mistrust. The government could launch by Christmas Day. "It is true that certain segments of humanity do not believe in the Event symbolized by the star of Christmas," McConnell admitted. "But there is no religion or no nation on earth ... that does not respond with hope and longing to the angel's song of Peace on Earth, Good Will to Men."[4]

McConnell's sentiments suggest an alternative interpretation for the early space age in the United States, particularly its social and political history. The swift and deafening public reaction to the sputniks in the press and in Congress has obliged historians to recognize the fear-inducing effect that Soviet accomplishments in space had on US institutions and psyches. Hurried American efforts to launch a satellite, establish a federal agency for aeronautics, and pass an education bill in the weeks and months after *Sputnik* led scholars to emphasize paranoia and national vigilance in describing the basic flavor of US politics and society after October 1957.[5] But the Soviet satellite, as McConnell's editorial testifies, catalyzed thoughts and feelings that contradicted the prevailing ethos. A host of scientists, engineers, journalists, and academics argued that the sputniks, though they seemed to have spurred the Cold War into a new domain, had actually opened a door to rapprochement. Viewing Soviet space feats not as a reflection of communist genius but of humanity's collective knowledge, optimists such as McConnell were hopeful that the quest to master space would unite human beings in a way that atomic energy had failed to do after Hiroshima.

The enthusiasm with which many Americans greeted the space age, and the salutary impact they thought space technology would render on international relations, revealed the continued intellectual currency of cosmic philosophies that had circulated in Britain, Germany, the Soviet Union, and the United States during the previous half-century. Post-*Sputnik* advocates of "space for peace," not content to predict or dream, felt responsible for ushering the earlier theories into practice. And indeed, whereas in the 1920s and 1930s the interplanetary project had constituted a loose philosophical and social discourse, in the late 1950s it hardened into a bona fide political campaign aimed at steering space exploration in peaceful directions. Idealistic notions about outer space moved from the meeting rooms of rocket societies, publishing firms, and the dimly lit offices of a handful of writers all the way to NASA, the State Department, and the UN General Assembly (UNGA). The conceptual links between space exploration and world peace that science-fiction authors and engineers had originally hatched were now advanced by lawyers, diplomats, and powerful politicians, up to and including the president. The torch passed.

Interplanetary ideas manifested themselves in early US space policy in myriad ways: leadership in the International Geophysical Year (IGY); cooperation with Europe in space science research; restriction of space weapons to preliminary study; and support for an inspection system that might guarantee the use of space rockets exclusively for peaceful purposes.[6] But at the dawn of the space age, two measures most captured the imagination of would-be sanctuary builders. The first was civilian control of the fledgling National Aeronautics and Space Administration. For those aspiring to prevent the Cold War from infecting humanity's future in space, competition among various institutions for control of the nascent space agency was a competition over the basic character of American space exploration. Having been beaten into space, the least the Eisenhower administration could do, many observers agreed, was to create a vivid contrast between its approach to space and that of the bombastic Nikita Khrushchev, first secretary of the Soviet Communist Party, who emphasized the new technology's translations to military power. Keeping NASA out of Pentagon hands meant the services could not run away with the robust space budget to pursue fantastical weapons projects that might escalate the arms race. James R. Killian, Eisenhower's newly appointed special assistant for science and technology, explained that the United States "must have far more than a program which appeals to the 'space cadets.'"[7]

Second, promoters of the sanctuary doctrine pursued international controls for space activities, particularly in the United Nations. Even before *Sputnik*, US officials determined that if space technology were "to be a blessing or a curse," developments in this unpredictable field would have to be brought "within the purview of a reliable armaments control system." Only the United Nations seemed to have the proper authority to ensure the peaceful development of space technology and thus guarantee the status of space as a weapons-free sanctuary.[8] Having failed to place the atom under international auspices in the 1940s, the United States was confronted, thought US Ambassador to the United Nations Henry Cabot Lodge, with "a similar opportunity to harness for peace man's new pioneering efforts in outer space." As he told the First Committee of the General Assembly six days after *Sputnik*, "We must not miss *this* chance."[9] Within a year the interplanetary mood had pushed the United States toward the establishment of a special ad hoc committee—the Committee on the Peaceful Uses of Outer Space (COPUOS)—that would explore UN resources for peaceful space activities, develop international cooperation, and study the legal problems that might arise from exploration. After the General Assembly established the COPUOS as a permanent body the following year, it became the principal venue through which UN member nations negotiated international space law, including the landmark 1967 Outer Space Treaty. In *Sputnik*'s backwash, one could glean not only struggle in space but also the beginning of the medium's governance.

Sputnik Globalism

Urgent as sober policies for space seemed, it was difficult for the new generation of interplanetary advocates to be heard amid the national cacophony after October 4, 1957—the *Sputnik* "shock." Eisenhower's enemies in Congress, the press, and everyday Americans inveighed against the president for allowing the Soviets to beat the United States into space and for reacting so nonchalantly. Particularly revealing were the hundreds of letters that poured into the White House over the following weeks pleading for increased spending on defense, science education, and the technological infrastructure necessary to compete with the Soviet Union. Many citizens begged Eisenhower to increase taxes in order to speed up the US space effort. Some simply wrote to scold the President. "Our error in every case," one citizen scribbled, "is that we have not set out to WIN THE COMPETITION but only to exert a limited effort. In a war or contest, the second best is a loser!" Another promised that if the United States did not soon surpass the Soviets in space, he would never vote Republican again.[10]

But panic and scorn were not the only reactions to *Sputnik*. As opinion polls conducted in the weeks and months after the launch revealed, indifference was another. According to one survey, 54 percent of Americans had never heard of Earth satellites and only 20 percent had any notion about the nature and purpose of satellites. When pollsters asked a sample of Baltimoreans what they thought to have been the most important thing to happen during the previous week, only 59 percent thought the world's first satellite preeminent. Many thought the crisis in Little Rock, the World Series, and "their own affairs" to be more important. Remarkably, although *Sputnik* had surprised more than half of Americans, fully 44 percent reported not to have been surprised. Gallup found, similarly, that in Washington and Chicago opinion on the threat of *Sputnik* were evenly mixed. Forty-six percent of respondents said the satellite had not dealt a serious blow to American prestige; only 43 percent said it had. Allen Hynek, associate director of the Smithsonian Astrophysical Observatory, told the Associated Press that laypeople he had met reacted casually. "Their attitude," he reported, "seemed to be that we had lost the ball on the 40-yard line but would surely win the ball game.... It was a shocking mixture of complacency and superiority." Measuring the pulse of the American people after *Sputnik*, the political scientist Donald N. Michael concluded that "opinions did not indicate unanimous psychological shock or national loin girding, as the press and many issue makers have insisted."[11]

Beyond apathy, many Americans conveyed enthusiasm, even elation, for the Soviet satellite, for they perceived in it an opportunity to deescalate the Cold War. Among the letters addressed to the White House were expressions of hope that the Soviet satellite—and indeed any future US satellite—would foster greater understanding between nations and become a force for peace. Church groups, peace advocates, and women's organizations implored Eisenhower to make the first US satellite "a positive expression of our desire for peace."[12] The Women's Prayer Crusade for World Order and Peace asked why a US satellite could not mean "something far different to the world, why it cannot be a symbol of World Unity and Peace and why, instead of fear, it can't bring God's blessing to all as it wings its way around the world."[13] Many citizens suggested names for the satellite that would inspire both admiration of the United States and recognition that any launch would not only be a national accomplishment but also a victory for all humanity. "Satellite for Peace," "Star of Hope," and "Freedom Sphere" were among the most popular recommendations.[14] "Think," wrote one, "what a far-reaching impression this would have on the peace-loving populace of the world, plus the immeasurable amount of prestige and dignity that would come our way as a result?"[15] If the first US satellite could be "made to look like a star," another suggested, perhaps the it would carry "not a sinister 'beep,' but a light of hope."[16]

In a matter of weeks, John McConnell's original proposal to orbit a satellite in the name of world peace grew into a nationwide education program—itself named Star of Hope—aimed at promoting space exploration as a "moral equivalent to war." Based in San Francisco, Star of Hope printed pamphlets, newsletters, and editorials dedicated to advocacy of a robust US space program completely divorced from the Department of Defense. Star of Hope would "serve international understanding" by pressuring the United States to launch a visible satellite symbolizing "world friendship." "It will be a civilian rather than a military satellite," the organization's Statement of Purpose declared, "and launched through the participation of the world scientific fraternity."[17] In a document simply entitled "Purpose," McConnell wrote a short pledge he planned for the satellite to carry into space and to which all Star of Hope members were to commit themselves: "I, a citizen of this planet, dedicate my friendship and knowledge to work for peace among all men. I will aid the efforts that heal, build, and unite."[18]

Compared to, say, the campaign for nuclear disarmament, as a peace movement Star of Hope was diminutive. Outside of San Francisco and the local towns surrounding Bakersville, McConnell enjoyed limited visible support. But his organization was not insignificant, either, largely because its chair promoted it so tirelessly. He wrote to the editors of more than 5,500 local newspapers urging them to tell their readers about the Star of Hope (figure 11). Many obliged. The *Washington Post*, *New York Herald Tribune*, *Chicago Herald*, *Baltimore Sun*, and *Christian Science Monitor* each ran stories about the activist's idea. McConnell appeared on the *Today Show*, the *Arlene Francis Show*, and dozens of radio broadcasts to spread the word. He received letters of support from numerous public officials and powerful personalities including Adlai Stevenson, Eleanor Roosevelt, Billy Graham, Wernher von Braun, and Senator Henry "Scoop" Jackson (D-WA). Both senators from McConnell's home state of North Carolina, Sam Ervin and W. Kerr Scott, also declared their support for the project. Chester Bowles (D-CT), later a foreign policy adviser

FIGURE 11. Star of Hope pamphlet. "By Laws for the Regulation, Except as Otherwise Provided by Statute or its Articles of Incorporation of Star of Hope, Inc.," n.d.; Star of Hope Pamphlet, Star of Hope proposals/ideas, box 33, JMP.

to John F. Kennedy, wrote approvingly to Star of Hope's secretary, Peter Hill, that "all of us who are looking for a world of peace and goodwill will sincerely hope that the 'race to the moon' can be turned into a more constructive exploration of and use of outer space."[19]

For a brief time between late 1957 and mid-1958, the donations that Star of Hope received allowed McConnell to travel widely in support of a goodwill satellite, including the 1958 Atoms for Peace Conference in Geneva, where he showed up with translations of his proposal. McConnell's primary goal at the conference was to acquire as many signatures as possible. Looking to gain the assent of Eisenhower, he was particularly delighted to learn that Glenn T. Seaborg, the US delegate and a frequent adviser to the president on science matters, was willing to support Star of Hope if McConnell could acquire a signature from the Soviet delegation. McConnell had written to Khrushchev the previous October, explaining that his endorsement "would add meaning to the declaration and inspire people throughout the world to do their part in building foundations for world peace." Though he had not received a reply, McConnell acquired the Soviet signature on the last day of the conference. By then, however, Seaborg had left.[20]

Although a US satellite launched in the name of "world friendship" gained minimal traction inside the White House, the idea that satellites might ease Cold War tensions permeated popular discourse in the weeks and months after *Sputnik*. Star of Hope, in fact, had prominent imitators. S. Fred Singer, then director of the Center for Atmospheric and Space Physics at the University of Maryland, penned a "Statement of Conscience" for the *New York Herald Tribune* urging US leaders to inaugurate programs that would advance space exploration "in a spiritual and moral way." Nothing could better symbolize the ethical foundation of US-led exploration than "a peace and goodwill satellite" that other countries would juxtapose to the secretive Soviet program. Congressional staff saw fit to include Singer's "Reply to *Sputnik*" in a special document collection for US representatives a year later. In it, Singer proposed that in addition to launching a goodwill satellite, the United States should collaborate with the Soviet Union and other nations on a manned lunar landing. He asked, ruefully: "Isn't the prospect of an interplanetary voyage much more appealing than [a] devastated Earth![?]"[21]

Singer partnered with noted French American industrial designer Raymond Loewy—well known in New York for his stylish trains and cars—who declared his company was teaming up with Singer and other satellite engineers to launch a "Star of Goodwill," what he hoped would be a "spiritual companion" to *Explorer 1*. Like many Americans, Loewy was disappointed that the United States had not been first into space, but not because being first would have proved the nation's technological superiority. Instead, he felt that the Soviet Union's achievement heralded only national greatness; a US satellite, on the contrary, would represent "this country's desire for world peace, for global fraternity and good will." As he wrote on December 10, "The Russian attitude allows for no such interpretation of the Sputniks." Like McConnell's proposal, Loewy's Star of Goodwill would have "absolutely no military connotations" and instead carry microfilms "of the flags of every nation and a symbol of every known religion."[22] Not only did NBC interview Loewy about his idea on television but on January 29, two days before the United States successfully launched its first satellite into orbit, Lyndon Johnson (D-TX), then immersed in national space policy, submitted the Star of Goodwill proposal to the congressional record while on the Senate floor.[23]

The quantity and prominence of individuals dedicated to space exploration as a humanistic and unifying adventure pointed to the rise of a new interplanetary figure in the United States, one that can fairly be called the "*Sputnik* globalist." These were individuals for whom the Soviet satellite stimulated hopes of peace. Humanity's first thrusts into the cosmos encouraged them to think more cooperatively, more diplomatically, and with greater recognition of the unifying potential of new technologies. As the term "globalist" implies, these actors believed that interdependency was the fundamental condition of modern life. While prizing diversity, they esteemed allegiance to humanity over religion, nationhood, or race. And, crucially, they argued governments should derive policy from considerations of the global. *Sputnik* globalists believed that the surrender of national sovereignty to international organizations was a necessary precondition for space exploration. The cosmos, a sacred zone to be reserved for the betterment of human civilization, and space technology, the means to reaching that zone, should be governed by all.[24]

Sputnik globalists were stitched together by competing emotions. The first, what they considered to be the antidote to Cold War politics, was empathy. If only the superpowers could truly appreciate the enemy's desires and insecurities, perhaps each could take a measured step away from conflict. McConnell recommended that US and Soviet leaders trade places for a time. "Why not move President Eisenhower and Congress to Moscow and then have Khrushchev and the Presidium conduct their business from Washington?" he asked Star of Hope members in May 1958. "This should give satisfactory assurances to both sides that there would be no surprise attack, for it is assumed that those who would have to press the button for an Atomic attack, would not do so if they were to be the first casualties."[25] After seeing McConnell on the *Arlene Francis* show, a woman from Los Angeles wrote to the program about a grand reconciliation in space. God, she surmised, was waiting for NASA to launch its first satellite "so that there may be a conference in the sky of the big three . . . God, America, and Russia." She anticipated that God would set aside a small planet on which the United States and the Soviet Union might work out their differences, one God might dub "Amussia Land" (a cross between America and Russia) to reflect mutual respect and cooperation.[26]

Fear, too, motivated *Sputnik* globalists, an emotion they marshalled against what they perceived as the endless perpetuation of the Cold War. Figures like McConnell, Loewy, and Singer feared that leveraging space technology for military power would feed the garrison state and increase the likelihood of nuclear war. Echoing the concerns of many commentators, one Star of Hope sympathizer wrote that he hesitated to associate with the goodwill satellite proposal given his "serious doubts as to whether the project could be freed of possible exploitation in militaristic or super-patriotic ways."²⁷ Peter Hill wrote to another member that fear of a general war with the Soviet Union had "paralyzed our moral capacity as a nation." Intercontinental missiles had made no room for "intercontinental kindness."²⁸

Alarm, in turn, morphed into acrimony. *Sputnik* globalists displayed a clear frustration with—even outright hostility toward—Cold War nationalism, an enmity made particularly intense by their sense that East-West competition had come to define the space age. For them, it was not national achievement and Cold War rivalry that had sparked the first satellites but rather the cooperation of dozens of nations in the IGY, an eighteen-month endeavor (July 1957 to December 1958) to explore the Earth's atmosphere, polar regions, and oceans. When planning for the IGY was just getting underway in mid-1955, a Russian-American citizen implored the government to pool its resources with the Soviets and build a space program that would reflect the accomplishments of "all nations," not merely those of the United States. Space exploration, in his humble estimation, should be "a United World project." He exclaimed: "Let us face the outside world as mature citizens of the planet Earth!"²⁹ In another letter to Eisenhower, a Massachusetts resident emphatically resisted all post-*Sputnik* calls to race the Kremlin in outer space. The best response to the Soviet advantage in technology, he insisted, "is not to strive frantically to beat Russia to the punch—as you and your supporters seem to be doing." This would only exacerbate tensions. To avert "disaster," the United States must immediately cease all development of armaments, as well as all nuclear weapons testing. "And the way to stop," he reminded the president's staff, "is to stop! Not talk! Not try to connive to gain some petty advantage over your adversary! Stop being an adversary! Become a friendly competitor!"³⁰ McConnell summed up his organization's attitudes toward the Cold War in a simple choice: "Grow up or blow up."³¹

Not confined to the scattered activists of Star of Hope or individual citizens outraged enough to complain directly to the president, one could find *Sputnik* globalists in the ranks of leading American opinion makers, who helped convey interplanetary ideas to a wider audience. In writing his editorial in the *Toe Valley Review*, for instance, McConnell had drawn inspiration from an article written two weeks before in the *Saturday Review* by its larger-than-life editor, Norman Cousins. *Sputnik* was an ironic event, Cousins mused, for the very inauguration of humankind's greatest and most exciting adventure threatened also to "pulverize" human societies. Regrettably, the significance of the satellite was "measured more in terms of space platforms from which intercontinental wars can be waged than in terms of man's new capacity to extend his citizenship to the universe." Such was the "unhappy meaning" of *Sputnik*. There was "no universal release or jubilation," only hard talk among the American public and in policymaking circles about the weakness of US science and technology and the need for a vigorous national response. Yet improvements in education, science, technology, and defense were not the essential requirements of the moment, Cousins insisted, but rather improvement "in our reasoning, in our judgement, and in our moral imagination." The United States should surrender its adherence to "unfettered" national sovereignty in favor of allegiance to the idea of the United Nations. Only world government, he argued, could ensure that missiles and satellites would not carry hydrogen bombs. Rather than merely "conjure up more effective ways of destroying the world," Cousins urged tapping "our intelligence and moral imagination to the fullest in creating a working design for a better tomorrow in which all the world's peoples can share."[32]

Malvina Lindsay, an insightful columnist for the *Washington Post*, agreed. She questioned whether humankind was yet ready to explore outer space, given its "current national and group squabbling." To become true "space citizens," Americans and Soviets would first have to abandon prejudice, hatred, and allegiance to their respective political, ideological, and economic systems. Suppose the Soviets first made acquaintance with foreign beings in the universe, she asked: "Is it possible these might be more highly developed than themselves—yet not Marxists!?" What if it were the Americans who were the Earth's first diplomats to space? "What if we there find beings more advanced than ourselves, who know nothing of people's capitalism, of earth's religions, of our Constitution, and who refuse to accept the American way of life?" Here was a plain critique of the status quo. Lindsay hoped that one day, a "Man from Mars" would appear, for he could tie the superpowers together in peaceful exploration of outer space.[33]

It was easy to be pessimistic, though. As it stood, the United States was not a responsible space citizen. In separate article Lindsay captured the anxiety of the times in a fictional conversation between "Mr. Hardpann," whom readers were to understand as the security-minded US government, and "Mr. Stellar, the indefatigable idealist." "Why think of a satellite only in terms of weapons?" asks Mr. Stellar. The United

States could choose to open a "new frontier of peaceful opportunity" by framing its first satellite launch as a hope for the future, rather than a tool of national prestige. But Hardpann scolds him. "Be realistic, Stellar. Outer space isn't the proper place to talk of peace." The only way to achieve peace would be to "beat the Russians to the moon... to bargain from strength." Stellar replies that world opinion had already turned against the United States because of its militaristic posture. "There's alarm everywhere," he noted, "because both we and the Russians are treating this new chapter in man's mastery of nature as just another development in the cold war." Despite Mr. Stellar's appeal to the better angels of human nature, Harpann gets the last word. The Cold War and its politics were an inescapable reality. The United States would be forced to race, Hardpann concluded: "It's no use."[34]

"Space for Peace" and the Birth of NASA

Assessments like Lindsay's appeared everywhere. Their ubiquity suggests that *Sputnik* globalists did not inhabit an intellectual fringe. In many ways, they reflected the national mood. Space for peace was "an unassailable position" in the wake of the Soviet satellite.[35] But how might thinkers like John McConnell actually influence space policy in an environment in which the government was likely to subordinate their alarms and aspirations to the more immediate national security requirements of the Cold War? Other than write to members of the administration or voice their concerns in the press, there appeared few options.

It was to their benefit, however, that the Eisenhower administration, ever mindful of the worldwide psychological impact of US policies, recognized the potential of Sputnik globalism for image making. The president observed that the Soviet Union's space program was shrouded in secrecy, tied inextricably to the military, and used by Khrushchev to boast about the superiority of communist organization and technology. In February 1958, Rand Corporation political scientists emphasized to the president's National Security Council (NSC) that acquiring superiority in space meant more than achieving dramatic firsts or developing more powerful ICBMs and sophisticated satellites. Maintaining a level of moral—that is, political—advantage over the Soviet Union was equally if not more important. The American program, therefore, must be kept open, peaceful, and scientific. "From now on," their report urged, "the US should recognize the need for restoring credibility in US superiority, stress our peaceful intentions and [the Soviet Union's] aggressive ones, and *disclose* and *publicize* US outer space activities according, first and foremost, to the effect on the US international position."[36]

Top US officials had already taken initial steps to equate American space exploration with peace and cooperation over the course of the previous year. In his State of the Union address that January, Eisenhower had proposed to "mutually control the outer space missile and satellite development." That same month, Ambassador Lodge submitted a proposal to the General Assembly that space be "devoted to exclusively peaceful and scientific purposes." And at the Four-Power Disarmament Conference in August, the US delegation called for technical studies for an inspection system to ensure that space vehicles were not intended for military use. In *Sputnik*'s wake, however, efforts of this kind had to be doubled if the United Sates was to fashion a favorable image of its space activities.[37]

Eisenhower attempted to do just that in a series of high-profile letters to Soviet Premier Nikolai Bulganin in January 1958. In a broader jeremiad on the development of ever-more destructive weapons, the threat of nuclear war, and the absence of support for the United Nations, the president warned his counterpart that there would soon be "powerful new weapons which, availing of outer space, will greatly increase the capacity of the human race to destroy itself." The two superpowers thus stood at a "decisive moment in history" in which they could realistically curb the arms race. Eisenhower recalled with regret the missed opportunities for international control of atomic energy during the mid- to late 1940s and considered the frontier of space technology an opportunity for rectification. "Let us this time, and *in* time, make the right choice, the peaceful choice."[38]

Following an address at the National Press Club on January 16, Secretary of State John Foster Dulles implored Bulganin to accept the president's outstretched hand. He suggested that "if they are at all sincere in their profession of peace," the Soviets should "jump at the chance" to control both space technology and the medium itself. In line with Eisenhower's rhetorical theme of lost opportunities, Dulles reiterated that the United States had offered in the spring of 1946 to put nuclear weapons under international control. "Never in history" had such a "great and generous a gesture" been extended by a country in such a hegemonic position. Given the Soviet Union's ostensible lead in satellite technology and missile propulsion, American offers of control may appear to be "sour grapes," he said; but the elder statesman hoped, "from the depths of my heart," that "platitudes about peace" could be transcended to ensure that space would be used in the service of "science and humanity and not in the interests of war."[39]

How genuine were these overtures? Were there not significant advantages to be gained from achieving superiority in space? Was the United States merely attempting to stall a weapons race in space because it was already behind? In the Kremlin's estimation this was certainly true: Washington was trying to put under international auspices that which it did not possess. Yet many US officials responsible for national space policy already maintained a genuine eagerness to preserve space as a demilitarized zone, both because space weapons were costly and impractical and because the benefits to American prestige would be greater if the United States explored space solely in the interests of science.

The President's Scientific Advisory Committee (PSAC), which Eisenhower endearingly referred to as "my scientists," embodied the administration's measured approach to space. It consisted of some the United States' most eminent scientists, including Hans Bethe, I. I. Rabi, and Lee DuBridge, and was chaired by James R. Killian, president of the Massachusetts Institute of Technology. Most of the committee's scientists harbored serious doubts about the capability of space technology to achieve military superiority vis-à-vis the Soviet Union and instead advocated, in Zuoyue Wang's phrase, "technological skepticism," an awareness of technology's limitations as well as its possibilities. Machines, they insisted, could not be divested from their social, political, and anthropological contexts. They could not solve every problem, but unchecked, could cause them. This cautiousness, like that of so many Sputnik globalists, came from the struggle to control nuclear weapons. Most PSAC scientists were liberals and moderates who had worked on the atomic bomb and radar during the war, had supported J. Robert Oppenheimer in the H-bomb debate, and had worked for Truman's Science Advisory Committee during the Korean War. They hoped after *Sputnik* that incipient developments in space technology would have a salubrious effect on foreign policy and thereby redeem science from its immoral associations with the bomb.[40]

Similar convictions, rooted in politics and belief more than in science, extended to the president himself. At the beginning of his first administration Eisenhower had positioned nuclear weapons at the center of US national security strategy despite his discomfort with the moral questions surrounding their use and development. Whereas in 1952 the president had been sure that the bomb would be indecisive in a future war with the Soviet Union—America's industrial capacity, its "Detroit deterrent," would remain the most important factor—by the mid-1950s, after nuclear crises in Berlin, Suez, and the Taiwan Straits, he was no longer so confident in the wisdom of "massive retaliation." *Sputnik* began its first orbit amid this reorientation in Eisenhower's thinking. Even this most ardent supporter of the "New Look" paused to reflect on the volatility of a deterrent balance in which each superpower possessed quantities of orbital bombers, antisatellite missiles, lunar bases, and manned space stations.[41]

In no area of policy were these considerations as consequential as the embattled government's deliberations regarding a federal agency for aeronautics. Over the spring and summer of 1958, Eisenhower and his science advisers determined that the institutional contours of what eventually became NASA would define the government's space efforts both at home and abroad. The elemental choice between military and civilian control was central to the debate over exploration and American space policy more broadly. An agency housed under the Department of Defense (DoD), many officials believed, would bestow undue influence and power upon the services, induce uncontrollable expenditures, and present an aggressive face to the US space program. Civilian control, on the other hand, would promote international cooperation, attract the best scientists, and mark the United States the leader in peaceful space exploration.

From the time the United States first began satellite research early in the 1940s, the struggle for space was intense. Because there was no agreement about where "air" ended and "space" began, the Army Air Force (AAF) argued that outer space was naturally its area of expertise. The "X-series" of jet aircraft were representative of this claim. The Bell X-1 broke the sound barrier over California in 1947, and the Air Force designed the X-15 to fly at fifty miles above the Earth, at the edge of the atmosphere. Conversely, the Army emphasized its control over ballistic rocketry in the immediate postwar years and, under Eisenhower, antisatellite ordnance. The fight over institutional claim to space peaked in 1958 as Congress debated the contours of the space bill. The would-be agency seemed at home in any number of bureaucracies: the Army, Air Force, Navy, Atomic Energy Commission (AEC), and National Advisory Committee for Aeronautics (NACA) each vied for a space agency under its exclusive control. Other recommendations included joint ventures between the National Science Foundation (NSF), the National Academy of Sciences (NAS), the universities, and a cabinet-level Department of Science, as well as a new independent agency. In the interests of expediency, too, some officials advocated giving responsibility for space to the DoD's Advanced Research Projects Agency (ARPA), which Defense Secretary Neil McElroy had just created under the leadership of General Electric executive Roy Johnson and physicist Herbert York to consolidate military space missions. Its care of several ongoing space projects, including satellite reconnaissance, made it as good a nest as any.[42]

Considerations of domestic and world opinion conditioned internal debates about the establishment of a space agency. Concerns that space technology would tread the same path that atomic energy had a decade prior ensured that the most important feature of the bureaucratic struggle was the proper balance between civilian and military authority. In a memo to Eisenhower three months after *Sputnik*, Killian expressed that although it was "entirely feasible" for the Pentagon to sponsor R&D for space, there were "deeply-felt convictions that the more purely scientific and non-military aspects of space research should not be under the control of the military." Such an arrangement would limit the new agency "to narrowly military objectives" and, crucially, put the United States "in the unfortunate position before the world of apparently tailoring all space research to military ends."[43]

Initially Eisenhower disagreed. He favored military control given the services' jurisdiction over most of the relevant technology and facilities, the "paramountcy of defense aspects," and most pertinently his desire to avoid undue duplication of space research. While it was a foregone conclusion that any new space agency would occupy itself with peaceful research, it was incumbent on the administration to capitalize on DoD's substantial experience and resources in space, the Army Ballistic Missile Agency (ABMA), Naval Research Laboratory (NRL), USAF's Ballistic Missile Division, and Cal Tech's JPL chief among them. "Defense could be the operational agent, taking orders from some non-military scientific group," he thought in a February 4 meeting with GOP leaders.[44] Later in the day, when he met with PSAC members Herbert York and George Kistiakowsky, Eisenhower remarked that DoD objectives in space should receive "the highest priority" and that he "did not think that large operating activities should be put in another organization." It was unwise, he thought, to concentrate talent outside the Pentagon.[45]

Others were not so sure. In the same meeting Killian and Vice President Richard Nixon, appreciating the impression a space agency would have on "our posture before the world," thought it wiser if nonmilitary research in space were carried out by an agency "entirely separate from the military."[46] S. Paul Johnson, director of the Institute for Aeronautical

Sciences, similarly argued that the exploitation of space fell "more nearly into civilian-scientific areas rather than into military areas." Probing outer space, he wrote in a memo to Killian, would be of greater interest "to the scientist than to the strategist." Referring to the ideas for military applications of space technology then circulating in the services' competition for the agency, Johnson remarked in jest that "we can discount at this time most of the 'Buck Rogers' type of thinking." Instead, the principal military interest in space remained improving surveillance, communications, and long-range weather forecasting.[47]

On March 5 Eisenhower received an important memorandum from Killian, Bureau of Budget (BoB) director Percival Brundage, and Nelson Rockefeller, chair of the president's advisory committee on government organization. The document recommended that the civilian space effort be housed under a revitalized and redesigned NACA. Despite budget cuts and mission rollbacks dating from World War II, the organization possessed an experienced technical staff and robust research facilities, had moved aggressively into space research, and notwithstanding its civilian posterior had a history of collaboration with the DoD. If not permitted to wade deeper into the space pool, what reason was there for NACA in the first place?[48] The Defense Department, by contrast, was "a military agency in law and in the eyes of the world." Giving the space program to the Pentagon, the memo stressed, "would be interpreted as emphasizing military goals." Nonmilitary aspects of the space program could fall into neglect, cooperation with other nations in space science could be made more difficult, and civil-military relations could suffer under stringent secrecy requirements. The law should be amended to allow NACA to tap DoD resources, provide for a single director reporting directly to the president, free it from civil service payrolls, and permit contracts with private industry.[49]

This advice seemed to have made an impression on Ike, who in a meeting with the memo's authors reversed his earlier enthusiasm for a space agency centered on DoD competency. Eisenhower agreed with Killian's assessments about the "limited scope" of military space activities over the near term. Although military use was "acceptable" in terms of "application of knowledge," the president remained "certain ... that discovery and research should be scientific, rather than military." Except for ballistic missiles, Eisenhower "felt that there is no problem of space activity that is not basically civilian." He immediately ordered BoB to draft a space bill based on NACA before Congress adjourned for Easter.[50] Budget moved quickly, so that by April 2 Eisenhower was pitching the new legislation in person before a joint session on the Hill, where he emphasized that "a civilian setting for the administration of space function[s] will emphasize the concern of our Nation that outer space be devoted to peaceful and scientific purposes." That same afternoon the president issued a memo to DoD and NACA officials ordering an expeditious transfer of all nonmilitary projects to the latter, civilian agency. While the Pentagon would retain all programs relating to "military weapons systems or military operations," NACA would take over the remaining work.[51]

The fledging space act received a substantial boost from the PSAC, which authored a best-selling pamphlet in line with the administration's growing consensus on civilian control. Admittedly the entire enterprise of space exploration had been inaugurated by the "military quest" for long-range weapons, the report began, and indeed there were substantial benefits of space technology for national defense. But the impetus for exploration, the scientists argued, derived at a more foundational level from humanity's "compelling urge" for discovery. The military exploitation of space was unavoidable given the circumstances but was necessary only "to be sure that space is not used to endanger our security." In its concluding paragraphs on military applications, the PSAC derided the copious speculation about satellite bombers, moon bases, and advanced space weapons then circulating in popular science magazines. "For the most part," it wrote, "even the more sober proposals do not hold up well on close examination or appear to be achievable at an early date."[52]

That the PSAC chose to address the more fantastical possibilities of military spaceflight revealed a crucial reality: the general public, to which the pamphlet had been addressed, associated the "military" use of space with the futuristic and encompassed the elemental questions of war and peace in space with which *Sputnik* globalists like John McConnell, Norman Cousins, and Malvina Lindsay concerned themselves in the aftermath of the Soviet satellite.

Nowhere was this association clearer than in the frenzied congressional debate over the BoB's draft bill, which took the military-civilian divide to near-theatrical lengths. On May 20, the Subcommittee on National Security and Scientific Developments testified before the House of Representatives. John McCormack (D-MA), chair of the subcommittee, considered militarized space to be a "very grim prospect." McCormack warned of "manned orbital bombers traveling at satellite speeds" and space-based military bases capable of destroying Earth-sites with ease. His colleague Kenneth Keating (D-NY) imagined "death rays from whirling mechanisms" and "missiles launched from outer planets." These possibilities should not be dismissed as "fiction," they argued. Those responsible for policy could not "afford to let such tragic works of human ingenuity take place," lest they risk "global suicide."[53] During the hearings, there was widespread agreement that the US space program should avoid such dark potentialities, and instead be devoted to improving life on Earth. The prospect of interplanetary travel or colonies on the moon made anything possible. Space technology, if devoted to peaceful pursuits, would unleash knowledge that the United States could use to "eliminate poverty everywhere," "bloom the desert," and even "change the climate of various regions of the world."[54]

Indeed it would be the United States, in particular, bestowing the fruits of new human adventure to the world. The reasons for a space program devoted to peace, McCormack reasoned, "lie in the direction of the American tradition of friendly world leadership." Adopting NACA as the operative agency was necessary to show "that we not only believe in peace, but ... are willing to take concrete steps with all nations and peoples of the world to achieve peace." Dalip Singh Saund (D-CA) argued that the passage of the bill should be expedited "in order that America may once again take the lead in this splendid approach" to space. Keating urged fast-tracking the legislation because it would demonstrate that the United States was aiming for "a future of vision[,] not vexation." "It will show the world," he continued, "that we have no desire to engage in swashbuckling among the stars ... that we choose the path of greatness, not oblivion." McCormack even suggested that if any other nation "got a decided advantage ... they would dominate the world and impose their will," but that the United States would "never do that." He explained, "Whatever advance the United States would make would not be for the purposes of imposing its will on others but for peaceful purposes."[55]

In addition to strengthening America's leadership in the promotion of peace, a civilian agency had the potential to foster greater trust among the nations of the world and protect outer space from long-standing hatreds. "We want to keep space as an area where mankind can put aside quarrels and there work for the betterment of all men through our understanding of the universe," McCormack preached. Keating seconded this utopian vision in his own statement. It was now possible for mankind, "freed from the ties of his earth-bound existence," to shake off "some of the jealousness and differences which beset human affairs." Those gathered at the hearing all agreed that these ideals were a real possibility, and that the United States should pursue them for moral reasons. Several representatives commented on the "unpolitical" and "nonpartisan" atmosphere surrounding these discussions. "There has not been one dissenting voice," McCormack remarked.[56]

Though there were in fact many dissenting voices, particularly in the Air Force, the bill had moved quickly through Congress in the spring. Johnson and Senator Styles Bridges (R-NH) introduced the Senate version of the space bill (S.3609) on April 14. McCormack introduced the House's draft (HR.11881) the following day. Both versions had passed by mid-June, and a bipartisan blue-ribbon panel chaired by Johnson met to iron out major differences between them. The final snag proved to be the agency's proper authority. Whereas the House bill had proposed a relatively weak seventeen-member advisory committee, the final bill retained the Senate's proposal for a much stronger seven-member policy board, which Eisenhower opposed on grounds that it would undermine presidential power and encourage bureaucratic logrolling.[57]

Johnson and Eisenhower met on July 7 to break the impasse. To assuage the president's concerns regarding the possibly diminished role of the executive in space matters, the Texas senator suggested that Ike serve as chair of the policy board. Eisenhower assented, and on July 29 signed into law the final, compromise bill. That autumn, the federal government would abolish the NACA, and subsume its infrastructure under a new agency, the National Aeronautics and Space Administration. A civilian agency, NASA's chief philosophy was for activities in space to "be devoted to peaceful purposes for the benefit of all mankind." Yet the Pentagon would still get its piece of the pie. Projects "peculiar to or primarily associated with the development of weapons systems, military operations, or the defense of the United States," the act's declaration of purpose read, "shall be the responsibility of, and shall be directed by, the Department of Defense." NASA was also to furnish for the Pentagon and its subsidiaries any "discoveries that have military value or significance." Henceforth, the president would decide whether a given program met this description; hence, the amorphous barriers between military and civilian space activities would permit the chief executive to justify transferring a given project to the services or the wizards at ARPA. The Space Act established a Civilian-Military Committee with seats reserved for DoD and at least one representative from each branch of the military.[58]

NASA was both a cornerstone in the Eisenhower administration's plan to equate the United States with the peaceful uses of outer space and evidence of how firmly entrenched the cultural connections between space exploration and international renewal had become in popular consciousness. The new agency reflected Eisenhower's vision for space as well as his desire to build contrasting images of Soviet and American science. He could now boast of an open, liberal-democratic space program controlled by civilians. The Soviet program, though it had "captured the imagination" of the world with its accomplishments in space, remained secret, autocratic, and controlled by the military. While the federal government could not ignore the space race—indeed it created NASA to address this—US officials emphasized the agency's civilian character to suggest that a nation's intentions in space were equally relevant to technological outcomes.

March to the United Nations

As the conflicted language of the Space Act indicated, it remained an open question whether an irenic American space program was sufficient to prevent the arms race and US-Soviet tensions generally from extending into space. As the law stipulated, DoD would continue to control all projects deemed relevant to national security, and as we shall see, this included a veritable mountain of programs ranging from obvious military applications like missiles and satellite reconnaissance to those that overlapped neatly with civilian goals like navigation and weather forecasting. At the same time, Soviet leaders marked NASA a manifestation of the space program's duplicity. They proclaimed—accurately—that programs for war-making and scientific research were inextricable, the USSR's classified military program dispersed among varied (and competing) Experimental Design Bureaus and the Strategic Rocket Forces. Would the newborn civilian program be strong enough to withstand the established interests of the Pentagon, aerospace defense industry, and cold warriors in Congress? Would the Soviet Union and, for that matter, the ambitious space programs of Britain and France, follow the American example? The military utility of space and the impetus for competition having already been established, the answers were far from certain.

As Eisenhower's exchanges with Nikolai Bulganin suggest, the lesson of the atomic revolution hung heavy over early deliberations about space exploration: it taught that once technology crossed a certain threshold of development under national auspices, for national purposes, control would prove impossible. In speech after press release, article after newscast, contemporaries expressed their fears that policymakers would "make the same mistake" with space technology as they had done in 1945. Cousins captured the original sin of the atomic bomb in the "Sense and Satellites" article that sparked John McConnell's Star of Hope campaign. "We never paused long enough to think through the meaning of the nuclear explosives we so ingeniously created," he had intoned. Regarding the bomb as "just another weapon," US leaders failed to appreciate the "vast intelligence and imagination . . . required to keep the new force under control."[59]

Now, when the same generation of cold warriors again held the reigns of technological revolution, it was imperative that new paths be charted, new initiatives explored, new prudence and patience called forth. Given ten years of discouraging disarmament talks between the superpowers, Norma Herzfeld, vice president of the Catholic Association for International Peace, and her husband Charles, a University of Maryland physicist, urged UN control of outer space. "Very soon," the couple warned, "it will be too late to do anything and the world will be exposed to cosmic terror and blackmail." Of course, international control of space would not stop nuclear weapons development or the ability of nations to use space research to build better ICBMs. Yet control would establish an important precedent for peaceful cooperation, and "divert the fascination of mankind from the power struggle of today toward a thorough conquest of the universe carried out in common."[60] Even before *Sputnik*, Henry Cabot Lodge urged the General Assembly to "take the problem [of space technology] in hand now before future developments complicate the problem of control in this field."[61]

Even among those analysts of the situation who advocated military vigilance in the face of Soviet space technology, UN control seemed a sensible possibility. In his best-selling book *War and Peace in the Space Age*, Lieutenant General James M. Gavin suggested that all military operations be conducted under UN auspices. Despite the organization's troubles in easing Cold War tensions, the United Nations, if given support by its member nations, could prove "a fully effective instrument of peace" in the space age. Colonel Martin B. Schofield of the Air War College warned that the "presence of a variety of devastating military forces, of many sovereign states, constantly deployed throughout international space may not be conducive to peaceful living. . . . It may be sounder for the United States, while it is an early contender in the exploration of space, to use its position of influence to the best advantage by strongly advocating a form of international control over the use of space."[62]

Donald Cox and Michael Stoiko, authors of *Spacepower: What It Means to You*, outlined four reasons to surrender control of space exploration to the United Nations that others often repeated. First, there were significant political advantages. A UN ban on weapons in space might prove "a crack in the door" to more comprehensive disarmament. If the world organization could successfully broker an agreement on space law, perhaps participating countries could then transfer their nuclear weapons, missiles, and satellites to a "UN Air and Space Force" that would police the use of outer space in the same way the United Nations was then overseeing the Suez region. Second, the United States and the Soviet Union could benefit economically, for if the United Nations alone could create missiles, then the cost to national economies would be reduced and wasteful redundancy eliminated. Third, if governments were to surrender space technology to UN control, then national missile programs would have to be centralized and streamlined, thus solving the interservice rivalry then plaguing the Pentagon. Last, the authors emphasized the "psychosocial" aspects of international control: "Man's moral conscience would be freed . . . he could look up at the heavens and know that he no longer need have any fear that some shooting star he might see in the dark night could actually be the glow of a deadly warhead plunging down to annihilate him and his family."[63]

In addition to commentators in the press, scientific community, and defense establishment, popular opinion also reflected a measure of enthusiasm for UN control. In his analysis of post-*Sputnik* opinion in Western Europe and the United States, Princeton political scientist Gabriel Almond showed that there was increasing support for "a more active role" for the United Nations. The notion that nuclear weapons could be hitched to satellites provoked "a widespread sense of personal and national vulnerability" to modern weaponry. Consequently, there existed "popular pressure to 'meet,' 'negotiate,' 'disarm,' 'disengage,' and 'try to settle,'" in dealing with thorny political problems. The success of *Sputnik*, Almond concluded, increased the appeal of the global body as a venue for disagreements as well as the expansion of cultural and economic exchange programs. UN Secretary General Dag Hammarskjold confirmed such thinking in his annual report to the General Assembly. "Rapid strides in scientific discovery," particularly atomic weapons and space technology, had strengthened "the tendency to link the United Nations with all aspects of international life." In many circles, he noted, the novel problems posed by space-age technology "cannot be handled without the help of world institutions. Accordingly, it is felt widely that since international machinery exists in the United Nations and its agencies, that machinery should be used in efforts to handle these pressing questions."[64]

As the atomic bomb had done in 1945, *Sputnik* triggered greater interest in and support for multilateralism and even world government. Political science journals, international law reviews, and other academic trade publications confirmed the reemergence of world federalism as a plausible political project. "The thinking of policymakers as well as of the general public in the West remain wedded to the ideology of global regulation," wrote Hungarian-born sociologist Paul Kecskemeti. Though regulatory regimes might fail, existing international organizations prove ineffective, and geopolitical realities enforce a Cold War balance of powers, "this neither diminishes the suggestive power of the ideology of regulation nor renders the principle of the balance of power more acceptable."[65]

Among academics, few were more interested in the international political implications of space exploration than MIT political scientist Lincoln P. Bloomfield. After graduating from Harvard in 1941, Bloomfield served in various posts in the Navy and OSS during World War II. After the war, he worked for the State Department for eleven years, before arriving at MIT's Center for International Studies in 1957. There, Bloomfield built a reputation for strong political analysis as well as a keen interest in the prospects for world order. A self-described realist, Bloomfield nevertheless thought that considerations of power had to be "tempered with moral authority and political imagination," an approach he brought to his scholarship on space. Bloomfield believed that although romantic ideas about cooperative space exploration were easy targets for political pragmatists, the cosmos, in truth, presented "both [a] need and [an] opportunity for the development of institutional forms of international cooperation." "All rhetoric aside," he suggested, "the fundamental long-range task facing the United States in its international strategy is to substitute processes of cooperation, order, and eventually world law for the anarchy and narrow nationalism that continue to endanger world peace and stability." If but thirty nations could transcend Cold War bipolarity, a "limited but meaningful community" would emerge. Like John McConnell, Raymond Loewy, and S. Fred Singer, Bloomfield advocated offering the first "moon-shot" to the international community: "A 'UN shot' of this nature," he wrote in *International Organization*, "could serve as a telling countermove against the spirit of nationalism which frustrates the quest for more genuine international collaboration."[66]

In his edited volume *Outer Space: Prospects for Man and Society*, Bloomfield warned that space exploration would introduce several vexing problems for bureaucrats and lawyers to sort out: traffic control, radio-frequency allocation, liability for damage caused by spacecraft, and security from space-based weaponry. Humankind, he argued, would inherit as many puzzles from space as they would solutions. But none could be resolved in space. They had to be "dealt with on earth, through statesmanship, in carefully thought-through military policies, in diplomatic negotiations, and in creative and imaginative planning in the fields of international law and international organizations." It would be up to flexible thinkers to develop a novel apparatus upon which the foundations of a "space-for-peace" regime could be laid.[67]

It is worth pondering what common assumptions undergirded both academic rationalizations for UN control, such as Bloomfield's, and the social ones that activists like McConnell forwarded in his Star of Hope campaign. One "space-shocked political scientist" at the Rand Corporation, Joseph M. Goldsen, offered three possibilities. First, enthusiasm for global governance of space derived from understandings of the "separateness" of space, not only as a tranquil, virgin area to protect from weaponization but also as an area particularly ripe for political and technological cooperation. In many political and professional circles, moreover, thinkers held up the universality of science as a *modus operandi* for the peaceful development of space exploration. Scientists and engineers could create a universally beneficial program for probing the cosmos "if the politicians would only let them alone." The scientific community especially feared that parochial national interests would, as they had in World War II, coopt the purportedly immutable values of science. Last, Goldsen observed the confluence of nineteenth-century rationalism, cosmopolitanism, and anxieties about nuclear holocaust in the development of ideas about outer space. The result was an antagonism toward traditional notions of sovereignty that had prevailed after 1914 and again after 1945 in the UN Charter.[68]

Proponents of UN participation in space governance were emboldened on March 15, 1958, the eve of the space act's submission to Congress, when the Soviet Union submitted to the UNGA a proposal to control outer space. It called for a ban on the use of "cosmic space" for military purposes, the liquidation of all foreign military bases, the establishment of a UN "framework" for international control of space, and a special UN agency to facilitate international cooperation in space exploration. Eisenhower, Soviet leaders argued, thought too narrowly about the demilitarization of outer space. Disarmament measures pertaining to space could not be separated from broader political disagreements between the superpowers. Should not banning ICBMs open the door for disarmament measures back on Earth? Should not foreign bases also be eliminated as forwarding zones for atomic destruction? After all, it was not ballistic missiles that threatened mankind, but rather the nuclear weapons that they could potentially deliver to enemy targets.[69]

Dulles rebuffed the proposal. The Soviets were "mixing up two things that are quite unrelated." For starters, the global system of US overseas garrisons depended on the "voluntary co-operation" of the host nations, which in many cases welcomed the presence of US troops out of fear of Soviet "imperialism." The Kremlin, he argued, was trying to kill the American plan by attaching an impossible measure to it.

Only five days later, however, a visit to the State Department by the UNGA president, Sir Leslie Munro of New Zealand, forced US diplomats to take the Soviet proposal more seriously. Munro referred both to Eisenhower's January 12 letter to Bulganin and Dulles's speech at the National Press Club, noting that his own public statements were consistent with their enthusiasm and optimism about using space for strictly peaceful purposes. Yet if Munro were to say anything more in his role as president of the Twelfth Assembly, he would need to confirm the official US position. Undersecretary of State Christian Herter reassured him that "the subject of outer space control was in the forefront of our thinking at the moment." The legal novelties of space exploration presented formidable obstacles, yes, but Herter had "reviewed the past US proposals for [the] peaceful uses of outer space, and affirmed that this was [a] continuing objective." Munro referred to the Soviets' new proposal and noted that although it was unlikely to pass a vote, it had nevertheless "captured the initiative." The world was "groping for ideas and answers," Munro implored. It was incumbent on the United States to "move soon with new and specific ideas."[70]

This advice came at a crucial juncture. Over the spring and summer of 1958, not only was Congress debating the National Space Act, but Eisenhower's NSC was busy drafting what would become the United States' basic space policy, NSC 5814/1. In this seminal policy study, the NSC contemplated the consequences of space exploration for science and technology in the twentieth century as well as its implications for international relations, particularly political competition with the Soviet Union. The report outlined the United States' most important goals in space, objectives that would guide US space policy into the late 1960s: developing and exploiting space capabilities to help achieve the nation's other scientific, military, and political aims; acquiring recognition as a leader in space; pursuing international cooperation in scientific space research and with allies in military research; building a system to develop and regulate space programs; and utilizing space technology and scientific cooperation to "open up" the Soviet bloc.[71]

Like Rand's recommendations for US space policy and Munro's directions to State officials, NSC 5814/1 kept world opinion central to its assessments and recommendations. The report underscored the significance of "psychological exploitation" by recommending that the United States "judiciously select" projects that would achieve "a favorable worldwide psychological impact." Given the perceived Soviet lead in space technology, the United States would counterpunch by "maintaining [its] position as the leading advocate of the use of outer space for peaceful purposes." To achieve its objectives, the United States needed to continue the type of international scientific cooperation the government had pursued during the IGY; invite foreign scientists, including those from the Soviet Union, to work in US laboratories; and propose multilateral arrangements for the regulation of satellite launchings and ownership of radio frequencies. NSC 5814/1 recalled the Kremlin's March 15 proposal to the United Nations to ban the use of space for military purposes and correctly surmised that international control would be a key topic for debate when the Thirteenth General Assembly met that September. If the United States were to maintain its status as the leading promoter of peace in space, then it would have to arrive at the United Nations with "an imaginative and positive position" of its own.[72]

While scarcely imaginative, the US "position"—to the delight of *Sputnik* globalists—was to recommend the creation, through formal resolution, of a UN ad hoc committee capable of steering nascent space law into maturity and preserving outer space for peace. "It is clear that the potentialities for good or evil that will arise from the exploration of outer space are enormous," Lodge wrote to Secretary Hammarskjold in a letter attached to the US proposal. As the world's leading multilateral institution, the United Nations should therefore spearhead the development of legal frameworks for space exploration and house the organs necessary to research the novel questions of the space age. The State Department's suggested language for the resolution reflected the government's attention to the emerging idealist consensus on space politics. The purpose of the committee would be to block "the extension of present national rivalries" into space and promote its exploration "solely for the betterment of mankind." However, it was "essential," the State Department urged its mission at the United Nations, to maintain a separation between the peaceful uses of outer space and broader disarmament goals involving missiles and nuclear weapons.[73]

When the General Assembly met in September, the American proposal for a UN space committee immediately hit a roadblock: its future composition. Valerian Zorin, Soviet ambassador to the United Nations, proposed that the new committee consist of two equally represented "sides": four Soviet-bloc countries, four "Western," and three "neutral." After all, the Soviet Union and the United States were the only two nations "practically" exploring outer space. But Lodge scoffed at this suggestion, reminding the General Assembly that "there are no 'two sides' to outer space. There are not, and have never been, 'two sides' in the United Nations." Membership on the ad hoc committee, he insisted, should reflect the composition of the General Assembly itself, and consist only of those countries best able to contribute to space exploration. Lodge's delegation thus proposed an expanded membership of eighteen states that restricted the Soviet orbit to a small minority. The US and Soviet delegations met several times to renegotiate the committee's composition, but neither seemed interested in compromise. Zorin emphasized that without the mutual consent of the superpowers, "there would be no cooperation." Lodge attributed his diplomatic failure to the apparent fact that the two sides simply "work from entirely different premises about the nature of relations between states, the structure of the United Nations, and the nature of the world." To Undersecretary of State Herter, it was clear that the Soviets wanted to "wrest initiative" from the United States in space-for-peace propaganda. It was an "indication of [the] political importance which [the] Soviets attach to leadership in this field."[74]

In the face of Soviet intransigence, top US officials trekked to Manhattan in quick succession to urge establishment of the ad hoc committee in the UNGA. Dulles went first. Reminiscing the lost chance for international control of atomic energy, the secretary of state insisted that in space exploration, "we should move as truly 'united nations.'" Four days later, Eisenhower asked whether outer space would be explored exclusively for peaceful purposes or whether it would become "an area of dangerous and sterile competition." He pointed to the recently concluded Antarctic Treaty, which forbade military activity on the continent and opened it up for scientific investigation by any nation, as a model for space. The president proposed that UN member nations agree that celestial bodies were not subject to national appropriation; that war be forbidden in outer space; that no nation, under strict verification, place into orbit any weapons of mass destruction; and that the UN develop an international register for all launchings and a permanent program for international cooperation. "The choice is urgent," he pleaded, "And it is ours to make." In their subsequent orations, Lodge and Senator Johnson, who chaired the Senate Aeronautical and Space Committee, seconded the president's proposals and noted the bipartisan support for the US resolution. "We can use this new dimension to destroy ourselves through the extension of national rivalries into outer space," warned Johnson, "or we can use this new development as a vehicle for international collaboration and harmony."[75]

Johnson, whose political career would be tightly bound to the US space program for the rest of his life, was particularly forceful before the General Assembly. The Senate majority leader was a true believer when it came to space exploration. He saw in space an opportunity for political, economic, social, and spiritual rejuvenation as well as an opening to thaw the Cold War. Perhaps, he suggested, conflict would disappear altogether as humankind ascended in unison: "Barriers between us will fall as our sights rise to space. Secrecy will cease to be. Man will come to understand his fellow man—and himself—as he never he has been able to do. In the infinity of space adventure, man can find growing richness of mind, of spirit, and of liberty." Johnson's anticipation stemmed in large part from a conviction that outer space was a special oasis untouched by humans and thus devoid of the Earth's "legacies of distrust and fear and ignorance and injury." Space was "unscarred by conflict," Johnson reminded the delegates: "No nation holds a concession there. It must remain this way." Such rhetoric was the culmination of cosmopolitan arguments about the sanctity and inviolability of outer space in the wake of *Sputnik*, and indeed Johnson's words presaged a generation of space-age idealism that would occupy his own administration and that of John F. Kennedy's as manned spaceflight took off (figure 12).[76]

Despite these appeals to future-mindedness, morality, and cooperation, the Soviets did not participate, convinced as they were that the proposed composition for the committee did not reflect political reality in the wake of repeated Soviet achievements in space. The US delegation pushed forward with its proposal for a committee composed

FIGURE 12. Lyndon B. Johnson frequently propounded the bipartisan consensus on the peaceful uses of outer space in the United States. *Washington Star*, November 19, 1958.

of eighteen states in which Soviet socialist republics would constitute a decisive minority. The General Assembly voted on November 24, 1958, and defeated the Soviet proposal, fifty-four to nine, in favor of the Western composition. The Committee on the Peaceful Uses of Outer Space was now an official UN body, but the Soviet Union—as well as Poland, Czechoslovakia, India, and the United Arab Republic—denied the committee's legitimacy and boycotted its future plans.[77] Three years later, in 1961, the Soviets would join the committee as a means to resist US space reconnaissance; for now, though, the COPUOS's mandate was decidedly limited. The United Nations ordered the committee to report to the General Assembly's Fourteenth session on the resources that the United Nations and other international institutions could muster to facilitate the peaceful uses of outer space; the continuation of IGY-style cooperation in space "on a permanent basis"; the organization of mutual exchange and dissemination of space research; coordination of national space programs; possible organs for facilitating international cooperation in space under UN auspices; and, perhaps most important, the legal complications that could arise from space exploration.⁷⁸

Reactions to COPUOS varied. US officials, for one, were ebullient. In a review of the negotiation process issued the following spring, Clarence Dillon, Herter's replacement as undersecretary of state, was enthusiastic about the committee's chances to "provide an opportunity for effectively continuing [the] leadership of the United States" in promoting "space for peace." He noted in the report that "while the United States currently suffers from technological disadvantages in the outer space field, it has a clear advantage in its willingness . . . to take constructive measures respecting international control." The Eisenhower administration, after all, had placed this prerogative at the center of its space policy. It was true that the United States could not achieve true control without the Soviet Union. Still, the Soviets had seemingly shot themselves in the foot. They, too, had proclaimed their support for international control of outer space. But when a clear opportunity to research the issue arose, the Kremlin instead clung to its "lead" in the space race. In this way, US space policy—premised as it was on images of American benevolence and Soviet deviancy—proved effective: if Soviet cooperation was the ideal outcome, then Soviet obstinacy was next best.⁷⁹

Space Sanctuary: A Matter of Policy

It appeared at the end of 1958 that interplanetary notions about international cooperation in space, the preservation of the cosmos from weapons, and global governance of the medium had not only survived the onset of the "space race" but had actually found expression in germinal US space policy. Scientists and engineers favoring a symbolic reversal of the government's approach to atomic energy had the ear of the president. The Congress and the public enthusiastically supported peaceful space initiatives such as the IGY and shunned extravagant weapons projects. The Pentagon, though it would retain all projects deemed relevant to national defense, lost control over scientific exploration. Creation of the COPUOS heartened proponents of international controls for the revolutionary technology as well as progressive laws to outlaw war and colonialism from space. Even without Soviet participation in the new UN committee, US officials appeared to have preempted the "specter," as Lodge called it, of a space-based arms race run wild.⁸⁰

As several scholars have pointed out, this effort was attributable in large measure to the Eisenhower administration's drive to legitimate satellite reconnaissance. Measures that guaranteed exploration "exclusively for peaceful purposes" helped distinguish the *weaponization* of space—which connoted all the "Buck Rogers" notions of space planes, satellite bombers, ASAT weapons, and moon bases—from the *militarization* of space, a term implying that although the services and the Central Intelligence Agency (CIA) were engineering spy satellites, these technologies served to monitor compliance with arms control and thus served the interests of peace.[81] Certainly the extent to which incipient space policy was based on uncertainty about Soviet intentions and capabilities is difficult to overlook.

But in explaining why the generation of policymakers responsible for navigating the dawn of the space age pursued the measures they did, it is unwise to dismiss considerations of ideology and culture. As suggested by the enthusiasm with which many Americans greeted *Sputnik*, interplanetary narrations of the coming superpower reconciliation in space continued to have popular appeal and hence real political influence. While it is unquestionable that the Soviets' "firsts" in space served to expand and intensify the Cold War, they also became an impetus to temper it. Refining the premises of the interwar period, *Sputnik* globalism came to encompass more than a resistance to the wastefulness of the space race or the imminent danger that as a technological choice the satellite would follow the path of the atomic bomb. It represented a clamor for fundamental changes in human nature and the ascent of those changes to the practice of international affairs. In many ways, it sought plainly to rewrite the history of the world.

These impulses, which two world wars had impressed deep in human consciousness, were readily apprehended by the crafters of early US space policy, who used the sanctuary paradigm to wage a propaganda battle, a war of national character, with the Soviet Union. They understood that the space race was a contest not only of science and technology but also of prestige and identity. The United States had to surpass its rival in terms of technical achievements as well as portray itself as a peaceful, cooperative, and beneficent space power. Civilian control of NASA and its leadership in the United Nations on governance of international space activities were but two pillars of this image-crafting strategy.

Indeed, the competition over which nation could more vigorously promote "space for peace" became a race-on-top-of-a-race, an adjunct to the technological rivalry between the superpowers already occupying immense political, intellectual, and economic resources. Attempts to garner world prestige by disparaging the enemy's intentions and authenticity in space harmed what might have been more comprehensive agreement and cooperation. The Soviets' condemnation of Eisenhower's overtures to international control, for one, damaged the spirit of goodwill that space rhetoric was intended to nurture, but so too did Dulles's accusation that imaginative Soviet proposals were purloined from American minds. Khrushchev's insistence that international control of ICBMs be tied to total nuclear disarmament and the elimination of foreign bases, while ambitious, was impracticable and deeply political; yet at the same time, the United States' refusal to consider any other disarmament measure alongside the banning of weapons in space revealed its obstinance in surrendering any military advantage.

In any case, it is reasonable in light of the evidence to question whether US evangelism regarding "space for peace" was effective. World opinion, for one, remained convinced that US activity in space was intended to produce military superiority. An extensive study conducted by the US Information Agency (USIA) concluded in July 1959 that the "reaction to space development, from all audiences, shows a clear tendency to equate achievements in this field with military power." Global interest in US initiatives for international control were decidedly "subordinate and unspecified" compared with the military implications of space. In fact, there was a "widespread conviction" that space projects were not truly for the sake of exploration but were "essentially military exercises." And despite talk of cooperation on both sides of the Iron Curtain, the public still viewed space exploration through the lens of Cold War competition. "The concept of a space race," the report concluded, "appears to be almost automatically injected" into the worldwide reaction to space-feats.[82]

Malvina Lindsay, who had conjured Mr. Stellar and Mr. Hardpann one year earlier, was equally incredulous. In another article published on the first anniversary of *Sputnik*, Lindsay described a report written by an "agent of outer space" named Rik. The alien document appraised the likelihood that the United Nations would "become an agency capable of deterring Earthlings from blowing up their planet and possibly throwing the solar system off base." While Rik extended hope that the United Nations could "lead Earth toward becoming an orderly and civilized planet," it faced daunting obstacles, namely, "petty human conflicts," "ideological bigotry," and an obsession with cultural difference. Humans, in other words, were in a "retarded" stage of development, too premature to explore space responsibly.[83]

Even US policymakers remained unsure if they had successfully staked a claim to leadership in the peaceful uses of outer space. Karl G. Harr, the president's special assistant for security operations, thought not. In a November 1958 memo to Eisenhower, Harr cited the administration's initiatives regarding "space for peace." He concluded that "we have not thus far been able fully to realize the potential of this issue." Although the "basic blame" for the UN deadlock "must rest with the Soviet Union," Harr explained, the United States had not done all it could have. Echoing the NSC's recommendations from July, Harr advised that "we must act boldly and soon, if we are to place the United States in the position of world leadership in this dramatic new field."[84]

Molding an image of the US space program as the peaceful antithesis to its secretive Soviet counterpart was a task made all the more difficult by the American defense establishment, which at times entertained a wholly different interpretation of what it meant "to lose the initiative." In the view of many defense intellectuals, politicians, and military leaders, arms control in space implied letting down one's guard. It indicated gullibility, naivety, and eventually surrender against a foe that had never let its eyes off the prize of "space supremacy" and, by extension, its quest for world domination. Beating the Soviets in space—or, as alarmists argued, merely keeping pace with them—required prodigious capital and personnel investments in military projects that would ensure US access to and maneuverability in space. Peace, many military leaders argued, relied on it.

McConnell later captured in his memoirs the dawning realization that a space sanctuary would not grow unimpeded within American space diplomacy. Reflecting on his activism, he regretted that the United States had never truly orbited a Star of Hope. The military, he insisted, had blocked his campaign. At the end of 1962, rather than a Christmas light in space, the Pentagon had launched a secret geodetic satellite on Halloween. Far from a Star of Hope, its name was A.N.N.A. While the second "N" stood for NASA, the other three letters stood for Army, Navy, Air Force.[85]

Truly, the battle for the soul of American space policy had just begun. *Sputnik* globalists began amid their measured successes to confront a countervailing vision of the future. It was a vision in which the historic animosities of states and empires extend into space indefinitely. Rather than serve as a cooling salve for international rivalries, space exploration only accelerates and intensifies them. Competing for new territories or natural resources on planetary bodies, nations ratchet up the destructive capacity of their weapons. Spacefaring planes and massive space stations attempt to control the most strategically valuable orbits, and lunar missile stations menace Earth around the clock. For every dream, it seemed, there was a nightmare.

Part Two
Nightmares

Chapter 4

Lunartics!

The images appear stripped from a science-fiction novel. There are cylindrical metal tanks, vacuum-sealed into the surface of the moon, equipped with air conditioning, cold storage facilities, and laboratories for physical and biological experiments. Construction vehicles raze lunar material to make way for launch stations, a nuclear reactor, and living quarters for a crew of up to twelve men. There are even solar-powered mechanisms for extracting oxygen and water from the natural environment. These were only some of the technical ambitions of the US Army's Project Horizon, a 1959 plan to install a military base on the moon. In two cumbersome volumes—more than four hundred pages altogether—the Army laid out in detail the manifold purposes of a lunar outpost. Project Horizon would help develop techniques for moon-based surveillance, Earth-to-moon communications relay, and scientific investigation in the lunar environment; serve as a launching platform for further space exploration; provide emergency staging areas, rescue capabilities, and navigational aids for astronauts; and "protect potential United States interests on the moon" generally. Construction could begin as soon as 1964 and take as little as two years to complete.[1]

Aside from the obvious scientific and political prestige available to the nation that first established a permanent base on the moon, requirements for national security also loomed large in the Army's considerations for

FIGURE 13. In the late 1950s, the US Army engaged in research for a lunar military base. In this drawing from its report, a lunar construction vehicle breaks ground. Courtesy of the US Army.

Project Horizon (figure 13). "Moon-based military power," the report suggested, "will be a strong deterrent to war because of the extreme difficulty, from the enemy point of view, of eliminating our ability to retaliate." Given the logistical difficulties of reaching the moon and a bona fide American presence there, the Soviets, who reportedly had begun to plan their own moon missions, might think twice about attacking the United States. But crucially, the reverse was also true: "if hostile forces are permitted to arrive first," the Army stressed, they could "militarily counter our landings and attempt to deny us politically the use of their property." Accordingly, the Eisenhower administration should make Project Horizon a crash program with "authority and priority similar to the Manhattan Project."[2]

Project Horizon revealed a particular way of thinking about space that gained considerable traction from *Sputnik* through the middle of the 1960s, one that contradicted the ideas the interplanetary movement had propounded in the first half of the twentieth century. Far from an apolitical sanctuary that governments should preserve from the Cold War, outer space, according to a host of US military leaders and defense intellectuals, would soon become a vast new theater in which to wage it. Project Horizon's architects put it best: "The extent to which future operations might be conducted in space, to include the land mass of the moon or perhaps other planets, is of such a magnitude as to almost defy the imagination," they wrote. "In both Congressional and military examination of the problem, it is generally agreed that the interactions of space and terrestrial war are so great as to generate radically new concepts." Indeed, a moon base was "merely a point of departure" for US research into military space technologies and new war-fighting strategies for the medium. To maintain its military competitiveness in space, the United States would soon need to explore space planes, manned orbiting battle stations, and perhaps even bombardment satellites. Together with NASA, the Pentagon would need to study the laws of celestial mechanics and their implications for military operations in space.[3]

Project Horizon and similar oddities are practically absent in historical and popular accounts of the space age, and for one simple reason: they were never realized. As Eisenhower's engagement with Soviet leaders, the creation of the COPUOS, and NSC 5814/1 each suggested, the United States had in the late 1950s moved in a direction at odds with the more colorful ambitions for space that many planners harbored within the Defense Department's scattered R&D offices. Subsequent US policies under Kennedy and Johnson—antiweapon UNGA resolutions, the OST, and the cancelation of Project SAINT (satellite interceptor), Dyna-Soar space plane, and MOL, just to name a few—further undermined the notion that space was destined to become a theater of armed conflict. Over the course of the early space age, the government struck down dozens of seemingly harebrained schemes akin to Project Horizon on budgetary, bureaucratic, technical, and political grounds. There would not be, as a S. Paul Johnson put it in the debate over NASA, "hordes of little men in space helmets firing disintegrators into each other from flying saucers."[4]

Considering these historical developments, it would be easy enough to dismiss some of the Pentagon's foolhardier plans for space. But in the late 1950s and 1960s, dreams of moon bases, satellite bombers, orbiting space stations, and other fanciful technologies comported fully with widespread understandings of what warfare would soon entail. Science-fiction authors, spaceflight advocates, and military analysts painted a picture of outer space thronged with advanced military hardware. Within a generation, they predicted, satellites capable of bombing any target on Earth would circle the planet continuously; astro- and cosmonauts would carve rudimentary installations into the lunar surface; and darting space planes would police cis-lunar space around the clock. Another one hundred years after that, colossal spacefaring battleships would defend fortresses in orbit and on the moon, and weapons capable of changing weather patterns, evaporating water supplies, and burning cities to ash would threaten the existence of anyone who opposed their wielders. Near-Earth orbit, these oracles claimed, would soon become an immense coliseum in which the fate of formerly earthbound politics would be decided in swift, kinetic exchanges.

Such imagery mattered not only because of its prevalence in American culture but also because of its interest to officials responsible for crafting military policy after *Sputnik*. The striking depictions of space war that pervaded public discourse convinced the US defense establishment that it stood on the threshold of a strategic revolution. Control of key points in space would be as decisive in twentieth-century international politics as control of critical waterways had been for the British navy in the sixteenth and seventeenth centuries. The geopolitical theories of Halford Mackinder, Alfred Thayer Mahan, and Nicholas Spykman enjoyed a renaissance as scores of journalists, academics, and military officers attempted to put the coming age of space into perspective. The adage that "control of space means control of Earth" found disciples in every corner of US military discourse, a trend duly noted by Soviet onlookers.[5] Some prognosticators went so far as to suggest that war on Earth would become obsolete: spaced-based military technologies would help transplant earthbound conflicts to the void, thus sparing civilian populations from violence. The sky no longer being enough, the stars became the limit.

Wars in space—or at the very least wars started *from* space—represented a future many found dubious. But powerful interests, particularly the Air Force, insisted that the future would arrive regardless of whether the civilian government decided to invest in new military space technologies. To justify its costly national campaign toward American "space supremacy," the USAF conjured the specter of Soviet space domination, or, perhaps more precisely, *world* dominion through superiority in space.[6] "The announced intention of Communism is the domination of the world," one general reminded Americans. There was "no reason to believe that this policy has changed" in light of incipient space exploration.[7] Given the Soviets' penchant for secrecy, deviancy, and aggression, the only way to preserve the peace was to be first in establishing command of space. The purportedly stabilizing influence of nuclear deterrence should be extended to outer space to prevent Soviet blackmail. "Operations in space will remain peaceful as long as peace-determined men keep it free," claimed USAF Secretary Eugene Zuckert in 1962. "It is our job to make sure that if anyone controls space from a military standpoint, it is the US that does so."[8]

Compelling images and technical descriptions of space war, their subversive implications for military strategy, and increasingly widespread convictions about the necessity of preventing Soviet domination of space propelled the Pentagon's very real interest in advanced military space systems like Project Horizon. As early as 1958 the Air Force Ballistic Missiles Division in Los Angeles commissioned several studies by defense contractors aimed at defining the contours of military spaceflight. These so-called Systems Requirements (SR) studies, most of which are still classified, demonstrate a keen interest in the hardware needed for war in space. In addition to lunar bases and nuclear weapons tests for the space environment, they included an analysis of a military test space station, a strategic orbital base, and satellite weapons systems. "Dr. Strangelove," historian Dwayne Day has written, "would have been orgasmic."[9]

Given the symbolic appeal of futuristic space technologies to early Cold War defense intellectuals and military officials, and the concrete steps taken by aerospace firms and the USAF to study the potentialities of "space war," it would be a mistake to dismiss plans like Project Horizon as curiosities of space-age paranoia. There are, it seems, several questions of interest to the serious historian. How did Americans envision outer space as a theater of armed conflict before the OST, and from where did these visions emerge? What benefits did promoters of space weapons articulate? With what drawbacks did their detractors counter? What did the new "experts" of space warfare believe to be the strategic implications of extending conflict beyond the Earth's atmosphere? And perhaps most important, in view of the support "space war" received in powerful circles: why did the USAF and its allies fail to achieve their vision during the late 1950s and 1960s?

Historians have spilt much ink explaining the significance of US military space programs to the Cold War, particularly reconnaissance satellite projects such as SAMOS (satellite and missile observation system) and CORONA.[10] Yet, curiously, scholars have largely ignored the *idea* of using outer space as a platform for military hardware before the Strategic Defense Initiative (SDI). In a 2006 survey of the field, Stephen Johnson pointed to the dearth of scholarship on space power doctrine, which first emerged in the late 1950s, referring to it as a significant "hole in the research" on space policy and technology.[11] Perhaps this is because, as noted, the "Buck Rogers' type thinking" typified by pursuits such as Project Horizon never materialized. Then again, perhaps historians have not written about outer space as a theater of war because, in the end, the Earth remained throughout the Cold War the most efficient and cost-effective platform for delivering nuclear weapons.

Whatever the reason for this lacuna, it is necessary to address "space war" as a discrete historical concept, one that, despite its failure to manifest in history, can nevertheless tell us much about not only American military imagination at the dawn of the space age but also the attitudes, fears, inclinations, and aspirations of US leaders, professionals, and citizens during what they considered to be a watershed in military history. Recently declassified documents and the writings of the space power theorists shed light on the United States' contested relationship with space militarization and the prospect for war in outer space.

The primary source record illuminates three broad historical phenomena. First, it marks the expeditious transition of "space war" from the realm of imagination to strategy, from the fantastic worlds envisioned by science fiction to the severe, arcane formulations of space age defense analysts. Second, it highlights the emergence in the late 1950s of a peace-centered discourse within the Pentagon, particularly the USAF, that military leaders hoped would justify space weaponization as national policy. Although advocates of space war were proposing an intense escalation of the arms race, they cloaked their proposals with the mantle of stability and peace—both would be bought by space power. And third, the documents show that while fiscal and technical obstacles helped drown the more provocative military space projects of the late 1950s and 1960s, political and ethical factors proved equally damning in the decade after *Sputnik*. The failure of the USAF to equate space weaponization with peace derived from countervailing opinions—widespread in and outside of government—that equated peace with the *preservation* of outer space from weapons and war. The power of these counterforces, even if the money and technology for advanced space weapons had been available, made "space war" anything but certain.

A New Theater of War

To understand how some nooks of the US defense establishment began to take an interest in moon bases, satellite bombardment, and space-based warfare generally, it is necessary to briefly consider popular culture. Before the mid-1960s, after all, what little knowledge Americans possessed about outer space derived from the era's copious science fiction, which often described space as a source of great danger and argued that military solutions to that danger were the most reliable. On other planets and aboard their spacecraft, human beings in contemporary films, comics, television shows, and novels come into frequent contact with hideous and violent creatures, but ultimately defeat them with bravery and ingenuity. A freakishly powerful Martian sucks its victims dry in *It! The Terror from Outer Space* (1958). The spacefaring crew of *Forbidden Planet* (1956) is stalked by an invisible monster. In movies such as *Earth vs. The Flying Saucers* (1956), *Invaders from Mars* (1953), and *War of the Worlds* (1953), viewers could feast their eyes not only on civilizational threats from cunning aliens and their often-superior technology but also on the military's heroic means of fighting back despite relative weakness. In the words of one film historian, militaristic imagery constituted the "basic flavor" of these stories; it was the "glue" holding them together.[12]

One important aspect of the era's science fiction, particularly in terms of the impact it would have on US conceptions of military possibilities in outer space, was its proximity to the conflicts, technologies, and politics of the day. There were, of course, light-speed spaceships, sexless aliens, and medicines that would instantly cure any affliction. But often such futuristic imagery comingled with more pedestrian depictions of technology, politics, and culture in space. Countless stories simply repackaged the Cold War in more futuristic wrapping. The inaugural issue of the comic *Major Inapak the Space Ace* (1951) is illustrative: it describes a world divided into two political spheres, the totalitarian "East" and the democratic "West," which builds "several rockets ... to circle the world like small satellites loaded with atomic death and ready to fall on the East zone if they started a war." The available technology is little more foreign. When the West learns of an alien attack on Denver, the brave Major Inapak (so named for a nutritional supplement children were to put in their milk) rockets to the moon to find a weapon powered by a crude electric relay that more closely resembles a Bell Telephone network than any future technology. In stories from other comic serials, characters often communicate through satellite radio and wear pressurized suits then under production at NASA.[13]

Fictional accounts of space war in the United States similarly relied on real-world politics, even racial hatreds. In creating alien characters to be conquered by American protagonists, comic book artists and science-fiction authors exhumed World War II-era stereotypes of Asians and Africans as sub- and superhuman, and often simply cast pugnacious and evil space aliens as aliens to the white American nation.[14] In a 1963 issue of *Space War*, a humanoid cat creature, caricatured to remind readers of Japanese militarism, attacks a rugged space-cowboy, but is predictably defeated. Another story in *Space Man* reflected anxieties of rising Chinese power in its portrayal of an alien government called "the Great Revolutionary People's Republic." Its leader is a Fu Manchu–like villain bent on galactic conquest. The cunning-yet-arrogant miscreant holds the US moon base hostage but is defeated by the US-led "Galactic Guard."[15]

Fictional stories, of course, did not by themselves shape factual conversations about the prospect for conflict in space. Although military and political leaders often referenced Buck Rogers or flying saucers in serious discussions about future military policy after *Sputnik*, science fiction was merely the cultural and imaginative gelatin for studies such as Project Horizon. These and similar projects required intermediaries to translate the fanciful into the plausible. Over the course of the late 1950s and early 1960s, advocates of spaceflight—frequently referred to as space "boosters"—proved themselves such able traffickers in ideas about the military implications of exploration. Equipped with knowledge of the technical aspects of space technology and feverish anticipation for the future, science writers—even "interplanetary" authors such as Willy Ley and Arthur C. Clarke—became, wittingly or unwittingly, conduits between the fantastical world of science fiction and the very real world of military policy.

Their greatest advantage was ubiquity. They were helped along by popular magazines including *Time*, *Collier's*, *Life*, and *Nature*, which ran special issues on space exploration, as well as popular trade journals dedicated to the topic, of which *Aviation Week and Space Technology* and *Space World* were only the most popular. Some of the early Cold War's most widely distributed books, too, dealt explicitly with the prospect of human spaceflight. Ley's *The Conquest of Outer Space* (1949), Clarke's *The Exploration of Outer Space* (1951), and Wernher von Braun's *Space Frontier* (1968) introduced millions of Americans to the technical prospects of exploration, as did a procession of marquee essay collections.[16]

Space boosters emerged as key actors in the ferrying of "space war" from fantasy to formulation because they were able to translate the technological abstractions of science fiction into digestible accounts of how given military systems might work. Most conspicuous in this regard was von Braun, whose dream of human spaceflight mixed with a troubled history of engineering implements of war.[17] Von Braun spoke across the country between the late 1940s and early 1950s, for example, on the feasibility of a military space station capable of bombing targets on Earth. Even to lay audiences, he spoke with great specificity about the technical prospects of such a weapon. "If we fire a small rocket missile from the station in the direction opposite to its orbiting around the earth, so that its final velocity becomes but 1,600 feet per second slower than that of the station," von Braun told a crowd at the University of Illinois, "the missile will tangentially enter the earth's atmosphere near the perigee of its elliptical path." Using radar observation and remote controls, the missile then could be guided to any target on the globe, a capability that would adorn the United States with "military omnipresence." Nearly a decade before the space age had begun in earnest, von Braun invoked what would, in the wake of *Sputnik*, become an axiomatic proposition: "It appears to me that that in the atomic age the nation which first owns such a bomb-dropping space station might be in a position to virtually control the earth."[18]

To be sure, enough naysayers existed to throw such ideas of spacefaring warships into serious doubt, a fact von Braun perceived as well as anyone. Would not such a station be a "sitting duck" for enemy countermeasures? Might a bomb in space simply fail to "drop" without the gravity of the Earth's atmosphere? Eisenhower's NSC would later dismiss the idea of orbital bombardment as inefficient and costly compared with traditional land- or submarine-launched ICBMs. Yet von Braun considered this shortsighted. There would always be people, he was fond of saying, who gazed upon a donut and saw only the hole. Rarely did the cynics consider the costs of too much restraint. Technology would not stand pat, after all, and more ill-intentioned powers were standing by, waiting to take advantage. "We've got mighty little time to lose," von Braun told a San Diego audience in 1952, "for we know that the Soviets are thinking along the same lines. If we do not wish them to wrest the control of space from us, it's time . . . we acted!"[19]

The stature, technical acumen, and wide readership of boosters like von Braun opened the floodgates of speculation. A caravan of writers—many attempting to make serious predictions but others simply trying to increase circulation—rushed to speculate what else the future might hold for the US military in space. As early as 1948, the astronomer Robert S. Richardson could anticipate a newfound capacity to launch nuclear weapons to targets on Earth from bases on the moon. In a special article for *Collier's*, Richardson surmised that launching nuclear-tipped rockets from the moon would be simpler than on Earth, for the moon's gravity was one-sixth of that on Earth, and there was "not a breath of air" to interfere with initial liftoff. "In an artillery duel between the planets," he explained, "the advantage would be all on the side having the lower surface gravity to cope with." Targeting Earth from the moon would be "like throwing rocks downhill." Such environmental conditions made the moon "the world's ideal military base." Two illustrations by the talented space-fiction artist Chelsey Bonestell Jr. accompanied Richardson's harrowing technical account. The first depicts a rocket lifting off from a military base on the lunar surface, suited military personnel and a second rocket—ready to launch—looking on. The second image is an aerial view of the rockets' target, New York City. Two enormous mushroom clouds billow up from midtown Manhattan and Queens. It was, as Richardson captioned the drawing, "the beginning of the end for New York."[20]

Subsequent renditions of the future were no less dramatic. In a clear adaptation of the superweapon from *The War of the Worlds*, space writer Robert Granville foresaw a "Death Ray" that could deflect the sun's thermal energy with inflatable mirrors. Such a weapon could melt warships, set towns ablaze, or boil seawater into hurricanes that would strike coastal cities. Jurist Stephen Gorove, who helped pioneer the study of space law in the late 1950s, was equally vivid; the "masters" of space, he foresaw, would have the capacity "to change the weather, to cause drought and flood . . . to control the tides and the levels of the sea, to alter the course of the Gulf Stream and change temperate climates to frigid." For Gorove, the geopolitical dictums of Mackinder and Spykman, who in the early to mid-twentieth century maintained that nations capable of presiding over a central "heartland" would reap political power, seemed inadequate in the space age: "*He who controls the Cosmic space,*" he prophesized, "*rules not only the Earth but the whole Universe.*"[21]

The business of divining future wars in space became so popular by 1962 that Otto Binder, editor of *Space World*, devoted a special issue to the topic. In "War in Space: Can Americans Win It?" commentators ranging from Willy Ley to Walter Dornberger, Bell Aerospace vice president (and former director of the German V-2 program), speculated about the shape of space conflicts unfolding just over the horizon. One booster anticipated the conversion of interplanetary space into "a vast battleground" in which space planes, rocket bombers, and orbital "dreadnaughts" duked it out for space supremacy. Another predicted that the Soviet Union might soon detonate hydrogen bombs in space and on the moon to intimidate Americans or perhaps someday would divert meteors from their trajectories toward US territory. Ley wrote of space planes whose machine guns, "doubly effective in the vacuum of space," would be more difficult to defend than larger weapons like rockets. Camouflaged with black paint and coated with a "radar-absorbing" substance, such craft would menace enemy outposts or defend the United States' own space "fortress." The fictional flavor of the special issue did not preclude comment by serious military officers. Major General Osmond J. Ritland, chief of special weapons at the Air Force Space Command, remarked that although the weapons featured in the report resembled "wild dreams" to most, he had no doubt that given enough time, these systems would prove technically feasible.[22]

Nor was Ritland's sanction an anomaly. When weapons analysts, political scientists, war strategists, and other, ostensibly more sober "experts" began outlining the contours of space war in the late 1950s and early 1960s, they did not leap far from the more reasonable expectations of conflict that science-fiction authors and space boosters had envisioned in publications like *Space World*. Some baldly compared the two. Donald Cox and Michael Stoiko argued that the power to police outer space should be vested in a special UN space force that would "patrol the interplanetary regions in a fashion similar to the modern science fiction heroes." While similar theories disputed the idea that such a collaborative, international force would determine space politics, all agreed that space would soon become heavily militarized, heavily weaponized, and perhaps even a new theater of war. The picture of future space conflict that Cox, Stoiko, and other defense intellectuals painted, the *New York Times* remarked late in 1958, seemed "no longer [to] belong to the Buck Rogers category."[23]

This assessment emerged from a growing sense that the technologies needed to establish space as an arena for warfare were already at hand. One could flip through British analyst M. N. Golovine's detailed study, *Conflict in Space: A Pattern of War in a New Dimension,* for a survey of the dizzying complex of space weapons that would soon occupy orbital space. In addition to the ICBMs that had already traversed space, several offensive systems were on the docket: Delayed Impact Space Missiles (DISMs), for example, would make programmed deviations from the normal trajectory of traditional ICBMs, and could be destroyed by the launching station during flight. The Positive Control Bombardment System, a recallable ICBM, consisted of three-ton nuclear-armed satellites in one-hundred-mile orbits. Nuclear-armed bombardment satellites (NABS) "virtually the ultimate in mobility and a powerful psychological deterrent," would remain in high, variable orbits and threaten to deliver hundreds of megatons worth of TNT using retrofire rockets to slow their payloads out of orbit. Gradually, ballistic missiles would give way to orbital weapons completely.[24]

A panoply of defensive systems lay just around the corner as well. Satellite Protection for Area Defense (SPAD), a system of hundreds of early warning satellites, would provide global coverage against enemy ICBMs; Random Barrage System (RBS) involved launching 20,000 to 100,000 satellites in random orbits to destroy enemy vessels using their own orbital trajectories; Project Needles, which the Pentagon successfully implemented with the help of MIT's Lincoln Laboratory, orbited hundreds of thousands of tiny copper needles to create an artificial ionosphere from which communications could be bounced. Project SAINT, though canceled in 1962, endeavored to engineer a satellite capable of inspecting and, if necessary, destroying an enemy space vehicle. Offensive and defensive systems alike would be maintained by agile repair craft (Project SMART) and supplied by small, manned vehicles (Project SLOMAR) that would also serve in rescue missions. Eventually, large manned space platforms would oversee the logistical interplay between offensive, defensive, and support systems that would, of course, include the seemingly numberless collection of reconnaissance, early warning, weather, communications, and mapping satellites already being developed and deployed.[25]

From novel technologies flowed novel strategic considerations. Analysts believed that the implications of these military systems would be nothing short of decisive. Journalists wrote of bombers, battleships, submarines, and Earth-based defense grids becoming "obsolete" in space-age warfare.[26] Spaceborne armies would replace the slow-moving brigades that had defeated fascism in World War II. Even jet power seemed retrograde. "Crack divisions will embark in rocket gliders," wrote James B. Edson, the Army's R&D director. "They will rise like flying fish above the atmosphere, re-enter, and glide to their destination." Within two hours such forces could respond to problems a half a world away. They would soon appear aboard large satellites, on lunar bases, and perhaps even on other planets.[27]

Flowing naturally from the idea that space technologies would eclipse weapons as recent as the B-52 Stratofortress and the Polaris submarine was the notion that warfare itself would transition from Earth to space. If the most important military technologies were housed in orbit, the thinking went, conflict would surely follow. Initially, of course, conflicts in space would involve a "mix" of Earth-based and space-based systems. Albert C. Stillson, a defense analyst at the Library of Congress, reminded readers of *Air Force Magazine* that "fighting in outer space would automatically involve earth-bound military power, and fighting on earth would automatically involve outer space military power." Space weapons would be essential to future wars, but not decisive on their own.[28] Over time, however, conflict would involve a greater percentage of space-based systems, become more automated, and shift the spatial burden of war from Earth to space.

"Cis-lunar" space, the broad swath of emptiness between the geostationary orbits of Earth and the moon's own gravitational field, would soon become the focal point of the most decisive military exchanges. To control this critical zone, the United States, the Soviet Union, and future spacefaring powers would have to inundate the region with space vehicles, both to stake a claim to as many orbits as possible and to defend them from enemies. Attrition in outer space would be decided not by the country that *first* accomplished military feats in space, argued Russian-American aircraft pioneer Alexander P. deSeversky, but the one that achieved "the fustest with the moistest."[29]

Emblematic of the importance to which US defense analysts ascribed cis-lunar space was the "gravity well," a celestial phenomenon—first introduced by Robert Richardson in a 1943 issue of *Astounding Science Fiction*—whereby the gravitational force that a large body exerts gradually widens and weakens as one moves farther out from that body into space.[30] As futurist and General Electric engineer Dandridge Cole explained, the Earth was positioned at the bottom of its own funnel-shaped well, roughly 4,040 miles deep. As the prodigious quantities of fuel necessary to propel rockets into orbit attested, it was far easier from an energy perspective to "drop" objects into the well toward Earth than it was to throw objects out of it. The military utility of Earth's gravity well was simple: the nation capable of taking possession of positions at the top of the well could attain a supreme advantage. G. Harry Stine, a rocket enthusiast and science-fiction author who later interviewed Cole, drew an easy analogy: "Put one person at the top of a well and another person at the bottom of a well. Give them rocks to throw at one another. Which person is going to have more time to see the opponent's rocks coming, more time to get out of the way, and more room to maneuver? Which person has the best opportunity to dodge rocks? Which person has the greater opportunity to do something? Which person stands the greatest chance of being hurt the worst?" As defense analysts would continue to proclaim into the twenty-first century, "a military space regime at the top of the gravity well would be a dominating position to initiate military activities for both offensive or defensive operations."[31]

In much the same way, military thinkers came to view the moon as a critical staging area for military operations in space. From both a strategic and tactical point of view, lunar outposts presented several stimulating possibilities. A missile base on the moon, many claimed, would strengthen nuclear deterrence, for any Soviet attack on such a base would require "about four and three-quarter days" to arrive, giving the United States ample time to retaliate not only with its Earth-based, second-strike forces but also with lunar-based, first-strike capabilities. Although enemies could feasibly destroy a moon base with their own space forces, such a facility would prove "impregnable" from Earth given the distance between them. In the case of a successful Soviet nuclear attack on the American homeland, a remote lunar base would vengefully respond in kind. Stillson thus predicted that a moon base "might cancel out the advantages to be gained by initiating total war." Edson suggested, moreover, that the military could make a moon base self-sufficient with the nickel, steel, and other minerals stored in meteor craters on the lunar surface. Moon bases, if self-supporting, could even "provide an important extension of the human habitat, and thus decrease the risk of self-extermination of the human race in a terrestrial conflict." Edson imagined a time when "the destruction of mankind on any single planet will be like the loss, in earlier times, of a city or a culture—a tragedy, but not the end of everything."[32]

The idea that a lunar military base would be decisive in a future war extended beyond the prognostications of experts like Golovine, Stillson, and Edson to the highest offices in the US Air Force. Brigadier General Homer A. Boushey, the first director of advanced research in the USAF, reasoned that because the moon lacked an atmosphere, a base there would not only make an ideal site for astronomical observation and communications relay but also an efficient platform from which to launch missiles. "From an energy standpoint, only one-fifth or one-sixth the energy is required to shoot a warhead from the moon to earth, as vice versa," he told the National Press Club in January 1958. Manned or unmanned stations could thus "catapult" missiles from shafts deep in the lunar soil, and those weapons could be then observed and guided from launch to impact. Any attack on a moon base from Earth-based platforms, moreover, would take forty-eight hours to arrive, giving personnel time to predict impact locations and to hide underground. The editors of *Air Force Magazine* seconded Boushey's recommendation that the United States install its base on the "dark side of the moon" to avoid detection by enemy governments. The base's "reception committee" for enemy weapons and landing parties could be built on the light side of the moon, while its offensive capabilities would remain harder to reach and monitor. Moon bases, therefore, provided significant advantages over Earth-based launching facilities, which required extensive tracking stations and were beholden to cooperative weather conditions. The lunar station appeared to represent "a retaliation base of unequalled advantage."[33]

From 1958, but particularly after Kennedy's September 1962 pledge that the United States would land men on the moon by the end of the decade, military leaders and aerospace enthusiasts insisted that despite their technical ambition, the construction of lunar military bases lay just around corner. At a symposium hosted by the Air Force Office of Scientific Research and the Aerojet-General Corporation, Boushey predicted that moon bases would be commonplace within twenty years (a US base would appear within ten), for while science and commerce were powerful motivations for exploration, "there is not the overwhelming urgency as for a military space capability."[34] The legal complications surrounding the use of outer space had to be tackled with rigor, of course, but if the superpowers could not come to terms on appropriate behavior in space, then it was "inevitable" that the moon would emerge as a strategic military asset. In its report on Project Horizon, the Army concluded that there were no technical obstacles to establishing a base on the lunar surface. Would-be space engineers could dig "holes or caves" into the moon and seal them with pressured bags. Early in 1958, the Martin Company, a prominent aerospace technology firm, proposed a "lunar housing simulator" consisting of inner and outer spheres of varying pressures, connected by vacuum-sealed airlocks. Westinghouse created plans for a moon-based nuclear power plant.[35] American Aviation Incorporated went on tour in 1962 with its model for a housing unit specifically designed for the three-man *Apollo* spacecraft that NASA planned to launch to the moon by 1969.[36] "Some persons will be quick to contest this concept [of a moon base] as 'Buck Rogerish thinking,'" wrote Army colonel Robert B. Rigg, "But they are wrong."[37]

One unsettling conclusion that contemporaries could reasonably have drawn from theories about the transplantation of warfare from Earth to space was that it would soon become more difficult to model. The tangled interplay of the weapons systems outlined in Golovine's book, for instance, represented a quantum escalation in the complexity of deterrence calculations: velocities increased, but so did warning times; delivery systems would be less susceptible to attack, yet so too would enemy targets; satellites, while improving reconnaissance, communication, and mapping, were delicate and hence vulnerable instruments. General Thomas S. Power, then commander of the Strategic Air Command (SAC), wrote that space warfare would be defined by "the operational relationship between space and time." The ballistic missile and its ancillary systems would create four-dimensional warfare, "a new regime of strategic operations in which utilization of the space medium will place a fantastic premium on action and reaction times." Future war seemed, if not more likely, then at least more difficult to predict.[38]

But the space-war oracles tried nonetheless. They conjured an image of space packed with advanced military systems interacting in lightning-fast but programmed trajectories. Tracking satellites would detect enemy ICBMs immediately on ignition, signal space-based defense platforms and ground stations, which in turn would mobilize ASAT ordnance. Satellite decoys and "barrages"—concentrations of orbiting debris—would provide suitable defense against enemy satellites and other space vehicles. Defense platforms in fixed positions would coordinate the entire enterprise. In reading accounts of how these electronic battles would unfold, one was tempted to wonder if human beings would eventually lose control of the process. With time, "remote-controlled" conflict would evolve into cool, dispassionate "robot warfare" whereby automated systems would respond reflexively to perceived threats.[39] In a letter to a colleague, USAF general Thomas White wrote that the United States might soon face "an era that is neither cold war nor hot war, but is characterized by a semi-overt-duel in the no man's land of outer space."[40]

Many strategists welcomed this eventuality as a civilizing one, for if wars would inevitably transition to outer space, perhaps they could be contained there. The constellation of orbital bombardment satellites, space vehicles, ASAT weapons, and interceptors, many theorists projected, would obviate the need to fight on the ground. Nuclear-tipped satellites could replace earthbound missile silos as the strategic targets of a major war. Offensive space vehicles could replace armies that had fought only to a bloody stalemate in Korea. Advanced technology might even enable militaries to engage in pitched space battles with remote, operational soldiers. But most important, space war could decide political outcomes without involving noncombatants. In the vacuum of space, argued economist Thomas Schelling, war "may involve few, if any, civilian casualties." Space war, in fact, could make large-scale conflict possible without initiating nuclear holocaust. Golovine proposed that orbital war could replace peripheral Cold War conflicts and inaugurate a strategic situation in which a preponderance of space weapons systems would decide the outcome of war. Dornberger wrote of the possibility for a limited "hide-and-seek" war in outer space in which the United States and Soviet Union would settle differences simply by destroying the other's spacecraft. Edson suggested that the moon be converted into "an agreed arena" for space weapons to decide political outcomes. All sides could avoid "terrestrial damage," and a clear winner would emerge to bloodlessly impose her will on the losers. "Full-scale orbital war," Golovine concluded, "might be the only human solution to the East-West ideological and political opposition." Though suspicious of the prospects for space war, Klauss Knorr, director of Princeton's Center for International Studies, agreed. Space, he wrote, was in many ways "an ideal theater for limited hostilities."[41]

Because the relevant technologies were so novel, and because their strategic implications were so radical, predicting how war would change in the age of space came to resemble a game of pin-the-tail-on-the-donkey. Political commentators and journalists with few technical skills gestured vaguely toward what they perceived to be the next generation of weapons. Engineers and defense contractors with little knowledge of bureaucratic politics or budget constraints simply guessed the extent to which practical decisionmakers would pursue their vision of the future. All seemed to ignore executive initiatives to preserve space from weaponry. None had complete information, and so all, in their own way, were stabbing in the dark. To compensate many writers reached back to similarly revolutionary moments in history. Some compared the inauguration of the space age to Columbus's "discovery" of the New World in the fifteenth century. Others went further back to the copper age invention of the wheel, or the Cambrian emergence of primordial life from rivers, lakes, and swamps onto dry land 430 million years before that.[42]

In the realm of military strategy, however, the space war oracles referred to texts that had been in circulation for less than seventy years. None seduced the space power strategists more than Alfred Thayer Mahan's "choke point" thesis, elucidated in the admiral's seminal 1890 treatise on geopolitics, *The Influence of Sea Power upon History*. In it, Mahan pointed to Great Britain's exceptional naval force to explain the growth and power of its empire in the sixteenth and seventeenth centuries and recommended that the United States build a force of similar strength if it were to compete as a world power. Mahan conceived of the open oceans as a "great highway" to be controlled at strategic points: the Strait of Hormuz in the Red Sea, the English Channel in the North Atlantic. Britain had controlled both at the height of its empire, and so had controlled trade flowing east and west across the Mediterranean and over the Indian Ocean to Asia. Mahan's theory came to have an enormous impact on Theodore Roosevelt, who surmised that if the United States was to control the Isthmus of Panama, it could enjoy a substantial share of all the wealth that passed through it.[43]

Space power theorists quickly adapted Mahan's thesis to outer space, what even discerning observers referred to as "the new ocean." In what they called the "Panama hypothesis," strategists articulated a vision of space geopolitics that substituted "space" for "sea" in crude fashion. "There are strategic areas of space," wrote Dandridge Cole in *Astronautics* magazine, "which must be occupied by the United States, lest their use ever be denied [to] us through prior occupation by unfriendly powers." The Soviet Union could claim "lunar Panamas" that would make US commercial or military action there impossible. Comparisons between outer space and the open seas were so widespread that New York publishing giant Hill and Wang reprinted Mahan's book in 1957.[44]

Extrapolations of Mahan's thesis to outer space took on greater weight early the following year when the US orbital *Explorer I* confirmed the existence of layers of ionized radiation in the upper atmosphere, what came to be known as the Van Allen belts (named for the University of Iowa scientist who pioneered the engineering behind the satellite). When the US government discovered that high-altitude nuclear detonations dispersed dense clouds of electrons, thus adding onto the particles occurring naturally in the Van Allen belts (see chapter 5), military planners began to speculate that by creating a thick layer of radiation in the upper atmosphere with nuclear weapons, one could control access to space through the Arctic and Antarctic poles, where, because of the Earth's magnetic field, radiation would be negligible. NATO planners surmised that the West could control access to space through these holes in the radiation cloak—a modern-day Panama Canal.[45]

The Panama hypothesis extended beyond Earth orbit as well, to points in cis-lunar space called libration points. In celestial mechanics, libration points (what astrophysicists refer to as Lagrange points for the French astronomer who helped discover the phenomenon) are distinct locations between two orbiting bodies at which the gravitational pull of each are in perfect balance. At such "L" points, an object will retain its position relative to the two larger heavenly bodies rather than enter its own orbit around either. In any given gravitational relationship between two large bodies, there are five Lagrange points. Within the Earth-moon system, because of the gravitational pull of the sun and the type of orbit that the moon performs, objects stationed at three of the Lagrange points—L-1, L-2, and L-3—would eventually drift away, making them "unstable." In contrast, the other two points, L-4 and L-5, are stable and therefore a capable home for a spacecraft. They are, in short, parking spots in space.

As with Earth's gravity well, the military significance of Lagrange points was apparent. Nations in control of the L-4 and L-5 points could regulate space traffic, deny enemies the use of particular orbits, and launch targeted strikes against the Earth, moon, and positions in cis-lunar space. Dominating stable Lagrange points, as Stine later put it, would permit a space power to "detect and take action against any threat originating anywhere in the Earth-Moon system." Like the moon itself, these points constituted invaluable high ground positions: "uphill" from both the Earth and the moon, the occupying force could monitor and control the entire spectrum; without a gravity well, it could "maneuver at will."[46]

The strategic lesson of "choke points" and of nineteenth-century geopolitics more broadly was clear: if the United States were to maintain its military and economic hegemony, and if it was to prevent the Soviet Union from acquiring a preponderance of power, control of space was nonnegotiable. In an essay titled "Reflections on Sea and Space," Peter Ritner, an editor for the *Saturday Review*, recalled the degradation of French military power in the seventeenth century and pointed to the empire's fateful decision not to build a state-of-the-art armada to match that of Great Britain. The United States should heed the lesson in space. "Admiral Mahan was neither the first nor the last philosopher to realize that new rules make a new game," he insisted. "We Americans are playing today with new cards and for new stakes. Only a generation of gamblers and pioneers—such as we must again become—can hope to enjoy the game. Or win it."[47]

Space Power Is Peace Power

By 1960 science-fiction authors, space boosters, and the new cadre of defense intellectuals had together built the intellectual scaffolding on which powerful interests attempted to advance substantive military programs like Project Horizon. When John F. Kennedy, who had criticized his predecessor for reacting so nonchalantly to the military implications of the Soviets' space feats, took office in January, these forces became more sanguine about the prospects of a more robust national effort to weaponize space. But their hopes proved illusory. Within months it was evident that the new president was little more interested in satellite bombardment, moon bases, or other weaponry for space than Eisenhower had been. Johnson would later disappoint the space war oracles still further by ratifying the OST, which, in addition to forbidding nuclear weapons from space, banned military installations, maneuvers, and weapons testing on celestial bodies (see chapter 6). Though advocates of a robust national defense effort in space would fight an uphill battle throughout the mid-1960s, they accepted their charge with urgency and conviction.

"Advocates" of a comprehensive defense program in space counted ordinary citizens, politicians, aerospace executives, and many others in their ranks, but the term can reasonably be said to be synonymous with the USAF, space war's most visible and powerful proponent. The Air Force's vanguard support of space power and the technologies needed to achieve it was a natural outgrowth of the branch's leadership. Between 1953 and 1965, each of the three Air Force chiefs of staff—Generals Nathan Twining, Thomas D. White, and Curtis LeMay—were vocal proponents of preempting Soviet influence in space through intense military modernization. Eugene Zuckert, secretary of the Air Force from 1961 to 1965, spoke in cities across the country on the emergence of near-Earth orbit as "the new high ground" of modern warfare. The USAF's R&D agencies, especially, were replete with forward-thinking engineers focused on achieving space dominance. Bernard Schriever, director of the high-profile Western Development Division; Homer Boushey, first director of the USAF's Directorate of Advanced Technology; and Donald Putt, director of R&D at the Office of the Chief of Staff, were among a spirited coterie of USAF personalities who propounded the necessity of American military supremacy in space.[48]

Aside from the USAF's top brass, the notion that the United States should attempt to control space unilaterally emanated directly from its incipient "aerospace" doctrine, which held that space represented merely an extension of the air medium and that given its doctrinal and technical experience, the Air Force should inherit the most relevant DoD space projects. That the USAF would be the military's branch of the future was signaled early on, when at the end of World War II the AAF's commanding general, Henry "Hap" Arnold, instructed the eminent scientist Theodore von Kármán to come to the agency's headquarters to study the next big weapon: missiles.[49] After *Sputnik*, when it was clear that space technology would be crucial to national security—and that space projects would attract the most federal dollars—the USAF's legislative liaison urged the branch's leaders to "emphasize and reemphasize" the USAF's claim to military space projects until "no doubt exists in the minds of Congress or the public that the Air Force mission lies in space as the mission of the Army is on the ground and mission of the Navy is on the seas."[50]

And emphasize they did. None insisted on the indivisibility of air and space more forcefully than the USAF's chief of staff, General Thomas D. White. In the national press, in military trade journals, and before Congress, White chipped away at external claims to the space pie. "I look upon the Air Force's interest and ventures into space as being as logical and natural as when men of old in sailing ships first ventured forth from the inland seas," wrote White in *Air Force Magazine*. "Similarly, ventures into outer space require men who know the air."[51] For White and other USAF leaders, space power was merely the next evolutionary step in the development of air power. The X-series of aircraft—with a mandate to go ever "farther, higher, faster"—reflected the service's purportedly inherent claim to space missions. Following the historic flights of the Bell's X-1 and X-15 aircraft, shortly after *Sputnik*, Boeing began development on the X-20 Dyna-Soar, which would reach space on an ICBM and remain there by "bouncing" on the atmosphere before gliding back down to Earth. Such technological and strategic considerations were effective in establishing "aerospace" as a legitimate military concept. Early in 1958 Air University revised its service manual to reflect the fact that air power had moved "naturally and inevitably to higher altitudes and higher speeds until now it stands on the threshold of space operations."[52]

By 1961, the Air Force had accomplished its goal becoming of the Pentagon's "executive agent" in military space technology. NASA had swallowed the Navy's Minitrack satellite tracking network, its Vanguard facilities, and the NRL. It had also absorbed the Army's JPL and the ABMA, including von Braun's Redstone team. And though it lost its manned space projects to the Mercury mission, the USAF emerged comparatively unscathed from the expansion of civilian space activities. It furthered its influence by contributing Thor boosters for NASA's lunar probes and cloud cover satellites; constructing new facilities at Patrick Air Force Base in Florida; and donating Atlas boosters and personnel to Project Mercury among numerous other activities.[53]

Emboldened by new their authority in the space field, USAF leaders called for a national drive toward "space superiority." In fact, argued Bernard Schriever, the United States should immediately create an independent military service for space operations, one that would serve as a "a protective umbrella" for the other branches struggling to coordinate projects, goals, and funding. Military space missions had "special requirements," he wrote, and therefore deserved special considerations—designated research departments, laboratories, testing facilities, and personnel.[54] In 1958 Eisenhower had created ARPA to collect and manage all military space projects under one roof and had created a new office for a director of Defense Research and Engineering (DDR&E) to advise the defense secretary on basic and applied research, development, and evaluation of new weapons systems. But these measures had not gone far enough. With a series of stunning "firsts" in space, after all, the Soviets had demonstrated an unambiguous commitment not only to exploration, it seemed, but also to establishing the foundations for a military presence in space. Their rocket boosters had much greater thrust, and Soviet researchers had achieved extraordinary progress in the field of space medicine, a sign that the Kremlin entertained high ambitions for manned flight. The Soviets, Schriever observed, regarded their accomplishments in space with "militant pride."[55] He demanded a similar commitment from his civilian government.

The purportedly nefarious intentions of the Soviets were, from the start, the USAF's principal justification for American military preponderance in space. Schriever, Boushey, Putt, and a host of others argued that given the USSR's aggressiveness, deviancy, and furtiveness, it could not be trusted to abide by international agreements to govern outer space in the interest of all. Observers in and outside of the USAF argued that at heart, the Soviet Union remained a "slave society" that benefited only Communist Party leaders. They agreed that the Soviet people, "joyless but dedicated," had been churned into missiles.[56] Those who accepted these basic characterizations quickly concluded that if the United States permitted the Communists to station military bases on the moon and orbit offensive satellites, they would use such technologies to either blackmail the West into submission or initiate a war to dominate the globe. Indeed for half a decade after *Sputnik*, Air Force leaders demonstrated a marked preoccupation—one could reasonably

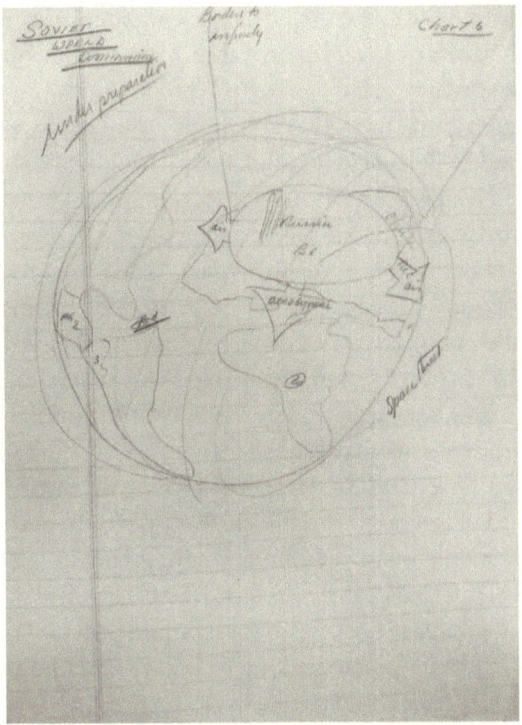

FIGURE 14. Bernard Schriever envisioned Russia extending its aerospace influence toward "borders to infinity." China, South America, and Africa all become "Red." "Soviet World Domination under Preparation," Statement by General Bernard Schriever before the House Committee on Armed Services, n.d., box 25, Bernard A. Schriever Papers, Manuscripts Division, Library of Congress.

call it an obsession—with the prospect of Soviet "world domination" from space. An internal USAF policy planning document fretted that Soviet preeminence in space would spell the end of the Cold War. Without sufficient ASAT countermeasures or offensive space capabilities of its own, the United States would be incapable of using space for either peaceful or military purposes. Americans would eventually have to cow to Soviet coercion.[57] Schriever appeared before the House Armed Services Committee equipped with maps charting the Soviets' step-by-step plans to achieve a "strategic aerospace envelopment" of the globe. The Soviet Union was "attempting to complement their significant progress in taking over Free Countries through political aggression by creating a global space capability to dominate the world" (figure 14). If the Kremlin were to achieve its aims, Schriever suggested, Americans' recent fascination with bomb shelters would be justified. Fenced in by the Communists, Schriever concluded, quoting the anthropologist Margaret Mead, Americans would be reduced to eking out meager lives in "screwed down garbage cans."[58]

The alternative outcome—*American* domination of space—would look quite different, USAF leaders insisted. For all tough talk of war in space, "peace" was their operative word. In an international political environment undergirded by US military primacy in space, peace and stability would reign. Schriever recalled that when SAC emerged after World War II as the nation's preeminent offensive force, Americans understood that despite the organization's capacity for raining hellfire onto enemy cities, "air power is peace power." The Cold War, he reminded readers of the *New York Times*, was kept cold by US bombers. Similarly, Americans should now understand that "space power is peace power."[59] The fact that both Eisenhower and Kennedy had repeatedly elucidated US policies intended to preserve space for exclusively peaceful purposes did not preclude the establishment of military supremacy in space, Schriever and other USAF officials agreed. In fact, US space power was consistent with the policy of "space for peace," for it would provide "the means for insuring that the policy is carried out."[60] Peace would be guaranteed by the threat of a war that could not be won.

This conception of peace differed markedly from others voiced in the State Department, the White House, and many congressional circles. For US officials whose policy orientations hedged closer to the sanctuary doctrine, "peace" implied international controls, disarmament, and diplomatic initiatives to strengthen cooperation in space. It meant the surrender of national sovereignty to international institutions, especially when the matter pertained to all humanity. Senator Henry Jackson (D-WA), for example, proposed that US satellites be launched "at the service" of the United Nations.[61]

Champions of US control of space, in contrast, equated peace with deterrence. Schriever, Boushey, and countless others envisioned a world in which US power ensured safety for all by threatening massive retaliation against any would-be aggressor. The United Nations, the International Criminal Court, and other international bodies could not be trusted to achieve global peace and justice, they reasoned, for such structures had failed historically to produce any enduring change. The enemies of peace and freedom could clandestinely develop space weapons, abandon treaties outright, or use the organs of international power to chip away at the foundations of global order. Only US military preponderance in space could guarantee peace and prosperity into the distant future. "We believe that space can be free to all for peaceful activity only if somebody keeps it free," General Zuckert intoned in an article for *Aviation Week and Space Technology*. "We are that somebody."[62] Chester Ward, judge advocate general for the Navy, recalled that previous periods of lasting peace had been undergirded by the unquestioned military prowess of a single political hegemon: the so-called Pax Romana relied on the prowess of the Roman Legion and the Pax Britannia on the Royal Navy. By extension, unrivaled US supremacy in space would produce an enduring Pax Americana.[63] In two decades' time, predicted another officer, the Earth's children would thank American "*men* on the moon" for world peace.[64]

The emphasis on "men" exists in the original document, and it underscores the masculine overtones of the military claim to outer space in the name of freedom and peace. Those who supported US control of outer space envisioned Americans as the stewards—the protectors—of Earth. They considered their nation to be the world's bastion for liberty, democracy, and peace, protecting the world from Soviet totalitarianism and militarism like Batman did Gotham from the city's numberless villains. In the USAF's rendering of the future, nothing short of Earth's survival was at stake in the absence of US control. "We cannot afford to be second-best in the conquest of space," wrote USAF Major General Dan C. Ogle, for anything other than space supremacy would "invite the binds of subjugation, the humiliation of inferiority, or the oblivion of destruction."[65]

Given their dogged pursuit of advanced weapons in space, one is tempted to cast these generals in the mold of a Buck Turgidson or Jack Ripper, the pugnacious air force commanders in Stanley Kubrick's 1964 dark comedy, *Dr. Strangelove*—insensitive to the politics of war, ignorant of technical details, and apathetic in the face of death on a massive scale. Yet the collection of military officers who promoted military missions in space were much more thoughtful than their zealousness may at first suggest. For starters, nearly all were well grounded in the technical aspects of the military space mission; indeed, they represented the vanguard of R&D within the Air Force. Boushey served as deputy director of R&D at USAF headquarters and was the first director of the Office of Advanced Technology under the deputy chief of staff for development. Zuckert was active in the technology field both before and after his tenure as Air Force chief of staff. In the early 1950s he served as a member of the AEC, and later was both chair of the board at the Nuclear Science and Engineering Corporation, a pioneer in radiation chemistry, and of

AMF Atomics, the atomic energy branch of American Machine and Foundry, Inc. Donald Putt held a master's degree in aeronautical engineering from Cal Tech, and served as director of both the Air Research and Development Command and the R&D branch of the Office of the Deputy Chief of Staff for Air Material. He was a member of the National Research Council, the Institute of Aeronautical Sciences, the Society of Navy Architects, and several similar groups. Schriever took courses in aeronautical engineering from Stanford and came to head NASA's most important development agency. James Doolittle earned one of the nation's first doctorates in aeronautical science—from MIT no less—and served on Eisenhower's PSAC and later as chair of Space Technologies Laboratories.

These men, World War II veterans all, were keenly aware of the profound impact new technologies could impose on warfare, and each was shocked and embarrassed by Soviet achievements in space during the late 1950s and early 1960s, achievements they were convinced had military implications. While many of these leaders applauded diplomatic overtures aimed at preventing the extension of the arms race to space, they argued the United States should hedge its bet against a foe it did not trust. For a brief window after *Sputnik*, these leaders could reasonably suggest that US military preparedness had few negative repercussions: it would boost the economy by creating jobs; encourage education in the sciences; generate new technology; protect the United States from space-based blackmail; and, in case the Kremlin had no interest in space weapons, an American military buildup in space would accelerate and protect the development of peaceful exploration. "You will note I have stated that the United States should win and maintain a capability, and I repeat the word capability, to control outer space," Putt reminded Congress in 1958. "I do not say that we must exercise control of space, but we must have the *capability* to do so. There is an important distinction between the two. We in the military fervently hope that all nations join together in whatever measures need to be taken to ensure that space is never used for any but peaceful purposes."[66]

Lunartics!

Disciples of space power made a compelling case for preemptive national efforts to dominate outer space. The very liberty and security of the so-called Free World, they argued, depended on it. It is in light of their substantive efforts that failure to achieve even a modicum of success is put in stark relief. Piece by piece, Schriever's vision of US military preponderance in space crumbled. McNamara's budget-conscious whiz kids "lowered the boom" on the most relevant projects.[67] The X-20 Dyna-Soar, after $400 million dollars, succumbed in 1963 to concerns about the lack of a clear goal for the space plane. Why, many asked, should the Air Force have a manned program alongside NASA's?[68] Later that same year DoD downgraded and then canceled Project SAINT, when the Soviets reversed position on the legality of spy satellites. Kennedy's campaign for a test ban treaty ended hopes for ASAT weapons equipped with nuclear warheads, as well as Project Orion, which sought to achieve nuclear rocket propulsion. The Dyna-Soar's follow-on, the MOL—a manned, stationary reconnaissance vehicle—collapsed in 1969 to concerns of redundancy and suspicions about its mission. Advanced projects including lunar bases, satellite bombardment, and armed space platforms, which had never left the stage of preliminary investigation under a skeptical Eisenhower, wasted away in neglect.

How did this happen? Considering the obvious military utility of space, the DoD's colossal research budget from the 1950s through *Apollo 11*, and the vision exhibited by USAF leaders, the defeat of "space war" as an American military paradigm is startling. The reasons for this defeat are multivariate, but three stand out. First, as the widespread support for a civilian-controlled space agency and a UN committee for space activities revealed, the political standing of the sanctuary doctrine was robust, and into the mid-1960s it continued to be a cultural force against which planners and theorists of space war found themselves competing for legitimacy.

Consider the Cold War, for a moment, from the widest possible lens: at heart, it was a struggle over whether the United States or the Soviet Union, democratic capitalism or revolutionary socialism, was better fit to lead humankind into the future. Crucially, by the time Army researchers began to draw blueprints for a manned lunar outpost, it was widely assumed in both countries, and in others around the world, that a significant part of that future was destined to occur in outer space. The idea that space technology would have profound implications for civilizational development on Earth was beyond doubt. Cold War competition in the space field thus necessitated more than a struggle over which nation's engineers could create the superior technology, or which could create it first. Bound up in the battle over the future were questions of war and violence, morality and ethics, technology leveraged for national power or technology for the common good. That governments should close off the cosmos—the human future—from warfare had become a near-axiomatic proposition even before spaceflight had become a practical probability. In the contest of images US officials waged with their Soviet counterparts, pursuit of advanced space weapons and military space infrastructure was, simply, a losing strategy.

The Cold War contributed to the declining appeal of space war in another way. As historians have amply demonstrated, the primary interest of US policymakers in military space technology was satellite reconnaissance. Through SAMOS, CORONA, and other spy satellite programs, defense planners would finally have reliable data on Soviet force levels, the location of airfields and launching facilities, and the ability to accurately detect nuclear weapons tests. Legitimizing these invaluable technologies required distinguishing them from "aggressive" and "active" military space projects akin to those favored by USAF leadership. "Passive" spy satellites would monitor arms agreements, provide early warning of enemy attack, and expand worldwide communications. In a balance of terror, these instruments were just what the doctor ordered. In contrast, controlling the pace and direction of futuristic space weapons seemed impossible. Although certainty about the coming military revolution propelled much DoD research on military space technologies, fear proved equally decisive in quelling that revolution. It was best, many agreed, to keep the genie in the lamp.[69]

And what if there was no genie in the first place? A third explanation for space war's failure is that so many educated observers rejected its basic premises. As swiftly as plans for advanced military space projects left the Bell Aerospace boardroom, appeared in the pages of *Aviation Week and Space Technology*, or left the lips of boosters like von Braun, they became subject to criticism in scientific, engineering, and political circles. Lee A. DuBridge, founding director of MIT's Radiation Laboratory and president of Cal Tech, referred to lunar outposts and other futuristic military technologies as "useless Buck Rogers stunts" and "utter nonsense." Why, he asked, would the United States attempt to launch missiles from the moon—over 240,000 miles away—when any conceivable target in a future war would be no more 5,000 miles from ground-based ICBM platforms? Moreover, the laws of physics precluded the use of the moon or satellites as platforms for bombs. Without gravity, bombs "just won't drop." At a conference in Los Angeles, DuBridge argued that occupying the moon had "not the slightest military value" and that constructing a lunar outpost would likely provoke conflict with the Soviet Union. Humankind's astounding technological achievements, he worried, rather than inaugurating a program of cooperative and practical research, risked transforming the United States "into a nation of space cadets" in which vast resources were squandered on "fanciful and fruitless" instruments of war. "There is plenty to do," implored the Manhattan-project veteran, "without trying to nail the American flag on the whole solar system by next week."[70]

Other prominent scientists and engineers agreed. Dinsmore Alter, a lunar scientist and retired director of the Griffith Park Observatory, said the moon was too valuable as an astronomical observatory to use as a military base.[71] Famed science writer Arthur C. Clarke remarked that moon bases "do not bear serious examination," for the difficulties of supplying such a base outweighed any potential advantages. Lunar-launched missiles could be more easily detected than those launched from Earth platforms, and would take longer to arrive, thus giving the enemy more time to prepare defenses.[72] "Outer space is new," Space Technology Laboratories President Simon Ramo observed, "but so is the bottom of the ocean. This does not mean we should put our retaliatory force there." C. C. Furnas, a former assistant secretary of defense for R&D, argued that ballistic missiles launched on Earth would always carry out their missions more efficiently, effectively, cheaply, and accurately. Another scientist insisted that even if the Soviet Union was able to control outer space, its new realm would be nothing but "a new Siberia," only more difficult to reach, maintain, and defend.[73]

Opposition to space weaponization extended beyond doubts of its technical feasibility. Commentators ranging from grassroots peace activists to executive-level officials questioned not only whether the United States *could* engineer a future of war in space but whether it should. Journalists often denounced the USAF and other promoters of advanced military space technology as "lunartics." These advocates, many observers claimed, threatened to extend the Cold War indefinitely; they were blindly selling the future in the name of continued military advantage. In response to *Space World*'s special issue on space war in 1962, a London-based reader remarked to the editors that he would have deemed the report "as the work of irresponsible fools" if it were not for the fact that so many US military officials were named as sources. The propensity to build armaments based not on realistic Soviet strength but exaggerated estimates amounted to "playing with matches in a gunpowder factory." Peter Ritner, convinced though he was about the geopolitical importance of outer space, nevertheless scoffed at traditional military claims to the cosmos. In particular, he lamented that the DoD would continue to monopolize the government's plans for space stations, which Ritner considered the most significant achievement since the Agricultural Revolution. "Is any sane man actually thinking of extrapolating these disputes beyond Earth's skin," he asked readers of the *Saturday Review*, "and waging political wars on battlefields of space and stars?"[74]

The USAF's vision suffered another blow early in 1964, when Lyndon Johnson received an important, long-range policy planning document on space technology and national security in the 1970s. The interagency group that authored the report took a measured look at the potential need for advanced military hardware, including space weapons, but cast a jaundiced eye on the generals' most fundamental assumptions. Much had been written—"probably too much"—about space as the high ground of future military operations, as well as the certainty that control of space meant control of the Earth. But space was neither predictably nor inevitably the key to future military power. The usefulness of space as a platform for military operations depended on the costs, technical feasibility, political favorability, and effectiveness of space-based technologies relative to existing, Earth-based systems. Moreover, the report stressed, shouting military voices were too certain of Soviet intentions in space. It was enough to speculate that the USSR "*may* try, and *may* succeed, in pursuing *some* promising avenues of development with important military applications. It is too much to say that they *will* do so; it is not a foregone conclusion." On balance, therefore, it was incumbent on policymakers to consider at what costs the United States might extend the arms race to space. Any technical problem the United States might pose to the Soviet Union in space would also complicate the US strategic posture as well. Introducing more complexity into the strategic environment did not necessarily make the world any safer; there was, the report stressed, "a point of diminishing returns."[75]

Notions that the future of armed conflict rested in outer space would continue to enjoy popular resonance in American film, television, and literature. Gene Roddenberry's *Star Trek* (airing first in 1966), Brian Aldiss's edited collection of short stories *Space Opera* (1974), and George Lucas's *Star Wars* trilogy (1977–1983) beamed images of fantastic war in space to millions of Americans. But the USAF's dream of American "space superiority" had by the mid-1960s suffocated from government neglect and often outright hostility from scientists, politicians, journalists, and ordinary citizens. Considering the pervasiveness of space-based warfare in US popular culture, the seriousness with which boosters and defense intellectuals attempted to define its contours, and the energy that the Air Force expended in trying to create a future conducive to it, the brevity of "space war" as a military construct is surprising. In tracing federal R&D of satellite bombardment, moon bases, and ASAT weapons in the months after *Sputnik* to the demilitarizing initiatives of the mid-1960s, the national debate over transplanting armed conflict to outer space seems to have lasted little more than half a decade.

Space war ultimately proved ephemeral because it was tied to other fleeting ideas of the early Cold War. One was the widespread conviction that terrifying new weapons, though a menace to humankind, would provide lasting peace. This idea was nothing new, but it enjoyed a powerful renaissance in the atomic bomb, which, many were convinced, would shock the world into pacifism.[76] Eisenhower's New Look strategy, in which national security would lean heavily on the threat of massive nuclear retaliation, was the central manifestation of this line of thinking. The USAF's leadership and several other corners of the DoD, particularly their R&D teams, considered advanced space weaponry to be simply the next iteration of deterrent force. The Soviet Union—and technology itself—would not stand still. Why should the United States?

A second transitory historical moment on which "space war" relied was the *Sputnik* shock. From 1958 to 1963, the Soviets had achieved a series of stunning space feats that seemed to foreshadow an uneasy period of strategic parity if not outright Soviet military supremacy. *Sputnik* was far heavier than the United States' first satellite, *Explorer 1*, for example. The R-7 Semyorka ICBM that boosted "fellow traveler" into orbit possessed far greater thrust than the Vanguard rocket that carried *Explorer*. The Soviet Union far outpaced the United States in space medicine; it orbited a human being before the United States; and, in 1962, it was able to return a satellite filled with biological specimens back to a prearranged target on Earth. Few questioned the USAF's charge that the Kremlin would attempt to leverage these accomplishments into bona fide space-based capabilities that would keep the United States out of space and menace it from above. But this, too, proved a temporary threat.

Ultimately, the appeal of converting outer space into a vast battleground replete with advanced weaponry could travel only as far as strategic conditions would take it. That support for "space war" was tied to contextual events at once comforted and concerned contemporary advocates of a space sanctuary. It was heartening, on the one hand, to know that brash calls for the next generation of weapons could prove fleeting as threats came and went. On the other, the revolving door of security and insecurity ensured that the quest for American preponderance in space always lurked just below the surface.

CHAPTER 5

The Cosmic Bomb

In mid-1959 the USAF completed a top-secret study on a question at once farcical and irresistible: What might happen if it detonated an atomic bomb on the moon? In its euphemistically titled report, "A Study of Lunar Research Flights," scientists from the USAF's Special Weapons Center proposed that a nuclear explosion on the moon could provide crucial details about astrogeology, lunar seismology, and planetary physics. It could illuminate the physical qualities of the space environment, the detection of explosions in space, and "the capability of nuclear weapons for space warfare."[1] Not least, there were political benefits to consider. The ten-man team that conducted most of the research for the study—which included a young Carl Sagan, then a doctoral student at the University of Chicago—concluded that "specific positive effects would accrue to the nation first performing such a feat as a demonstration of advanced technological capability." A large, bright explosion in the middle of the night sky would dazzle the world and remind Soviet leaders of American military supremacy. If global opinion could be properly prepared, the United States would swiftly wash away the embarrassment of *Sputnik* and forcefully reclaim the initiative in the space race.[2]

But the scheme failed to garner support within the Eisenhower administration. As debates over moon bases, space planes, and similarly ambitious projects revealed, technical constraints, tight budgets, and popular opinion possessed the necessary weight to drown the Pentagon's most ambitious plans to weaponize space. And yet, that the Air Force devoted resources to even studying the possibility of a nuclear detonation on the moon reflected the stubborn allure of integrating nuclear technology with space technology. Their mutual status as futuristic, powerful, and dynamic implied that the two should be married. It was no coincidence that in the wake of Hiroshima and Nagasaki, some journalists, groping for words that might capture the awesome power of the new weapon, dubbed it "the cosmic bomb."[3] Nuclear science, for its promise to unlock the secrets of the universe through study of its smallest units, seemed a natural corollary to space science, which promised to do the same through study of the universe's largest and most distant ones. Astrophysics and nuclear physics asked the biggest questions, created the grandest spectacles, and offered the possibility of unbounded energy and military power. It would become one of the greatest technological temptations of the Cold War.

In the late 1950s and early 1960s, this temptation manifested in two distinct but interrelated possibilities. The first was the use of nuclear explosions in space for military advantage: to disrupt communications, jam defense systems, and even destroy incoming missiles. That outer space was to become a new theater of armed conflict dictated that governments develop new methods for engaging the space-based technologies that would shape the future of war, a challenge for which the revolutionary power of nuclear weapons seemed ideally suited. Beyond their centrality to deterrence, the Pentagon had already laid plans to use nuclear weapons to destroy navies, poison food and water supplies, irradiate dense city populations, and eradicate enemy troops on the battlefield.[4] Scientists also proposed that "peaceful nuclear explosions" could be used to create harbors and canals, etch roads into mountainsides, and even excavate natural gas.[5] The fitness of nuclear weapons for such myriad tasks suggested their utility for war in the space age.

The second possibility was to hitch nuclear weapons to satellites, using them as bombardment platforms to reign fire on earthly targets. The idea was an archetypal example of what cultural historian H. Bruce Franklin has called "the Superweapon." Brewed from a mix of available technology and the imaginative paranoia of the early Cold War, the bombardment satellite, like all Superweapons, promised to make war obsolete by threatening overwhelming and unstoppable destruction. Depending on the number of bomb-tipped satellites available at a given time, and depending on the orbits of those satellites, their wielders could target every acre of the globe. Only nations with advanced ASAT weaponry or ballistic missile defenses could hope to mount respectable countermeasures. Bombardment satellites, like the hydrogen bomb, the ICBM, the nuclear submarine, and MIRV weapons, would enforce a Pax Americana.[6]

Yet for all their military appeal, both detonating bombs in space and dropping them from satellites proved ephemeral in US and Soviet policy. While the superpowers did initiate programs for exospheric nuclear tests, their programs were decidedly short-lived. Eisenhower and Khrushchev agreed to a testing moratorium late in 1958, and when testing restarted abruptly three years later the United States and the Soviet Union managed to test only a handful of weapons in space before their diplomats in Geneva successfully negotiated the Limited Test Ban Treaty (LTBT), which forbade tests not only in space but underwater and in the atmosphere as well. At the same time, the superpowers signed a seminal UN resolution calling on states to refrain from orbiting nuclear weapons, installing them on celestial bodies, or stationing them in space "in any other manner."[7] And though the Soviet Union had by the mid-1960s developed the controversial fractional orbital bombardment system (FOBs), it abandoned the program within twenty years having never orbited a warhead. In the United States, satellite bombardment failed even to escape the design stage.[8]

The fleeting terror of cosmic weapons proved another victory for the sanctuary approach to space first conceived by interplanetary thinkers after World War I. This was true not only because of moral and ethical appeals to preserve space as a realm of peace and cooperation but also because of rational military and political assessments about the consequences of extending the arms race to space. The Arms Control and Disarmament Agency (ACDA), born amid the contest over nuclear weapons in space, warned of a "grim game" in which governments, attempting to eclipse each new military space program with still more engineering, would lose control of the wheel. After satellite bombers would come space platforms to control them, both land- and space-based ASAT systems to destroy the platforms, and new countermeasures to stop those weapons in turn. "Missiles will bring anti-missiles, and anti-missiles will bring anti-anti-missiles," warned the former chair of the Joint Chiefs, Omar Bradley, shortly after *Sputnik*. "But inevitably, this whole electronic house of cards will reach a point where it can be constructed no higher. . . . And when that time comes there will be little we can do other than settle down uneasily, smother our fears, and attempt to live in the thickening shadow of death."[9]

"The Greatest Experiment Ever Conducted"

Late at night on July 9, 1962, people all over Hawai'i gathered hurriedly outside their homes and businesses so that they might witness something spectacular. In Pearl Harbor, sailors ordered themselves on the dock alongside their submarines. Organizers of the Miss Hawai'i pageant in Honolulu stopped the event so that the audience could scramble to nearby rooftops. Tourists packed hotel courtyards. College students brought their dates to the beach. Parents woke sleepy children. All rushed to whatever vistas they could find to gaze on what promised to be a fantastic lightshow: the US government was poised, after two previous failures, to detonate a 1.4 megaton hydrogen bomb roughly 248 miles above the Earth, in outer space.

Despite reports that weather might again delay the test, at 10:45 p.m., Honolulu time, a massive Thor rocket lifted into the sky from Johnston Island, roughly nine hundred miles southwest of Oahu. Fifteen minutes later, at exactly 11:00 p.m., spectators got what they were waiting for. In an instant, the rocket's W49 warhead exploded, igniting the Pacific sky in a flash of bright white light. Suddenly, "the blue-black tropical night . . . turned into a hot lime green," a color so unexpected that many observers gasped. The viridescent aurora gave way to a "lemonade" pink, then, slowly, to a terrifying red. "It was not the familiar orange of the tropical sunset but a deep, solid red, and the people afterward groped for words to describe it," reported *Life* correspondent Dick Stolley. "The glow bubbled aloft and boiled in the sky . . . It was as if someone had poured a bucket of blood." In New Zealand, the Fijis, and Samoa, observers stood in awe as swirls of green, yellow, red, blue, and orange moved along a north-south axis across the heavens. "It was like a watercolor sunset," recalled one witness. For six minutes, the night had become a radiant and prismatic day.[10]

This was Starfish Prime, the largest space-based nuclear weapons test of the Cold War. The blast was the climax of a yearslong effort by scientists and engineers to determine the impact that high-altitude nuclear explosions might have on military systems, including ICBMs, satellites, and ground-based communications. In the tense years between *Sputnik* and the Cuban Missile Crisis, the United States carried out five nuclear tests in space ranging from roughly two kilotons to the one-and-a-half *mega*tons that Prime released over the Central Pacific. Not to be outshone, the Soviet Union exploded four of its own weapons from 1961 to 1962. These tests, hastily conceived and in many cases dangerous, revealed that the political campaign to preserve space as a weapons-free sanctuary was vulnerable to the perceived military requirements of the Cold War. If warheads detonated at orbital altitudes might neutralize enemy missile attacks, what use was there for a space sanctuary at all? Still, the brevity of the tests also proved that the arms race and the Cold War generally were vulnerable to sanctuary politics. Starfish Prime had endangered satellites used for communication, navigation, weather prediction, and scientific data collection. Its emitted electrons raced endlessly in magnetic bands of high-density radiation circling the planet. The very ability of astro- and cosmonauts to escape the atmosphere was in jeopardy. By late 1963 the fragility of the space environment and the technologies inhabiting it proved ample reason to close the door to cosmic testing, one that the arms race had opened merely a half-decade before.

The notion that nuclear detonations in space might have military value grew directly from the *Sputnik* "crisis." Though many scientists, including Eisenhower's science adviser James Killian, considered the Soviet satellite a relatively benign technological achievement, others viewed it primarily a reflection of long-range missile capacity. One such scientist was Nicholas Christofilos, a Greek-American physicist then working at the Lawrence Livermore National Laboratory. He had never worked on a military project for his adoptive government (he had emigrated to the United States in 1953), but *Sputnik* spurred the eccentric-but-talented scientist to devise a solution.

An answer to intercontinental missiles, Christofilos argued, could be derived from his civilian research at Livermore. Late in 1957 his primary objective at the lab was to design a fission nuclear reactor called Astron. Theoretically the reactor would create enormous volumes of energy by superheating fusion fuel in a dense layer of high-energy electrons that had been trapped in a magnetic field. Engineers could fabricate such a field by constructing a magnetic "mirror" from which electrons would repeatedly bounce, increasing in intensity as time passed. Christofilos proposed to defend the United States from enemy missiles by extrapolating Astron's electromagnetic reactions to a planetary scale, for Earth possessed its own natural mirror, the magnetic Arctic and Antarctic poles. He proposed that if a nuclear bomb were exploded at the appropriate altitude, the emitted electrons would become trapped in the Earth's magnetic field and concentrated in a shell of radiation capable of disrupting or even destroying any missile that passed through it. In essence, an "Astrodome" around the Earth.[11]

"The Crazy Greek," as he became known, took the idea to his boss at Livermore, Herbert York, who initially thought there "was simply no place to take an intervention like Nick's." Any experiment would require numerous satellites; the United States had not yet successfully launched even one. The AEC, a natural sponsor for Christofilos's idea, did not have the capacity or the authority to take on a project at such a "grand scale." Neither could the services devise such a sophisticated experiment on their own.[12]

But two fortuitous developments intervened to keep the shell experiment afloat. The first was that activities associated with the IGY would soon confirm the existence of natural belts of radiation circling the Earth. With NAS funding, physicist James Van Allen and his small team at the University of Iowa attached Geiger-Müller counters to two of the first US satellites, *Explorer 1* and *Explorer 3*, as part of the Navy's Vanguard program. By April 1958 it was clear that those instruments had encountered some "real physical phenomenon." As Van Allen explained to his NAS audience on May 1, the two satellites had been bombarded, at particular points in their orbital trajectories, with radiation a thousand times more intense than cosmic rays (like "bees in a hive," he reported). Having conducted experiments with rockets continuously from 1946, the Iowan had made the first great discovery of the space age: the donut-shaped torus of energized particles soon dubbed the Van Allen belts. In less than a year, he had gone from a dusty basement laboratory to the cover of *Time*.[13]

Second, and perhaps more important, in January 1958 Herbert York became chief scientist at ARPA, which Eisenhower and Defense Secretary Neil McElroy established to house all military space programs on an interim basis. "Once in ARPA," York later recalled in his memoirs, "I found myself with both the responsibility and the authority for carrying out the experiments" his colleague had envisioned only weeks before. Invigorated, Christofilos circulated his proposal just as ARPA was taking shape. His classified paper—"On the Possibility of Establishing a Plasma Shield of Relativistic Electrons in the Exosphere of the Earth as a Defense Against Ballistic Missiles"—made the goal clear to the senior military officials who read it. Eisenhower's science advisers invited Christofilos to present his idea to the PSAC on January 3 and endorsed his experiment after two weeks of study. Killian and York briefed Eisenhower on March 6, emphasizing the tests' military implications. An electron shell could erode radio and radar transmission, damage or destroy arming and fusing mechanisms on ICBMs, and threaten crews of orbiting spacecraft with intense radiation.[14] The president sanctioned the proposal and instructed ARPA to launch a satellite to observe the results. What became known as Project Argus became one of the new agency's first tasks.[15]

And it had to act quickly. On March 31, Moscow announced a unilateral nuclear test moratorium to begin that October, triggering intense discussion among US and British leaders about whether they should reciprocate. Eisenhower and British Prime Minister Harold Macmillan agreed with Khrushchev to host an international science conference aimed at the problem of monitoring and verifying a test ban. This Conference of Experts met in Geneva over the summer and concluded that through air sampling, seismology, and electromagnetic measurements, governments could feasibly detect nuclear detonations down to one kiloton. On August 22, the day after the conference issued its report, the United States announced its own moratorium, also to begin at the end of October.[16] Planners of Christofilos's experiment—which now included the AEC, Defense Nuclear Agency (DNA), CIA, Armed Forces Special Weapons Project (AFSWP), and JPL, as well as all three branches of the military—had just a few more months to execute it.[17]

Secrecy was paramount. "At the time," York remembered, "we all wanted to keep the whole idea to ourselves." Christofilos "was doubly bothered when we talked about [Argus] in the Pentagon offices with only an ordinary door between us and the main corridor where just anybody might be listening."[18] Within Van Allen's Iowa team—responsible for engineering satellite instruments to measure electron density, diffusion, and decay rates—only a select few were privy to the true nature of their project.[19] ARPA developed cover stories for most of the civilian and military organizations involved, tying most Argus activities to scientific investigations related to the IGY. Most of the DoD personnel involved in the tests were also in the dark.[20]

On August 27 a Navy task force consisting of nine ships and more than 4,000 men successfully executed the first test, *Argus 1*, far beyond the Kármán line (roughly sixty-two miles) separating the Earth's atmosphere from the icy void beyond. A "great luminous ball" ensued, followed by a bright aurora that extended along the magnetic field lines where the detonation had occurred. Airplanes swept along the margins of the explosion range taking photographs. Sounding rockets launched from Patrick Air Force Base in Cape Canaveral measured the resulting radiation. *Explorer 4* passed by, day after day, week after week, collecting vast quantities of data. Argus's planners repeated the routine twice more on August 30 and September 6, just seven weeks before the test moratorium was scheduled to begin. *Argus 2*, like its predecessor, failed to reach its intended altitude, and onlookers were unable to see an aurora. But *Argus 3* ascended more than three hundred miles, supplied relevant data, and created a brilliant light show, "a red crown" and an "'X' of electric blue."[21]

The results were compelling and conclusive. Killian briefed Eisenhower in a memo on November 3. Just as Christofilos had predicted, the radiation created by each explosion had become trapped in the naturally occurring bands of radiation that Van Allen had discovered just the year before. "The action of the field," he wrote, "resembled that of the barrel of a recoilless rifle." Half of the radiation dipped into the atmosphere near the launching site, causing the auroras; the other half arched 4,000 miles above equator along the magnetic field lines. The phenomena observed, Killian stressed, contained much of military importance. "These electrons might damage electronic equipment in space, render the space above the earth temporarily lethal to man, generate worldwide radio-noise, especially in the HF (high-frequency) and VHF (very high-frequency) bands, and produce strong localized disturbance of the ionosphere at great distances from the explosion." DoD would have to modify its design requirements for electronic components aboard its ICBMs and IRBMs, ballistic missile and air defense radars, and especially short-wave radio equipment.[22]

Keeping Argus under wraps proved difficult. For beginners, the United States had engineered *Explorer 4* with the understanding that it was to be part of scientific data collection for the IGY; its findings would have to be made public. Over the course of late 1958, as the Eisenhower administration presided over the establishment of NASA and as US officials negotiated in New York and Geneva for a UN committee on space, the Pentagon contented itself by classifying Argus's results and releasing other *Explorer* data related to scientific exploration. The world would be none the wiser.

But Hanson Baldwin, a Pulitzer Prize–winning military correspondent for the *New York Times*, had caught wind of Argus in August, before the United States had conducted a single test (ARPA Director Roy Johnson traced the leak to Van Allen's team at Iowa.) Unsure of what to do with such an important story with potentially hazardous implications for national security, Baldwin consulted his colleague, the science reporter Walter Sullivan, another giant at the *Times* who had already polished his reputation covering the IGY. The two journalists reasoned that if they broke the story prematurely, the tests would never occur. Their suspicions were confirmed when Sullivan received a call from ARPA's head of security pleading the newspaper to hold the story.[23]

Baldwin and Sullivan obliged until they felt their exclusive story slipping away. In October, after the test series was complete, Christofilos gave a talk on the creation of an artificial electron shell at a gathering of the American Physical Society in Chicago.[24] Aside from the use of a nuclear device to supply the electrons for the experiment, he relayed many crucial details about the physical theories that had underlaid the Argus experiments.[25] Rumors of nuclear tests in outer space pervaded the annual meeting of the American Association for the Advancement of Science (AAAS), particularly Fred Singer's paper on "Artificial Modification of the Earth's Radiation Belt." When Singer and Van Allen gave a joint press conference at the meeting, a *Newsweek* reporter asked them about the relationship between the radiation belts and high-altitude nuclear explosions. Worse still, Soviet scientists had caught the scent. In March 1959, I. S. Shklovskiy and V. I. Krasovskiy reported the detection of high-energy particles in the lower Van Allen belts in an article for *Izvestiya*. "It is not to be excluded," they wrote, "that this zone has, if we may say so, an artificial origin."[26]

With its contacts at ARPA unforthcoming about when the government might release information on the test series, and with other news outlets hot on their tail, the *Times* broke its Argus story on March 19, 1959. "US ATOM BLASTS 300 MILES UP MAR RADAR, SNAG MISSILE PLAN; CALLED 'GREATEST EXPERIMENT'" was emblazoned on the front page. Though US officials would later emphasize the important scientific data extracted from the Argus experiment, particularly its confirmation of the "Christofilos effect," Baldwin hovered over the series' profound military implications. As one Pentagon official put it, Argus showed that a "teaspoonful" of radiation in the "sea" of electrons already circling the planet might prove integral to American security, especially if the US-Soviet test moratorium, to begin that month, should prove a failure.[27]

The paper's exposé did not come as a shock to US officials who had kept Argus a secret, but they struggled to defend the operation from charges, particularly in Congress, that the public had been lied to and that nuclear tests in outer space jeopardized the safety of the entire planet.[28] Baldwin had noted that at least with regard to radio frequencies, Argus had "made its effects felt almost globally."[29]

Who was to say that the tests did not have similarly universal ramifications in other areas? Argus's entire rationale, after all, had been to discover how nuclear weapons might behave at high altitudes and ascertain the effects such weaponry might have on the physical environments in which they were detonated. Because answers to these questions were still unavailable, Argus, many argued, had been at best risky and at worst massively irresponsible. "The Earth is so minute on the cosmic scale and its environment is controlled by the delicate balance of such great natural forces," thought British astronomer Bernard Lovell, "that one must view with dismay a potential interference with these processes before they are investigated by the delicate tools of the true scientist."[30] "By what right can any nation or group proceed to envelop our earth in a band of radiation, the harmful effects of which are still open to debate?" wrote one concerned citizen in a letter to the *Times*. "What will be the verdict of future generations, if any remain, on the irresponsibility of conducting nuclear tests without more concern for human welfare and moral issues?"[31]

Such sentiments were not uncommon in the wake of the *New York Times* leak, for they came at a time when fears about nuclear fallout approached their Cold War zenith. In the five years that had transpired since the *Lucky Dragon* incident—in which fallout from the March 1954 Castle Bravo thermonuclear test at Bikini Atoll had contaminated the crew of a Japanese fishing vessel—apprehension over radiation poisoning and environmental degradation had reached a fever pitch. Fallout had been the subject of extensive congressional inquiry in the spring of 1957, particularly the spread of radioactive isotopes such as cesium-137 and strontium-90. Scientists from the US Weather Bureau warned Americans about the probability that fission products would become enmeshed in natural weather patterns and descend to Earth as rain or snow. That same year, physical chemist Samuel Glasstone published a revised edition of his popular book *Effects of Nuclear Weapons*, which warned of radioactive, "stratospheric" debris carried thousands of miles through the upper layers of the atmosphere. Novelist Nevil Shute told of humanity's extinction north of the equator in *On the Beach*. Christofilos himself admitted that fallout at extreme altitudes "probably did come down." Within the band of radiation created by the Argus tests, he added, ominously, "a man on a satellite ... would be dead in less than 3 hours."[32]

The Conference on Discontinuance of Nuclear Weapons Tests that began on October 1, 1958, in Geneva assuaged these fears, at least temporarily. Eisenhower stated that if the Soviet Union refrained from testing any weapons during the ensuing negotiations, the United States would halt its own testing for one year, possibly more if a proper inspection system emerged and the talks proved fruitful. Great Britain, the only other nuclear power besides the Soviet Union, reluctantly agreed. Backed by the encouraging prospects for monitoring a test ban that the Conference of Experts had put forth that July, US negotiators arrived hopeful that the high-level statements and moral posturing about a test ban might evolve into something more.[33]

But from the start the moratorium was an uneasy, jittery animal. Just a day after negotiations began, the Soviet Union carried out a ten-kiloton nuclear test from Kapustin Yar, effectively nullifying the freeze. Another occurred at the same test site on November 3. Viewing the blasts as both a test of nerves and a delayed execution of tests months in the planning, Eisenhower kept the American moratorium intact. A week later, Khrushchev demanded that the United States, Great Britain, and France withdraw their forces from West Berlin within six months, precipitating a three-year crisis over the fate of the city that spanned nearly the entire moratorium.

From a historical perspective, it is surprising how long the pause in testing actually lasted. For peace activists, Cold War-weary politicians, and much of the public, it promised safety by finally closely the lid on radioactive fallout. But for almost as many military leaders, Cold War-*waging* politicians, and laboratory scientists, it promised to undermine safety by bringing the technological bases of deterrence to a standstill. How could one ensure nuclear deterrence, asked Edward Teller, the "father" of the hydrogen bomb, without nuclear development! The moratorium depended entirely on the success of the test ban negotiations underway in Geneva, yet those were mired by seemingly irreconcilable differences over the need for inspections. Since *Sputnik* Khrushchev had touted the Soviet nuclear arsenal as the slingshot that would restrain the American goliath, but at the same time he sought to reap propaganda rewards from the testing pause. Critics in the Soviet Union viewed the moratorium as freezing US strategic superiority in place, and those in the United States argued that it gave the Kremlin a free window in which it could catch up in secret.[34] Throughout, the Pentagon kept the AFSWP on high alert, ready to reignite testing at a moment's notice. The Soviet Union never stopped planning for new tests.

The wobbly edifice finally gave way in August 1961, when the Kremlin announced its moratorium was over. Over the next three months it conducted more than thirty tests, including the infamous "Tsar

Bomba" blast that set off fifty-eight megatons—4,000 times more powerful than the bomb that destroyed Hiroshima—over the Arctic Circle on October 30. The official justification for ending the moratorium was that the French had broken the atomic barrier and would not abide by the freeze. In reality, the resumption of tests had more to do with (1) applying pressure on the West; and (2) catching up with the United States in the development of strategic weapons. As Khrushchev told his top nuclear scientists in a special meeting on July 10, it was imperative to restart nuclear tests "because the international situation had deteriorated and because the USSR lagged behind the US in testing.... We would have to add to our nuclear might and show the 'imperialists' what we could do." In his memoirs, Andrei Sakharov recalled that during the meeting "it was perfectly clear that the decision to resume testing was politically motivated." Khrushchev's son Sergei later recalled that his father thought it "impossible to catch up with our rivals ... but he was trying to gain everything that our technology could provide."[35]

To be sure, the Soviet Union had not stood idle. The tit-for-tat logic of the arms race dictated that the Kremlin match US developments in strategic weaponry lest it fall behind in some small but decisive qualitative advantage. Megatonnage. Throw weight. Guidance systems. Early warning. Even projects that might be technically dubious or expensive demanded attention so long as the Americans were making strides. To wit: having followed the Argus tests closely, particularly their implications for missile defense, the Soviet Union initiated its own program of space-based nuclear tests shortly after it broke the moratorium.[36]

The test series originated in an effort to develop a prototype antiballistic missile (ABM) system called "System A." In their early studies on ABM, the Ministry of Medium Machine Building and the Academy of Sciences concluded that systems using conventional explosives would simply not suffice. But in December 1956 Yuri Khariton, one of the Soviet Union's leading nuclear scientists, supervised a test—designated "K"—that proved the utility of nuclear weapons for the ABM role. The explosive power of a nuclear warhead would enable System A's engineers to substantially reduce the number of ABM launchers in their proposed ring around Moscow. Ironically, the power of nuclear weapons also threatened the budding system. Their destructive effects, particularly electromagnetic pulse (EMP), would endanger the electronic components of the ABM warheads and the guidance systems of the missiles carrying them. To measure these dangers, the Central Committee sanctioned a high-altitude and space-based nuclear test series—also named "K"—that would put System A through the wringer.[37]

K-1 and K-2, bearing 1.2 kiloton warheads, exploded in the exosphere above Sary Shagan on October 27, 1961. Colonel Yuri K. Tsukov remembered a "poisonous green" cloud forming to the northwest of the test site. Radios produced only static. For months afterward officers found blind rodents on the steppe nearby. Officials finished the series a year later, almost to the day. On October 22 and 26 respectively—smack dab, it is worth noting, in the middle of the Cuban Missile Crisis—K-3 and K-4 exploded three-hundred-kiloton warheads over central Kazakhstan, where the resulting debris was supposed to intersect with two ICBMs that had launched from Tyuratam minutes before. K-3's EMP induced a powerful current that fused 350 miles of overhead telephone lines. Its low frequency enabled it to penetrate the ground, where it overloaded circuit breakers and ignited a fire that burned down a nearby power plant.[38]

These were unnerving consequences, but by that time the Starfish Prime explosion had already proven the awesome—even planetary—effects that nuclear tests in space could render. On Oahu, fuses and circuit breakers blew out. Hundreds of streetlamps extinguished while others suddenly lit. Local radio stations and telephone service failed. Church bells ringing out the 11 o'clock hymn were muffled by the sudden clanging of burglar alarms and air-raid sirens. For twenty minutes, communications between the United States and Australia completely died out. The same was true for communications between Hawai'i and Midway Island and for the Tokyo's links to Honolulu, California, and Buenos Aires.[39]

Before long it was clear the blast had reached even farther afield. The explosion had inundated the lower Van Allen belt with high-energy particles, adding a two-hundred-mile-deep layer to the Earth's natural geomagnetic radiation. According to one science magazine, in the South Atlantic over Brazil, where the belt dipped closest to the Earth, the detected radiation was ten times more powerful than the highest readings ever recorded in the Van Allen belts and about one hundred times the intensity. The test had dispersed a cloud of kinetic electrons that chaffed and burnt out the solar batteries of at least six satellites, including the Soviets' *Kosmos 5*; the United States' *Traac* (scientific), *Tiros 5* (weather), *Telstar 1* (communications), and *Transit 4* (navigation) satellites; and Britain's first satellite *Ariel 1*, which soon stopped broadcasting altogether. The belt forced DoD to postpone its launch of *Anna*, a geodetic mapping satellite, in the hope that the intensity of the radiation would weaken by the final quarter of the year. In a few cases, the useful life of the satellites affected had been cut to one-fourth or one-fifth.

NASA and the Defense Atomic Support Agency quietly organized a scientific symposium at the Goddard Space Flight Center in Greenbelt, Maryland, to discuss the startling findings. The gathered scientists estimated that in the belt's core, 2,000 miles above the Pacific Ocean, "one square centimeter of a satellite's skin would be bombarded by a billion bomb-produced electrons every second."[40]

These results unsettled NASA scientists and inflicted considerable political harm to American pretensions about leadership in the peaceful uses of space.[41] Yet from a military perspective, Prime delivered much of what its planners had hoped for. The AEC and the Pentagon had intended for the tests of 1962 to be bigger, bolder versions of Argus and to settle, once and for all, whether nuclear explosions could render effective defenses against enemy missiles and satellites.[42] Though the Dominic series did not include a test of ballistic missiles through the blast zone, the impact on satellites was apparent almost immediately. Telstar, which had begun beaming television broadcasts to Europe the day before Starfish Prime, soon began dying as it passed through the emitted electrons. On September 10, Britain's Science Minister Lord Hailsham wrote floridly to Harold Macmillan of *Ariel 1*: "Although badly wounded in his solar paddles, he is not quite dead."[43] The Air Force would soon parlay the Dominic tests at Johnson Island into the nation's second ASAT project, Program 437, what many in the military viewed as a chronically overdue response to the threat of bombs *stationed* in space.

The Weapon of the Future

Before conservative paragon Phyllis Schlafly became a household name for her campaign against the Equal Rights Amendment, she was, at heart, a defense intellectual. With Rear Admiral Chester Ward, she coauthored best-selling books on the culpability of American liberals in weakening US military strength; the strategic failures of Robert McNamara; the psychology of Henry Kissinger; and Gerald Ford's arms control policy.[44] Among the first of her many books was *Strike from Space*, also written with Ward. In it, they warned that the United States was on the verge of another Pearl Harbor. Whereas the Roosevelt government had ignored key intelligence about a surprise Japanese attack on the Hawaiian base, Johnson's wisemen now buried their heads in the sand rather than face

Soviet development of space-based nuclear weapons. "All the evidence screams the terrible conclusion that the Soviet strategy is *Make Noise in Vietnam, but STRIKE FROM SPACE*," they warned. "The Soviets deliberately trapped us in Vietnam to divert our attention and resources to a guerilla-conventional war and away from the nuclear threat in space."[45]

Such fears were not unreasonable given the times. The Soviet space program was shrouded in secrecy. The Cuban Missile Crisis had shown that the Kremlin was willing to assume substantial risks to achieve geostrategic parity with the United States. Gigantic rockets affiliated with a new orbital bombardment program appeared in Red Square parades. Yet by the time *Strike from Space* hit bookstores in 1965, the threat of space-based nuclear destruction had all but disappeared. Two years earlier, the United States and the Soviet Union had signed a UN resolution calling on states to forego stationing weapons in orbit, on heavenly bodies, or in space "in any other manner." The weapons on display in Moscow had never been tested. And Khrushchev's rivals, enraged by the premier's nuclear "adventurism" in Berlin, Suez, and Cuba, deposed him in October 1964.[46]

So why the alarm? One answer is that Schlafly and Ward had seized on recent space activities to create a narrative of military vulnerability that might justify greater defense expenditures and a more aggressive foreign policy. Indeed, red-blooded Goldwater conservativism was at the core of each of their joint projects. At the same time, it is impossible to ignore the extent to which orbital nuclear weapons had seized the national imagination by the early to mid-1960s. Satellite bombardment cropped up in front-page news articles, best-selling science-fiction novels, and nonfiction tracts like *Strike from Space*.[47] In the half-decade after *Sputnik*, too, the Pentagon engaged in a number of secret feasibility studies on satellite bombardment, details of which leaked into popular books and trade journals.

This outpouring of writing contradicted the failure US military leaders experienced garnering any meaningful support for orbital bombardment. Some observers, both at the time and since, dismissed arms control in space as low-hanging fruit for cynical cold warriors looking to score a propaganda coup.[48] Because so many scientists and engineers discounted orbital weapons on technological grounds, it was easy to argue that a ban on nuclear warheads in space would forbid weapons that neither the United States nor the Soviet Union were inclined to pursue in the first place. Given their disadvantages compared with traditional means of delivery, many argued, bombardment satellites failed to change the balance of power.⁴⁹ But barring nuclear weapons from space was not merely a political issue. American arms controllers relied on practical assumptions they deemed critical to national defense: the centrality of satellite reconnaissance to deterrence, the exorbitant costs of space weapons, and uncertainty about the goings-on of the Soviet space program. Some US officials, though they considered orbiting warheads a near-term feasibility, argued that unilateral restraint might oblige their Russian counterparts to follow the same path and that arms control in space would "lend impetus to the whole disarmament field."⁵⁰ These considerations helped create a gulf between the pervasiveness of space weapons in popular culture and political debate on the one hand, and their defeat in US space policy on the other. That divergence kept the space race from becoming something more.

Orbital destruction had been a subject of public fascination since the end of World War II. In July 1945 US Army intelligence regaled journalists from the *New York Times*, *Life*, and *Time* with details of a massive *Sonnengewehr* (Sun Gun) that Nazi scientists had modeled for use in combat. Their blueprints called for a gigantic mirror—three-and-a-half square miles and orbited roughly 5,000 miles up—that would harness the sun's solar rays and redirect them onto enemy cities and armies. The intense heat created by the focused rays could "make an ocean boil" or "burn up a city in a flash." Even under the skeptical cross-examinations of British and US intelligence, the scientists coolly insisted that had their work gone unimpeded the Third Reich would have been able to field the weapon within another fifty to one hundred years.⁵¹

Only months later, after Hiroshima and Nagasaki, the American physicist Louis Ridenour immediately connected the devastating power of the atomic bomb to satellite technology in a short story for *Fortune* magazine. "Pilot Lights of the Apocalypse" opens in an underground command center beneath San Francisco, where high-ranking military officials give the president a tour of the facility. The commanding general explains that there are more than 5,000 bomb-equipped satellites in orbit above the Earth, owned by a host of different countries, ready to strike enemy cities in the event of major hostilities. Because such a strike would descend from outer space, however, determining from where an attack originated is impossible. The command staff must therefore rely on "political" data—an ever-shifting list of political agitators—to determine which enemies might have the greatest motivation to initiate a war. No sooner does the president's tour end when the bunker shakes violently and pieces of the ceiling crumble to the floor. The command center loses all communications, and a series of flashing red dashboard lights show that major capitals around the world have been destroyed. Assuming this is "the real thing," one reckless colonel initiates a space-based strike against Denmark, top on the current list of antagonists. But it is only after this rash action precipitates an all-out nuclear war that the command staff realizes their mistake: the city's communications and relay lights had been affected by an earthquake, not a Pearl Harbor-like attack. Seconds later, a *real* strike on San Francisco destroys the bunker, ending the play.[52]

Was orbital bombardment truly feasible? Was Ridenour an oracle of wars to come? Not likely, many agreed, at least when the federal government began examining the possibility after V-J Day. When the Douglas Aircraft Company engaged in feasibility studies on the potential military utility of satellites, its analysts largely dismissed the notion that orbitals might be useful as platforms to launch missiles or drop bombs. In its seminal 1946 study, *Preliminary Design of an Experimental World-Circling Spaceship* (to which Ridenour contributed), Douglas concluded that although satellites could be deorbited over specified targets and thus serve as missiles themselves, their principal utility resided in reconnaissance, weather prediction, targeting, navigation, and communications.[53]

The limited capacity of satellites as weapons was reaffirmed when the Rand Corporation, which split with Douglas to become an independent, government-funded think tank, hosted a three-day conference on "unconventional weapons" in 1949. James Lipp, head of Rand's Missile Division, remarked that satellites were "qualitatively different from other weapons, as their primary purpose is not to destroy things." Lipp and other conference-goers agreed that the weight of atomic warheads made them poor candidates for satellite payload, given that rockets of the day lacked the power to boost them into orbit. Rather than engineer satellites as weapons of Cold War competition, Lipp argued, nations should collaborate on the effort given satellites' "universal significance."[54] Eisenhower's advisers were equally skeptical. In a seminal 1955 report the NSC argued that satellites would likely "constitute no active military threat to any country over which they might pass." Although the United States could conceivably use large satellite-guided missiles at Earth targets, it would "always be a poor choice for the purpose."[55]

The opening of the space age did little to change these perceptions. As Congress began debating the contours of the National Space Act in the spring and summer of 1958, the PSAC scoffed at the notion of weaponized satellites. Much had been written, it wrote in a special report to the public, about the possibility that space would soon become a theater of war, one replete with satellite bombers, lunar bases, and space planes. But "even the more sober of these proposals," the committee stressed, "do not hold up well on close examination or appear to be achievable at an early date."[56] That same year, when USAF generals Curtis LeMay and Donald Putt argued that the Pentagon should immediately begin research on orbital bombardment, Donald Quarles, deputy secretary of defense, demurred. He strongly objected to "the inclusion in the presentation of any thoughts on the use of satellites as a [nuclear] weapons carrier," and declared that the Air Force was "out of line" in advancing bombardment as a possible satellite application. Indeed, he forbade any consideration of bombardment in the USAF's future plans. And while both LeMay and Putt voiced their opposition to the directive on grounds that the Soviet Union might explore the nuclearization of space, Quarles remained "adamant." The Air Force stopped using its Weapons System designation for the military satellite program at the behest of ARPA, which had begun oversight of US military space programs in February. Its director, General Electric executive Roy Johnson, intended the move to "minimize the aggressive international implications of overflight." The change, he noted, would "reduce the effectiveness of possible diplomatic protest against peacetime employment" of passive satellites.[57]

Although it was increasingly clear that incipient US space policy would not include a bona fide program for bombardment satellites, Eisenhower did permit ARPA and the services to conduct preliminary research on space weapons.[58] The Rand Corporation devoted a full two years to the issue beginning in 1958.[59] In a special report prepared for the House Select Committee on Astronautics and Space Exploration, the think tank explained the feasibility of bombing targets from space. Its author, Robert Buchheim, admitted that if a warhead were to descend straight down from above its target, the propulsion requirements would be prohibitive. "However," he wrote, "the problem quickly becomes an entirely reasonable one if distances of several thousands of miles are allowed for accomplishing descent." Technicians could initiate and control unmanned satellites directly from stations on the ground, or flight crews aboard manned orbitals could direct warheads to their targets: "For such a case, the guidance operation could include direct line-of-sight steering of the bomb-carrying missile to the target—even a moving target."[60]

At the same time, the USAF Ballistic Missile Division in Los Angeles, just an hour's drive down I-10 from Rand headquarters, began a series of system requirement studies on space defense projects, many of which included orbital weaponry. Most of these studies are still secret, but several of the titles allude to a clear interest in bombardment: Strategic Orbital System (SR-79814), Earth Satellite Weapon System (SR-79821), and Advanced Earth Satellite Weapon System (SR-79822), for instance. The Air Force's Strategic Earth System Study (SR-181), Strategic Interplanetary Study (SR-182), and Lunar Observatory Study (SR-183) also included plans for nuclear weapons. Together these analyses imagined the hardware necessary for an "Earth Military Orbital Space Force," as a long-range planning document phrased the Air Force's ambitions. "The Soviets will put man on space platforms in cislunar space and on the moon as expeditiously as possible," wrote the USAF's diviners. "For both reasons, this Nation must have man in space, and man probably will be utilized eventually in space offensive weapons."[61]

Defense intellectuals, journalists, and military technologists brought the debate over orbital bombardment into the open. For many thinkers in the late 1950s and early 1960s, bomb-tipped satellites simply reflected the "next logical step" of deterrence.[62] Indeed, on first glance, they appeared to offer a number of significant military advantages: warheads aboard satellites could be armed immediately and launched to targets within minutes; because of their altitude and velocity, satellites in variable orbits were harder to detect and less vulnerable to enemy countermeasures; such weapons would not require extensive ground installations; and the number of personnel needed for effective operations was small compared with aircraft and missiles.[63] Then, of course, there were subjective political benefits. M. N. Golovine emphasized that recallable ICBMs (he called them nuclear-armed bombardment satellites, or NABS) were "virtually the ultimate in mobility and a powerful psychological deterrent." The *Los Angeles Times*' Frank Bristow thought that this weapon "of particular psychological horror" might even mitigate the risk of an accidental nuclear war. "Having a space-based weapon relatively secure from surprise attack," he wrote, "would give us the freedom to wait and thoroughly evaluate any alarm before giving the fateful order to release our city-smashing weapons."[64]

Even Thomas Schelling, dean of the arms control intelligentsia, considered the public outcry against bombardment satellites the product of "intellectual laziness," a failure to consider such weapons on their merits. Space-for-peace advocates, he noted, often considered the ability of nations to prohibit such weapons a barometer for how seriously those nations regarded broader disarmament. If the United States and the Soviet Union could not come to terms on an area that was technically and fiscally difficult to weaponize in the first place, then what hope was there that the superpowers would achieve reductions in missiles, submarines, bombers, or nuclear weapons? Although the compunction to protect the pristine wilderness of outer space was compelling in moral terms, Schelling warned against maximalist approaches to the disarmament of outer space. Bombardment satellites were unlikely to replace or outmode any arm of the US strategic triad. Adding satellite bombers to the existing weapons complex might "keep Khrushchev from being reasonably sure that he could get away with a surprise attack." If satellite bombers strengthened deterrence, therefore, the international community should not close the door. "If we have made it a test of our resolve to give up smoking for the sake of our health," Schelling analogized, "we may consider the test more important than our health, and stick by our resolve even when we receive evidence that smoking is good for us."[65]

What about feasibility? Cost? Majors Paul V. Bartlett and Relf A. Fenley, two USAF engineers, wrote that although manned satellite bombers required much greater thrust—and thus money—than bomber aircraft and ICBMs, these costs should not deter Americans from committing to the task. "The ancient Egyptians spent a large percentage of their national income over a period of years in building the pyramids," they reminded readers of *Air University Quarterly*. "It has been said that if we were to spend a like proportion of our national income for a like number of years, we could put the pyramids into orbit."[66] Air Force general Thomas White and ARPA director Roy Johnson seconded their assessment before Congress as it debated the contours of the National Space Act. The latter rejected the notion that it was "ridiculous to put a man in a satellite to drop a bomb because a bomb wouldn't drop."

> Actually, we do not know what the weapons of tomorrow are going to be. Work over the next 20 years might lead to a death ray, and if you have a death ray, that would be the weapon of tomorrow, and then obviously a man in a satellite up in the sky would be in a far better position to use judgement, to exercise control of that ray. So what I am saying is: let's not look at the problem of tomorrow in terms of weapons of today and just automatically say that there will be no military uses of space way out, including the moon. If we think in terms of present weapons, that is probably right. The bomb today is considered the ultimate weapon. I suspect 20 years from now the bomb will be passe.[67]

Whatever strategic and tactical benefits bombardment satellites might lend to US military power, and whatever their cost and viability, the most compelling rationale to pursue such weapons remained the possibility—the *likelihood*, as the Air Force argued—that the Soviet Union was diving headlong into development. When Khrushchev alluded to a "fantastic weapon . . . in the hatching stage" to a national radio audience in January 1960, Manhattan Project veteran Ralph E. Lipp speculated that the Russians were working on an orbital missile that could be stationed in space and recalled back to Earth on demand.[68] He was on to something. The US intelligence community determined that the Premier had referred to the construction of a one-hundred-megaton hydrogen bomb, but the Soviet Union soon began preliminary development of Sergei Korolev's massive *Globalnaya Raketa*-1 (Global Rocket 1, or GR-1), which served as the basis for one of the most frightening weapons of the Cold War: the fractional orbital bombardment system.[69]

FOBS, as the system came to be known, would allow the Soviets to orbit a nuclear missile and deaccelerate it out of that orbit onto Earth targets. Unlike traditional ICBMs, which follow an arched trajectory roughly six hundred to twelve hundred miles above the planet, fractional orbit missiles could trace a "depressed" trajectory as low as 125 miles. This lower flight path would dramatically reduce the fifteen minutes of warning time US ground stations could typically provide for missiles launched from Soviet territory. Because they used Earth's naturally occurring orbits, FOBS missiles could enjoy an unlimited flight range—a space bomber that need not refuel midflight. Its continuous orbit also made it impossible to precisely determine the missile's intended destination. Most spine-chilling was that FOBS weapons could deorbit along a polar axis, from south to north, thus bypassing the comprehensive system of radars the United States had established along stations in Alaska, Greenland, and England: the vaunted Ballistic Missile Early Warning System (BMEWS). "We can launch missiles not only over the North Pole, but in the opposite direction, too," Khrushchev boasted. "As the people say, you expect it to come by the front door, and it gets in the window."[70]

In a seminal 1962 National Intelligence Estimate of the Soviet space program, the CIA speculated that the Kremlin was unlikely to pursue such weapons given the technology available but warned that the Agency's ability to identify military programs was admittedly "poor." Although the accuracy, reaction time, targeting flexibility, and vulnerability of FOBS weapons made them unfavorable compared with

ICBMs until the end of the decade—the report stressed that "in the near term its military effectiveness would be minimal"—the Soviets might be interested in demonstrating a bombardment satellite as an act of propaganda. A developmental system could appear as early as 1965.[71]

This proved remarkably accurate. The Soviet Union initiated the first of three separate FOBS programs in March 1961, rocket engineer Vladimir N. Chelomei's design for a near-orbital missile, a modified UR-200A ICBM. In April 1962, the Experimental Design Bureau (OKB-1), led by Korolev, began development of the GR-1. The three-stage rocket weighed more than one hundred tons and could boost a 2.2-megaton-yeild warhead. The Kremlin would eventually scrap both programs in 1965 in favor of Mikhail K. Yangel's much larger R-36-O, a modified SS-9 superheavyweight ICBM. Fully fueled, it weighed 180 tons and packed between two and three megatons worth of TNT. As part of a broad propaganda campaign, the Kremlin paraded both Korolev's and Yangel's rockets through Red Square in military celebrations. Radio Moscow bragged that "the main property of missiles of this class is their ability to hit enemy objectives literally from any direction, which makes them virtually invulnerable to anti-missile defense means." NATO and US analysts were convinced that the missiles were operational and designated them the SS-10 SCRAG and the SS-9 SCARP.[72]

Whatever estimates US intelligence provided about Soviet bombardment from space, public stunts and pronouncements amounted to writing on the wall. In August 1960 the Vostok program achieved a stunning victory when it orbited a 10,000-pound satellite—replete with dogs, rats, mice, and other biological specimens—and successfully returned the "flying menagerie" to Earth within seven miles of the intended target.[73] Donald Putt reasoned that the weight of the satellite, coupled with its recovery so close to the landing zone, reflected a new capacity to deliver nuclear weapons from satellites to any place on Earth.[74] Following the inaugural flights of Soviet cosmonauts Yuri Gagarin and Gherman Titov in April and August of 1961, the danger seemed to have become acute. Khrushchev implied that the balance of terror had suddenly tipped decisively in the Soviets' favor: "We placed Gagarin and Titov in space and we can replace them with other loads that can be directed to any place on earth.... You do not have fifty and one-hundred megaton bombs. We have bombs stronger than one-hundred megatons." He repeated the message on December 10, indicating that the Soviet Union, if it could land Vostok capsules on prearranged targets, could "send up 'other payloads' and 'land' them wherever we wanted."[75]

"This Grim Game"

As swiftly as the cosmic bomb had taken hold of the US military imagination, it proved abortive in both its manifestations. Merely five years after the United States conducted its first nuclear weapons test in space—four since the Air Force completed "A Study of Lunar Research Flights"—the LTBT ensured that there would never be another. At the same time, US and Soviet officials cooperated on a landmark UN resolution banning the stationing of nuclear weapons in outer space and on celestial bodies. Just months after *Sputnik*, Roy Johnson had predicted that bombardment satellites would be the weapon of the future; by October 1963, two years before the Soviets tested a single FOBS missile, they had become a political dead end. Prevailing winds blew in favor of arms control.

During the 1960 campaign Kennedy had censured Republicans for the absence of a concrete plan for disarmament and promised arms control would be a top priority under his administration. Once in office, though he initially speculated that disarmament was "really just a propaganda thing," the new president took proactive steps to mitigate the arms race, including the one budding in outer space.[76] He appointed Jerome Wiesner, a vocal advocate of arms control, as his special assistant for science and technology. He oversaw the elevation of the US Disarmament Administration to agency status. And despite his criticism of Eisenhower, he continued many of his predecessor's initiatives: pursuit of a test ban; continuance of the moratorium begun in 1958; and the convening of the Committee of Principals, a group of high-ranking government officials the State Department created to discuss concerns about nuclear testing.[77]

The extent to which Kennedy himself subscribed to notions about the inviolability of space as a sanctuary from the arms race and from Cold War politics generally is difficult to determine. He authorized systems—both Air Force and Army ASAT missile programs, for example—that would target space assets while at the same time squelching broader USAF ambitions to weaponize space. As Paul Stares has noted, the Kennedy period was one "of false hopes and false starts" for advocates of robust military space development.[78] Yet one thing is certain: if not a true believer in the Interplanetary Project in the mold of an Arthur Cleaver or Arthur C. Clarke, Kennedy still spoke like one. "The new horizons of outer space must not be driven by the old bitter concepts of imperialism and sovereign claims," he told the General Assembly in September 1961.

"The cold reaches of the universe must not become the new arena of an even colder war." To this end, Kennedy proposed "keeping nuclear weapons from seeding new battlegrounds in outer space" as part of the broader US program for general and complete disarmament.[79]

The most significant prong in this campaign, and that which made nuclear tests in space a relatively short-lived feature of the Cold War, was pursuit of a test ban treaty. The LTBT resulted from a multitude of complex processes that took nearly a decade to unfold. In May 1955 Soviet delegates to the UN Disarmament Committee proposed discontinuing all nuclear tests. A dozen UNGA resolutions that followed addressed the possibility and urged a ban under a system of international controls. Though the other members of the Subcommittee of Five—the United States, Great Britain, Canada, and France—were generally supportive, negotiations proved laborious, halting, and vulnerable to obstinate Cold War politics. Should a test ban emerge in isolation from the goal of general and complete disarmament? The Soviets originally thought not, but reversed course after Khrushchev's 1961 meeting with Kennedy in Vienna. Should a ban be predicated on progress in other measures of arms control—namely, a suspension of fissionable material production for weapons and safeguards against surprise attack? The United States and Great Britain were willing to compromise, but France proved intractable. Most obstructive of all, would a ban require on-site inspections and other compliance measures? US negotiators were adamant that agreement was impossible without them; yet, fearing verification protocols would reveal its relative weakness to the American arsenal, the Soviet Union balked. In the General Assembly, in the UN's disarmament committees, and in the three-power talks that had produced the test moratorium, these issues prevented any serious daylight until mid-1963, when British, Soviet, and US leaders agreed to conduct trilateral negotiations on a test ban in Moscow. Kennedy announced the talks on June 10 and pledged that the United States would not be the first to resume atmospheric tests.[80]

From there, a cakewalk by comparison. On July 2 Khrushchev made a speech advocating a test ban exempting underground tests, the subject of a similar Anglo-American proposal the Soviets had rejected just the year before. The ban would forbid all detonations in the atmosphere, in outer space, and under water, environments in which both US and Soviet negotiators felt confident in national technological methods to verify compliance. The ensuing talks that began on July 15 lasted only ten days. Leaders of the three nations signed the LTBT on August 5.

After three weeks of hearings and floor debate, it passed Senate ratification, 80-19, on September 24, 1963.

When the LTBT entered into force the following month, commentators celebrated the agreement primarily as an easing of the arms race and a cork, long overdue, on radioactive fallout. Few dwelled on the significance of the test ban to one or another environment. But outer space—both as a physical medium in need of protection and a political imaginary with implications for national image-making—loomed large in the Kennedy administration's support for a treaty. From the revelations about Argus in 1958, but particularly in the wake of Starfish Prime, it was clear that exospheric tests jeopardized the civilian space program, contradicted US rhetoric about the use of space for peace, and tarnished the US claim to be the more rational, responsible, and irenic nuclear power.

The challenge of balancing military rationalizations for space tests with the dangers they posed to the civilian space program and US propaganda came through in a classified discussion Kennedy held with senior officials less than a month after Starfish Prime. Having reluctantly supported the resumption of high-altitude tests earlier that year, Kennedy now expressed serious reservations about Dominic, particularly space-based tests that might imperil civilian space missions. He had hitched his political star to the moon landing, after all, and with Mercury astronaut Willy Schirra set to take flight within the month, Kennedy wondered whether further tests should occur at all if they would "make a lunar journey prohibitive." When an unidentified participant in the meeting suggested simply launching spacecraft at sufficient distances from the nuclear tests, the president joked that he would "have to move the whole space program up to New England." Kennedy recommended instead "that we will not [conduct] any tests that raise any reasonable prospect of interfering until Schirra goes." "And," he added, "let's try to decide which of these tests we can throw out. We don't want to do them all, if we can help it." Further testing complicated not only the ability of astronauts and spacecraft to reach space but also the "political side," as Deputy National Security Adviser Carl Kaysen put it, of polluting space with harmful radiation. "There's not much use [in] our going to the Russians and telling them about the problem of electrons [in low-Earth orbit]," Kennedy warned, "and then going ahead and doing it ourselves and adding more electrons." Secretary of State Dean Rusk agreed, admitting that "some of these shots are creating a problem for us in space."[81]

This understated the matter. Though the Soviet Union had already conducted its own tests in space, it seized on Starfish Prime to throw US morality, and by extension US Cold War leadership, into doubt. Soviet ambassador to the UN Valerian Zorin labeled the Dominic series the extension of the arms race to space.[82] Soviet scientists issued reports that exospheric tests would irreparably alter weather patterns, damage crops, and poison Pacific fisheries. Platon Morozov, Soviet representative to the UN COPUOS, accused the United States of toying with the lives of astronauts.[83] Indeed the most damning rebuke came from the cosmonauts themselves, who spoke with authority on the danger of radiation to human spaceflight. Gherman Titov called Kennedy's resumption of testing a "real act of sabotage in outer space." The United States, he told reporters, "had no moral right" to conduct tests in space, where representatives of all nations were supposed to enjoy freedom of travel. "Some Americans," he chided, "apparently have forgotten that they are not the only inhabitants of our planets."[84] Yuri Gagarin likewise called on his counterparts Scott Carpenter and John Glenn to join him in condemnation of Prime. The test, Gagarin charged, helped complete the cosmos's transition from an oasis of peace to an arena of covetous military powers—that is, "people who would spoil the barrel of honey with one spoon of tar."[85]

The potential political liability of space tests had been apparent before the Dominic series even began. Physicist Georgi Pokrovsky scolded the Kennedy government in an article for Moscow's *New Times*. "America's official silence is eloquent proof that it has no valid arguments in justification of what can only be regarded as a criminal act," he wrote. The United States' "reckless adventure" in space reflected its indifference to "world-wide indignation." "Regardless of whatever military advantages it hopes to gain from this reckless experiment," Pokrovsky concluded, "one thing is certain: the United States and its policymakers are sustaining a heavy moral and political defeat."[86]

Blustering hypocrisy perhaps, but true, for outrage extended well beyond the Soviet Union. Three thousand Tokyoites protested after the first Dominic test in April, while more than three hundred demonstrators were arrested at the US embassy in London. Antinuclear activists in the Pacific sailed boats into the quarantined zone surrounding Johnston Island. Soon picketers appeared outside the White House.[87] The CIA caught wind of radio broadcasts and news reports from every corner of the globe criticizing the tests, Starfish Prime in particular. "The Americans are pursuing strategic and military aims exclusively" alleged one Egyptian daily. An "unpardonable crime," wrote a Sundanese editor. One Indian official thought space-based tests qualified as a "crime against humanity."[88] Even U Thant, often a voice of impartiality in his role as UN secretary general, viewed the confluence of the space age and the atomic age "a manifestation of a very dangerous psychosis."[89]

In Britain, more than seven hundred top scientists made a written appeal to the minister of science that governments openly discuss future "experiments" in space, nuclear and otherwise, with those whose research might be affected. Nuclear tests in space "might be harmless," their report admitted. "They may destructive. [But] to move ahead is to stake the future of mankind in an ill-considered game of chance."[90] Noted astronomers Bernard Lovell and Fred Hoyle inveighed against what they perceived to be the recklessness and irrationality of atmospheric testing. Lovell called Starfish Prime "one of the most clumsy and dangerous experiments ever devised—an affront to the civilized world." The scientists, engineers, and policymakers responsible for Dominic showed nothing less than an "utter contempt for the grave moral issues involved." Hoyle warned that the explosion's infusion of radiation into the natural Van Allen belts would make scientific experiments there more difficult for decades, possibly centuries. "The morality of making what might possibly be a long-term change in our environment," he admonished, "can very properly be questioned."[91]

Starfish Prime and its global denunciation had an important political consequence: they hastened the transformation of "outer space," only recently an abstract frontier in which technological innovation was shaping the future and winning (or losing!) the Cold War, into a real physical environment susceptible to human error and pollution (figure 15). The test helped introduce contamination and environmental danger into a space discourse dominated by ideas about adventure, ingenuity, and prestige.[92] Discovery of the Van Allen belts had yet to reach its fifth anniversary; how could US military leaders sanction such a "sledge-hammer blow" to the orbital environment when scientists had only begun to understand it?[93] "Cosmic madness!" cried British astronomers, "a fearful shot in the dark."[94] Certainly it did not help the test's organizers that less than one month after the blast, Houghton Mifflin released Rachel Carson's *Silent Spring*, which alerted the public to invisible pollutants and indeed helped spark the modern environmental movement.

Flowing from environmental degradation, of course, was the need for environmental protection. That nuclear tests in space menaced both astronauts and the orbital infrastructure upon which modern life

FIGURE 15. The US space-based nuclear weapons test, particularly the Starfish Prime explosion, caused international censure that further motivated the Kennedy administration to negotiate the Limited Test Ban Treaty. This cartoon was reprinted in a Soviet physicist's article on the tests titled "Crime in Space." *London Evening Standard.*

was beginning to depend dictated that outer space enjoy special protections. Although a by-product of arms control negotiations already decades old, the test ban treaty came to be understood as a central pillar of that protection.

A similar combination of popular opinion, cultural assumptions about the sanctity of space, and strategic concerns about extending the arms race to space contributed to the defeat of satellite bombardment in US policy. As the Kennedy administration reeled from its March 1962 announcement that the United States would resume atmospheric testing, it began, at the same time, to seriously address the threat of "bombs in orbit." On May 26 Kennedy issued a National Security Action Memorandum (NSAM 156) directing the State Department to assemble an executive committee that would consider space negotiations "with a view to formulating a position which avoids the dangers of restricting ourselves, compromising highly classified programs, or providing assistance of significant military value to the Soviet Union and which at the same time permits us to continue to work for disarmament and international cooperation in space." The potential for an agreement on nuclear weapons in space was among the most urgent priorities for the new task force, dubbed the NSAM 156 Committee for the memorandum that had created it.[95]

Only a month before the committee began its work, US negotiators in Geneva had been caught by surprise when Canadian Foreign Minister Harold Green proposed a ban on nuclear weapons in outer space, notably without the advance counsel of the United States. Both the United States and the Soviet Union had expressed interest in such a ban during previous talks in the Eighteen Nation Committee on Disarmament (ENCD) and in the UNGA, but only as part of an agreement on general and complete disarmament, what one official aptly referred to as a "pie-in-the-sky" arms control objective seeking total elimination of weapons of mass destruction (WMDs). Green's proposal was for a *separate* ban that might be negotiated without considerations of broader disarmament questions. Despite internal disagreements, the NSAM 156 Committee initially recommended that US negotiators "make clear their firm opposition" to a separate ban because it did not provide for inspections.[96]

Kennedy was ready to accept this recommendation, but first wanted to know: Had the NSAM 156 team considered a ban relying on unilateral, or "national" means of verification, whereby each side would police the other? In practice, such an agreement would be *declaratory*, a statement of intentions that, while strongly worded, would not qualify as binding international law. When the committee's executive secretary, Raymond L. Garthoff, explained that a declaratory option had not received a "full appraisal," Kennedy ordered further study. "If nothing better can be achieved," he maintained in a memo to Secretary of State Dean Rusk, "such a declaratory ban might be in our interest." Must the administration's insistence on inspections be upheld in every area of disarmament, "regardless of our own possible interest in declaratory agreements in special cases?"[97]

By August it was clear that Kennedy had increasingly come to understand the ban as a tool capable of projecting a favorable image of the US military space program and as well as a pillar of his broader arms control goals. The result was another action document, NSAM 183, which instructed officials to "forcefully" explain and defend the US space program. The primary aim of such a defense would be to "show that the distinction between peaceful and aggressive uses of space is not the same as the distinction between military and civilian uses." Reconnaissance satellites, though engineered by the CIA and the military services, contributed to peace. Orbital delivery systems were decidedly aggressive.[98]

CHAPTER 5

The administration thus began a rhetorical offensive in which officials openly declared that the US government had no intention of placing nuclear weapons in orbit. The first act was a speech by Deputy Secretary of Defense Roswell Gilpatric in Sound Bend, Indiana, on September 5, 1962. Although there was little doubt that either the Soviet Union or the United States could station nuclear weapons in orbit around the Earth, Gilpatric noted, "such an action is just not a rational military strategy... in the foreseeable future." He acknowledged that the United States was then pursuing "extensive" military activities in space but affirmed that it had no program to orbit WMDs. There would be "no greater stimulus" for a Soviet thermonuclear arms effort in space than "a United States commitment to such a program," he insisted. Yet the deputy secretary also reminded his audience that if the Soviets *did* choose to militarize space, the United States would not hesitate to follow suit. It was an age-old proclamation: US weaponization could only result from Soviet weaponization; US military activity in space was necessary primarily because of secretive Soviet activity there.[99]

Kennedy seconded these themes a week later when he delivered his famous "moon speech" before a sweltering crowd of 40,000 at Rice University in Houston. Remembered mainly for its promise to land men to the moon by the end of the decade, the address said as much about national security in space, as Kennedy urged both vigilance in the face of purported Soviet aggression and a pacific approach to exploration. He pointed to the supposedly unspoiled nature of outer space to argue for a race to the moon that would not stoke international tension but rather foster international cooperation and discovery. "There is no strife, no prejudice, no national conflict in outer space as yet," Kennedy reminded his audience. "Its hazards are hostile to us all. Its conquest deserves the best of all mankind, and its opportunity for peaceful cooperation may never come again." But to ensure that space became a "sea of peace" rather than a "terrifying theater of war," Kennedy asserted that the United States should be the international leader in space exploration. The president maintained that his government was fit to preside over the early space age because it had begun pursuing a vision of space equipped not with instruments of destruction, "but with instruments of knowledge and understanding." Kennedy carefully balanced American desires to whip the Soviets at their own game against widespread aspirations for the Cold War to subside and eventually fade away. "I do not say that we should or will go unprotected against the hostile misuse of space any more than we go unprotected against the hostile use of land or sea," he remarked, "but I do say that space can be explored and mastered without feeding the fires of war, without repeating the mistakes that man has made in extending his writ around this globe of ours."[100]

Soaring rhetoric, but it concealed both internal disagreements about a nuclear-weapons ban in space and continued Soviet truculence in arms control. The Joint Chiefs maintained that progress in orbital weaponry might soon make space-based delivery systems competitive with submarine- and land-launched missiles. A ban on satellite bombardment would (1) hamper US military space programs; (2) set an unwanted precedent for arms control without inspections; and (3) give the Soviets a "free period in which to pursue clandestine development of an orbital weapon system with no fear of being overtaken by the United States." The CIA worried that Soviet negotiators would attempt to attach a ban on spy satellites to any agreement dealing with bombardment. Even when Kennedy dismissed these considerations in favor of the ACDA's recommendations for a declaratory ban on October 2, the Kremlin rebuffed US overtures on a resolution. When ACDA's Adrian Fisher reached out to Foreign Minister Andrei Gromyko and Ambassador Anatoly Dobrynin on October 17, they replied that any ban on orbital bombs would have to accompany the removal of US foreign bases in Europe and Asia. It was 1957 all over again.

But note the date. The disparate agencies of the government had finally agreed to reach out to the Soviets on a ban at the height of the Cuban Missile Crisis. Khrushchev's gamble convinced US military officials of the need to purchase "technological insurance" to hedge their bets against Soviet militarization in space. The NSAM 156 Committee would spend the following spring preparing contingency plans for Soviet space bombs, and in August 1963 Kennedy would assign top priority to the USAF's ASAT program, Project 437.

At the same time, however, the razor-thin margin by which the superpowers had evaded an apocalyptic nuclear exchange lent urgency to the entire arms control agenda. The LTBT certainly flowed from Cuba, as did the Hotline Agreement in which Washington and Moscow agreed to establish full-time telegraph and radiotelegraph communications between the two capitals. The ban on nuclear weapons in space was no exception. Under Secretary Walt Rostow, the State Department revived the idea in its policy-planning paper on *Post-Cuba Negotiations with the*

USSR, which the "Ex-Comm" that had navigated the missile crisis considered as part of its debrief of the affair. Amid Soviet attacks on spy satellites at the United Nations and sustained propaganda about the aggressive intentions of the American space program, the ban reassumed its value as a measure that might at once protect US spy satellites, prevent the Soviet Union from achieving an orbital bombardment capability, and promote the image of the United States as the more peaceful space power.[101]

By the time Mexican delegates at the United Nations tabled a draft treaty forbidding states from stationing nuclear weapons in space on June 21, 1963, the chances for an agreement had opened considerably. Although US officials neither supported nor opposed a *treaty*, they did find significant receptivity for the declaratory ban among their Soviet counterparts. On September 19 Gromyko informed the UNGA that the Soviet government was prepared, "here and now," to ban nuclear weapons from space.[102] Kennedy addressed the same audience a day later: "If we fail to make the most of this moment and this momentum," he intoned, referring to the LTBT, "if we convert our new-found hopes and understandings into new walls and weapons of hostility—if this pause in the cold war merely leads to its renewal and not to its end—the indictment of posterity will rightly point its finger at us all."[103]

US officials cooperated over the following week with delegates from the Soviet Union and a handful of other nations to draft the resolution, what Mexican Ambassador to the UN Padilla Nervo submitted to the General Assembly on October 17 as Resolution 1884, "Stationing Weapons of Mass Destruction in Outer Space." The document called on states to refrain from stationing WMD in space or on celestial bodies, and from facilitating the placement of such weapons in space by other countries. Stevenson spoke before General Assembly with great relief, extending hope that by "avoiding a nuclear arms race in space we will have taken one further step on the road to disarmament."[104]

It is worth examining why restraint won out in an area of technology that had always been dictated by a philosophy of aggregation, of more, more, more. The failure of orbital bombardment, both as a technology and a strategic construct, resulted from much the same political, cultural, and pragmatic military considerations that drowned the DoD's broader ambitions for space in the late 1950s and 1960s. First, what Arthur C. Clarke had in *Sputnik*'s wake called "the morality of space" still held enormous sway in public opinion. Few commentators then referred to outer space as a "sanctuary"—that term would take another decade to enter popular discourse—but they generously deployed its analogues. Lyndon Johnson called space an "island of peace"; the House Committee on Science and Aeronautics named it the "road to peace," and the *New York Times*, anticipating the UNGA's unanimous vote on Resolution 1884, dubbed it a "zone of peace."[105] And though clergymen had always stressed the sanctity of "the heavens," the term took on new meanings more closely associated with secular humanism and political utopia. Stationing nuclear weapons in space would belie this rhetoric and destroy the dream of cultivating the cosmos as a preserve in which human societies might enter true maturity.

Flowing from the continued inviolability of space in the political imagination was the intrinsic value of "space for peace" to Cold War image-making. Eisenhower had determined (rather, he had discovered) that the character of the US space program mattered at least as much as its technical prowess compared with the Soviet Union. Kennedy, though he pretended to be blazing a new frontier in space, learned from this discovery and made it a mainstay of his space politics and, by extension, his space policies. The most crucial benefit of the "bombs in orbit" resolution was that the US government had finally succeeded in fashioning a political distinction between "military" satellites and "aggressive" ones.[106] Orbitals that monitored nuclear weapons stockpiles and troop deployments, provided navigation to ships, or facilitated instantaneous military communications were passive and therefore served the interests of peace. By comparison, the bombardment satellite appeared little more than a Doomsday Machine.[107]

Tacit Soviet assent to spy satellites was the biggest prize of all. From the 1955 Geneva Conference—at which Bulganin had rejected Eisenhower's "Open Skies" proposal—the Kremlin had insisted that overhead reconnaissance, whether with balloons, satellites, or spy planes, constituted an egregious violation of national sovereignty, an objection that proved quite serious once Soviet air defenses felled Francis Gary Powers's U-2 in May 1960 and that of Rudolf Anderson two years later during the Cuban crisis. By the summer of 1963, however, Soviet attitudes toward space-based reconnaissance had suddenly reversed, a fortuitous development attributable to the increasing usefulness of the Soviet Union's spy satellite program, as well as the government's slow realization that secrecy was doing more harm than good given frequent American overestimations of the USSR's military strength.[108]

Finally, the Soviet and US governments agreed to close off space from the nuclear arms race because they determined that restraint benefited national safety. Their most immediate consideration was that orbital weapons represented a costly, unpredictable, and potentially disastrous upheaval in the balance of terror. Proponents of the resolution understood that within the context of continually evolving technology, deterrence would unfold with great volatility. Borrowing from economics, the ACDA warned of a "multiplier effect" in armaments: one nation develops a new weapons system; its rival introduces a system intended to counter it; the first system is modified to evade the countermeasures; and the countering system is updated in turn. "And so it goes." Defense intellectuals at the Rand Corporation had enough difficulty modeling a nuclear exchange given IRBMs and ICBMs alone. If either superpower found itself unable, or in an untimely manner, to engineer the countermeasures necessary for space-based nuclear weapons, the delicate balance of deterrence—already precarious given the multitude of weapon types, early warning systems, delivery vehicles, and ancillary programs—would be undone. A JPL study published the previous year had stressed the same thing. Stationing nuclear weapons in space could only produce instability and render a "destructive effect on an already depressed world opinion." The ACDA report's concluding statement summarized what would soon become official administration policy: "We do not think it desirable for either side to embark on this grim game in outer space.... The type of arrangement that now seems attainable [the ban on warheads in space] would not relieve us of the necessity of defensive precautions. However, it may assist in avoiding the opening of a new dimension in the arms race while we continue efforts to bring the race for existing types of weapons under control."[109]

In space, 1963 proved an eventful year. In June cosmonaut Valentina Tereshkova became the first woman, and the first civilian, to traverse the cosmos. The following month witnessed the birth of both the spaceplane and the geosynchronous satellite. Transit, the world's first navigation satellite system, began beaming coordinates to Polaris submarines before the year was out. The space race was proceeding apace. Yet the most consequential developments unfolded not in technology but in diplomacy and foreign policy. The LTBT and Resolution 1884 had closed off space to the nuclear arms race. Having begun to see the first fruits of their satellite reconnaissance program, Soviet officials finally ceased their opposition to spying from space. Kennedy even reached out to Khrushchev regarding a joint lunar mission. On December 13, these milestones abutted a new agreement at the United Nations: the Declaration of Legal Principles. Among several other stipulations, the resolution declared that states would explore and use space for the benefit of all countries; that space, free for all to explore, was not subject to sovereign claims or occupation; and that astronauts were the "envoys of mankind" and therefore deserving of assistance during times of emergency. The flurry of activity in Geneva and New York all pointed to the maturation of a new animal: international space law.

Part Three

Waking Up

CHAPTER 6

A Celestial Magna Carta

For those who supported the ban on bombs in space, combating the version of the future envisioned by men such as Bernard Schriever, Thomas White, and Homer Boushey was a process akin to toppling a vending machine. In the decade after *Sputnik*, there was little doubt that the weight of the USAF's ambition for a new high ground in space—that is to say, the practical military utility of the medium and the allure of futuristic "control" of space—was substantial. Given Soviet firsts in space, Khrushchev's chest-thumping about the lethality of his rockets, and the unknowns of the Soviet space program, hedging one's bets with advanced weapons projects for space seemed a plausible, even advisable, course of action. The obvious utility of satellite technology and of orbital positions for war fighting made the militarization of space a heavy obstacle to lift. Accordingly, the job required a sequence of small but momentum-building thrusts: high-level exchanges between US and Soviet leaders on arms control for space; the winning-out of civilians for control of NASA; the establishment of the COPUOS; bilateral negotiations to initiate cooperative space projects; the LTBT; and the space-related UNGA resolutions. By themselves, each of these shoves was insufficient to overthrow aspirations for space-based deterrence or indeed for "space supremacy." Together, however, they helped form a consensus about appropriate behavior in space and constituted

FIGURE 16. Seated, from left to right, are Anatoly Dobrynin, UK Ambassador Sir Patrick Dean, US Ambassador Arthur Goldberg, US Secretary of State Dean Rusk, and President Lyndon Johnson as they affix their signatures to the OST in a ceremony on January 27, 1967. Courtesy of the United Nations.

a determined and principled campaign to preserve the cosmos from the politics and weaponry of the Cold War.

The climax of this campaign—the push that finally grounded the notion of "space war" and "space control" as political and strategic constructs—was the Outer Space Treaty, for which leaders from Washington, Moscow, and London packed into the East Room of the White House to attach their signatures on January 27, 1967 (figure 16). Formally titled the Treaty on Principles Governing the Activities of States in the Exploration and Use of Outer Space, including the Moon and Other Celestial Bodies, this landmark agreement laid out, in binding fashion, seventeen articles that now form the backbone of international space law. The OST declared that outer space could no longer be subject to claims of national sovereignty; that military installations, military maneuvers, and weapons testing were henceforth forbidden on the moon and other celestial bodies; that states were bound to refrain from stationing nuclear weapons in space and on heavenly orbs; that space activities would be conducted in accordance with the UN Charter; that astronauts were "envoys of mankind" and therefore subject to assistance in the event of an accident; and that states would engage in space activities for the benefit of all humanity, regardless their of economic, scientific, or technological development. The agreement established outer space as a realm akin to the high seas, a zone free for use and exploration by all, "the province of all mankind."[1] When the UNGA convened to vote on the accord late in 1966, it won unanimous approval. The same was true of US ratification in the Senate, where the treaty earned an 88-0 "yes" vote.[2]

The OST represented the apotheosis of American space diplomacy. It succeeded in codifying the legal principles that the United States, the Soviet Union, and dozens of other UN member nations had laid out over the previous decade in academic debates, in the declarations of their political leaders, and in the COPUOS's deliberations. Space law's transplantation from declaration to edict was a profound occasion, one that reflected a political consensus about the future of exploration and, not insignificantly, the waxing power of international organizations in global space politics. The United Nations, and the COPUOS in particular, had claimed expansive influence over national activities in an area of technology with serious military and economic implications.

Despite the space treaty's apparent success, a consensus about the long-term efficacy of the agreement has been elusive. Some scholars consider the treaty to have been a prescient if measured step in the demilitarization of outer space and in the deescalation of the Cold War more broadly. Historian Hal Brands has portrayed the OST as part of a successful offensive against the arms race forwarded by the Johnson administration. Although the OST never garnered the attention of the Strategic Arms Limitation (SALT), Anti-Ballistic Missile ABM or Nuclear Nonproliferation (NPT) Treaties, he argued, its ratification created an atmosphere in which these ambitious accords became possible. Arms control advocates Helen Caldicott and Craig Eisendrath similarly concluded that despite worrisome loopholes, the OST has proved a highly durable document, one seriously capable of limiting the arms race to Earth.[3]

Others have their doubts. One popular interpretation is that the OST was merely low-hanging fruit in broader US efforts at disarmament in the 1960s, a political play to demonstrate détente could endure the shockwaves of the Vietnam War. Outer space, in this rendering, was easy to protect from weapons and warfare because weaponizing space was then technically difficult; because in the end, the earth remained throughout the Cold War the most efficient and cost-effective platform for delivering nuclear weapons. Had the technology to weaponize space been available, many scholars agree, the OST would surely never have been signed. Walter McDougall concluded that while the OST succeeded in opening space to programs in developing countries and the orderly development of space technology around the globe, the treaty was "all show and no substance" in terms of creating a space sanctuary. Writing amid the global debate over the SDI, he was justifiably glib in the face of US diplomatic triumphalism following the treaty: diplomats celebrated the agreement "as if in the absence of a treaty squads of astro- and cosmonauts, armed with flags, ray guns, and theodolites, would ascend on the moon in colonial warfare." The political scientist Everett C. Dolman has been even more skeptical. He regards the OST as merely a "reaffirmation of Cold War realism and national rivalry, a slick diplomatic maneuver that bought more time for the United States and checked Soviet expansion." The negotiations constituted nothing more than "a perverse competition of who could out-cooperate whom."[4]

Why is commentary on the OST, both in contemporary discourse and subsequent academic debates, so polarized? It is possible for the treaty to have been at once "an inspiring moment in the history of the human race" and merely a coup for US and Soviet propaganda?[5] Such attitudes, it is worth emphasizing, were not merely the product of predilections toward realism or idealism, glass-half-full versus glass-half-empty worldviews. Doubters derived their assumptions about the treaty from an appreciation for the frequent divergence of Cold War rhetoric from US and Soviet foreign policy, from skepticism about the United Nation's ability to enforce international agreements, and most of all from a conviction that revolutionary technologies could not and would not be called to heed the law. Promoters identified the treaty as a key component of the multilateral arms control process that included the LTBT, the Hotline Agreement, and the NPT. They placed the OST at the center of a teleology in which the Cold War would, piecemeal, dissolve as new areas of the globe were closed off to weapons and war. The OST's true believers and its detractors told different halves of the same story: the business of the Cold War was proceeding as usual—or it was in decline; nothing in the international system had changed—or else everything had changed.

If examined in broader and deeper historical context, both narratives break down. Not merely an immediate legal solution for contemporary international space activities, nor simply a political gambit to salvage détente, the OST both resulted from and reflected a series of broader phenomena. First, the treaty was the culmination of trends in the nascent field of space law going all the way back to *Sputnik*. In particular, a "negative" or "prohibitory" approach to governing space exploration prevailed in international legal debates in the decade after 1957. Even before *Sputnik* shocked the United States into taking seriously legal problems in space, international lawyers sought to define the future of human activity in space by outlining not what nations *should* be able to do in the medium but rather what should be *forbidden*. While the origins of space law as an academic and pragmatic discipline were rooted in precedents established by the law of the air and from concerns about national defense, they also sprung from the intellectual and cultural milieu that singled out the cosmos for special protection. Much like the interplanetary discourse of the 1930s and 1940s, discussions about how the international community should govern outer space and the technologies that would inhabit or pass through it diverged into two distinct vocabularies. On one hand, commentators employed a language of peace and science for what they considered admissible human activity in space: words like "freedom," "discovery," "cooperation," and "research" abounded in speeches, articles, and proclamations on the matter. The key phrases of negation, on the other hand, connoted war and violence: "missiles," "bombs," "weapons," "rivalry," and "conflict" were to be kept out. In an earlier time, thinkers such as David Lasser and Olaf Stapledon had envisioned outer space as an empty page onto which humanity would write a more peaceful future; now, in *Sputnik*'s wake, space lawyers conceived of their budding field of study as the means to guarantee the safety of the interplanetary vision.

Space law and space politics, in other words, emanated from space culture. From the earliest musings about the possibility of orbiting an artificial satellite after World War II through the mid-1960s, lawyers and policymakers wrote about the inadequacy of "terrestrial" or "anthropocentric" law in formulating effective rules for outer space. In the rapidly proliferating space law literature and in the OST negotiations, they warned against "the temptation of pressing available juridical and diplomatic material" into the cosmos.[6] On the contrary, they argued, the revolutionary enterprise of space exploration required a revolutionary set of legal principles that would expunge the corrosive by-products— imperialism, war, genocide—of historic juridical practice. "To extend our existing systems of law, with their imperfections and ideological limitations, their inherent conflicts and inconsistencies, under the guise of an 'international law,'" wrote the amateur-rocket-engineer-cum-attorney Andrew G. Haley, a founding member of the discipline, "would be to spread our terrestrial conflicts and intolerances wide and far through a universe that potentially offers tremendous vistas of a new age for man."[7]

That the OST reflected more than superpower posturing was also evidenced by the signal contributions of developing countries to the final contours of the agreement. Far from being a purely East-West confrontation, the OST negotiations represented prevailing, and indeed more substantive, tensions between North and South. Whereas the US and Soviet delegations arrived at the United Nations confident in the fitness of existing political and legal arrangements to outer space, leaders from the Global South harbored serious misgivings about "the weaknesses of classical international law." They insisted that far from applying traditional methods and practices to activities in space, the international community should adopt new, more maximalist approaches to this new field of law. While the North privileged the codification of rules that would allow for the orderly exploration and exploitation of space, the South attempted to enshrine a set of legal principles whereby nations would expunge imperialism, war, and economic inequality from human activity in the cosmos and by extension from the future of international relations.[8] Their lofty ambitions, their anxieties about technological and economic competition with the industrial world, and their memories of imperial exploitation dictated the terms by which they engaged their spacefaring counterparts at the United Nations.

A second conclusion cuts in the opposite direction. If the OST was not the hollow shell that its detractors had claimed, neither did it prove a reliable bulwark against the extension of the Cold War—and against new conflicts between the "haves" and "have nots"—to the cosmos. The soaring language with which the treaty's architects sought to inspire new political expectations was ultimately capacious enough to allow human history to pass through the back door. Just two days before Anatoly Dobrynin signed the treaty on behalf of the USSR, for instance, the Soviet military successfully tested FOBS, the weapon that first pushed the superpowers toward a ban on nuclear weapons in space. Although the treaty represented a legitimate milestone in the disarmament process, it did little to slow the development of conventional space militarization in the 1970s and 1980s: ASAT weapons, kinetic-kill vehicles, and particle-beam weapons among other high-technology projects. Contradicting the spirit of the poorer nations' contributions to the OST, moreover, both the United States and Soviet Union in subsequent years upheld traditional "rights" to exploit natural resources on the moon and other heavenly bodies, a conflict that would torpedo the 1979 Moon Treaty twelve years later. And though the OST contained strict *limitations* on space activities, questions abounded about what it *prescribed*, particularly the services it obliged the small club of space powers to perform. At heart, the OST represented a sober compromise between the Cold War prerogatives of spacefaring nations and the pacific, postcolonial version of the future envisioned by spacefarers yet to be. Within a decade it was clear that the agreement should neither be abandoned nor taken for granted.

The Ethical Roots of a Discipline

Governing space, as it had been for spaceflight itself, was initially a matter of speculation. The question of space law first appeared in a 1910 treatise by the Belgian lawyer Emil Laude, who predicted after the Wright brothers' historic flights that new codes would eventually govern the "ether" beyond the layers of breathable gas through which the first airplanes were then passing. If not "aerial law," Laude asked, then what? "It may be hazardous to predict," he concluded, "for the term Ether itself only hides our ignorance and we dare not propose the term Ethereal Law. But certainly it is a question of the Law of Space."[9]

Interest in the legal implications of the "ether" increased as transAtlantic engineers began experimenting with liquid-fuel rockets in the 1920s. When the Soviet Union held an air law conference in Moscow late in 1926, a senior official from the USSR's Aviation Ministry, V. A. Zarzar, observed escalating debates about a "theory of zones" that divided the air (subject, the theory went, to national jurisdiction and control) from the regions above (which many agreed should transcend such jurisdiction). Presaging Soviet objections to satellite reconnaissance three decades later, he asked if, in the event that aircraft were powerful enough to fly "at such tremendous altitudes, at improbable speeds," what would prevent one nation from flying high over the territory of another? "We only point out that the 'theory of zones,'" he wrote in a published version of his talk a year later, "is not so 'stupid' as would appear at first glance."[10]

He was right. By the early 1930s German leaps in rocketry propelled human activity beyond the threshold separating the "zone" of the air from that of space. Closely following these developments and similar advances in Europe and the United States, Vladimir Mandl, a Czechoslovakian lawyer, writer, and engineering professor, perceived the onset of a new realm of legal practice and in 1932 published the world's first monograph dedicated to it, *Das Weltraum-Recht: Ein Problem der Raumfahrt* (The Law of Outer Space, a Problem of Spaceflight).[11] In this short treatment,

CHAPTER 6

Mandl proposed a clear delineation, based on the unique capabilities of spacecraft, between legal rules for space and those for other shared spaces such as the air or the high seas. Because the medium possessed many unique and challenging qualities, space law would be "quite a different phenomenon than is the present law of jurists," Mandl predicted. Above all, traditional claims of national sovereignty would have to make room for more flexible regimes, for beyond Earth there began a vast area "independent [of] State power." In language that resembled his contemporaries at the British and American space societies, he suggested that spaceflight would diminish the importance of national territory and erode the power of state governments. No longer having "subjects," national governments would exist in a state of equality with their environs, and new communities based on personal ties would come to replace traditional loyalties.[12]

By 1957 most practitioners of space law, which by then had coalesced into a lively and expansive subfield, were concerned less with the existential problems of "the future" than the immediate legal consequences of the Soviet satellite and the onset of practical space exploration. There existed, all agreed, a glaring void in the available body of international law to deal with satellite communications, the allocation of radio frequencies, liability for damage caused by fallen spacecraft, and sovereignty over objects launched into space, celestial bodies, and the medium itself. "If you find yourself in command of a satellite in the ionosphere and you encounter a Soviet satellite," warned Philip C. Jessup, a Columbia University law professor, "you had better send back for instructions because the law books will not help you any" (figure 17).[13]

Over the course of 1957–1960, the pace and novelty of US and Soviet space feats prompted lawyers to urge international agreement on a set of basic principles that would guide spaceflight away from the Cold War and toward its intended applications in scientific research and economic growth. The Soviet Union successfully orbited the first living organism, a dog named Laika, on November 3, 1957, though the animal died of respiratory failure within hours. The United States followed suit at the end of the following January with its inaugural satellite, *Explorer 1*, which confirmed the existence of the Van Allen radiation belts. *Luna I* made the first heliocentric orbit around the Earth in January 1959. It was the first to fly by the moon (it was intended to impact the lunar surface) and succeeded in taking the first measurements of solar winds. The world's first communications satellite, its first weather satellite, and the first photographs of the Earth from orbit followed quickly behind.

FIGURE 17. "Isn't it time you were getting me in orbit?" Amid rapid technological achievements by the superpowers, many asked if legal constraints for space activities and for the medium of space were falling perilously behind. "Isn't It Time You Were Getting Me into Orbit?," *Christian Science Monitor*, July 30, 1961, 14.B Outer Space 14.B.11 International Space Law 196, box 344, RG 59, General Records of the Department of State; Office of the Secretary, Special Assistant to the Secretary of State for Atomic Energy and Outer Space, General Records Relating to Atomic Energy Matters, 1948–1962, USNA.

Amid these accomplishments the example of the atomic bomb conditioned early deliberations about how governments should proceed in space: failure to place nuclear weapons under international control in the mid-1940s, many contemporaries argued, taught that if legal frameworks lagged behind technological developments, the capacity for governments and international law to control events would dissolve.[14] "Good intentions," wrote one observer, "may be overtaken by events."[15] The pace of accomplishment threatened to launch the superpowers beyond "a point of no return" at which political rivalry and likely another arms race would become inevitable. The gap between ever-progressing technological capabilities and the formulation of rules to govern those capabilities, many lawyers agreed, "had widened to the point that the peace of the world is threatened." To avoid the mistakes of the bomb, space lawyers "must become eagles and stop being turtles," implored one New York attorney as late as 1965. The current pace was simply much too slow.[16]

Urgent as the need for a new coda of space law appeared to be, an equally voluminous body of legal thought, particularly in the halls of decision, advocated a more restrained approach emphasizing precedent, the requirement of facts, and the necessity for law to follow political realities. This "positivist" school of space law held that governments should "make haste slowly" in constructing a new legal order for space.[17] Chester Ward, judge advocate general of the Navy, put the matter colloquially: "It would be futile for a state legislature to attempt to draw up a state highway code without knowing the performance characteristics of modern automobiles and trucks," he wrote in the *JAG Journal* seven months before *Sputnik*. "The legislators would have to know the ability of modern drivers and the driving hazards of the highway system." Efforts to write laws for the space environment would be "just as futile" without sufficient knowledge of physical conditions in the medium and the particular and ever-changing qualities of spacecraft.[18] Space law, in brief, "should be based on the *facts of space*."[19] The State Department's counsel, Loftus E. Becker, thought attempting to move too quickly on codification would invite conflict, not prevent it. No nation, including the United States, had objected to *Sputnik*'s overflight of sovereign national territory. Tacit approval of overflight had continued with subsequent launchings on both sides of the Iron Curtain. Given this fortuitous legal development, why not wait for technological and political events to determine the shape of the law? NASA counsel John A. Johnson made a similar conclusion in perusing the mountain of new material on laws for space. "The fact that the recommendations of eminent legal scholars on this subject have been revised several times during the past few years," he explained to Congress in March 1959, "indicates the impracticality and, I might suggest, the imprudence of any comprehensive effort to settle this question with finality at the present time."[20]

Positivist approaches to this profound new legal field existed in constant tension with another body of interpretation to which the OST would eventually owe its greatest intellectual debt: the "naturalist" school of space law.[21] Naturalists held that international law flowed from human intuition about morality and ethics, that justice was rooted in "conditional and absolute" truths about human nature. They believed people possessed an innate knowledge of right and wrong from birth. The law was not "intangible and nonexistent" as the positivists believe; it did not derive from kings or legislatures. Rather, it predated organized attempts to codify the law. Indeed, natural law did "not exist in code form and has never existed in statutory form."[22] Haley, the natural school's leading light, frequently referred to the theories of natural law developed by the Spanish renaissance scholar Francisco de Vitoria, who held that human conduct existed before courts and statutory law and that the law constituted "a state of mind" controlling human behavior. Vitoria had aspired to extend individual principles of morality to international law upon receiving sobering reports from the New World regarding maltreatment of its native populations. Indian communities, he thought, should be afforded the same sovereign rights that nations in the Old World had enjoyed.[23]

Human beings, Haley and his colleagues maintained, must predicate space law on the same moral percepts with which they, as individuals, were naturally endowed. Crafters of space jurisprudence should operate well outside the boundaries of contemporary international relations and reach for rules that all could agree were "beyond terrestrial disagreement."[24] Such precepts could be found, for example, in the writings of Aristotle, Epictetus, Seneca, Mohammad, Ahikar, Abdullah Ansari, Sadi, Confucius, Mahabharata, Rabbi Hillel, and the New Testament. Haley was particularly taken with the Enlightenment philosopher Immanuel Kant.

Haley's cosmopolitan and humanist approach to outer space was most evident in a novel legal concept he introduced in the mid-1950s called "metalaw," a forward-thinking field of legal practice that sought to establish ground rules for how human beings should relate to other intelligent life in the universe. He derived the word from the Greek *meta*, which connotes "transcendence," and from the Old English *lagu*, a body of binding customs. Metalaw therefore implied erecting codes that would operate "beyond our present frame of reference."[25] In a reversal of the Golden Rule, metalaw's central tenet was that if human beings should come across other life-forms in the universe, those beings should be treated as *they* wished to be treated, not as human beings had traditionally treated each other. "To treat others as we would desire to be treated," Haley wrote in his most significant work, *Space Law and Government*, "might well mean their destruction."[26]

Haley's writings on metalaw are remarkable for their simplicity, imagination, and moral clarity. He insisted on four basic rules: that landing on a planet upon which life was assumed to exist first required a reasonable determination that no harm would come to that planet; that human beings must be *invited* by the life-forms on a given planet before a landing could occur; that astronauts decontaminate themselves both before landing on a new planet and before returning to Earth; and in attempts to communicate with foreign planetary systems, lines of communication must be kept perpetually open.[27] Anticipating his detractors, Haley wrote that even if a given planet was home only to primordial life-forms—even "an amoeba!"—humans would still need to respect their right to live in isolation from "the bleak and devastating geocentric crimes of mankind." Creatures ranging from single-celled organisms to sentient bipeds must first evolve into beings capable of understanding and relating to humans in ways that would promote peace and mutual habitation. This fundamental rule, Haley argued, was so central that "it would be better to destroy mankind than allow its violation."[28] Here was the kernel of what Gene Roddenberry, creator of *Star Trek*, would later introduce as the Prime Directive, the guiding principle of the United Federation of Planets prohibiting interference with the internal development of alien civilizations.

Haley contrasted metalaw with what he called "anthropocentric law," the historically bound legal customs of human beings, both on an international level between states and at more local levels between businesses and individuals. Despite the civilized language and tradition of legal practice, Haley thought it to be fundamentally "self-serving": it protected property, wealth, and the interests of those with power in their dealings with those with less.[29] While having provided a measure of order in human societies, the legacy of anthropocentric law was too crooked to justify its transplantation to activities in space, where Haley assumed the human future would unfold. To guard against "galacticide," space lawyers should "take a look again at what happened to the Indians."[30] If brought into the cosmos, traditional legal practices would produce results "similar to the Spanish destruction of Native Mexican and Peruvian civilizations in the 16th century or to the American trampling on native . . . tribes during the time of the westward push."[31] Continuous wars, colonialism, and strident nationalism had warped the bases of international law that figures like Vitoria and Hugo Grotius—both heroes of Haley's—had established as "fathers" of the discipline. "The indefinite projection of a system of anthropocentric law beyond the planet Earth," Haley wrote after *Vostok* and *Mercury* spacecraft had already catapulted humans into space, "would be the most calamitous act man could perform in his dealings with the cosmos."[32]

Alternatives were available. Haley's recommendations for novel legal structures were each predicated on guaranteeing that imperial greed, violence, and jingoism were "never permitted in space."[33] He suggested establishing a sovereign, independent authority to which "every human being would be a citizen by virtue of his existence." Every country on Earth would surrender its jurisdiction over space technology, such that no nation could launch, or even own, a space vehicle without a license. This supranational authority, working from premises of natural law, would administer every aspect of human activity in space, from public safety and health to immigration. "All of these regulations," Haley explained, "would conform to the most universal and enlightened principles of freedom."[34] Though mainly focused on the behavior of states and international organizations large enough to weaponize or pollute space, these reforms aimed at earthly poisons existing at every level of society, all the way down to the individual. Haley proposed banning from space any person "who would exclude others for any political reason" and the "nefarious (or impiously wicked)." The "moral make-up" of such persons was "inimical to the concepts of fundamental justice on which space law must be founded."[35]

The circulation of these ideas through trade journals, law conferences, and congressional committees marked the increasing currency of what many legalists referred to as "negative" or "prohibitory" space law. Negative space lawyers sought to curtail national space activities, to set limits on what the nature and goal of space exploration should be. As the name "prohibitory" suggests, lawyers like Haley wished to bar certain activities from space, particularly those that might be interpreted as militaristic, imperialistic, or those advanced in the interests of blind profit. Negative space law emphasized the importance of forecasting, of predicting what directions new technologies would push the law, and how the law might preempt changes that might rock the political boat. Accordingly, it privileged theory over practice. It placed future possibilities before present needs. And, most centrally, it valued morality over pragmatism.[36]

Consider, for instance, the "Magna Carta of Space," a declaration of legal principles for the cosmos promulgated by the Inter-American Bar Association at its annual meeting in Bogota, Colombia, on February 2, 1961. This draft treaty proposed that space and celestial bodies become the common property of all people. It forbade nuclear experiments in space and decreed that "war, in, by, or through space is hereby banned forever."[37] The document's principal author, New York lawyer William A. Hyman, was "an angry, provoked idealist." [38] He was angry because superpower competition threatened to inaugurate a frightening prelude to space-based warfare capable of snuffing out life on Earth. He was provoked because the inaugural flight of Soviet cosmonaut Yuri Gagarin had taken place at an altitude bordering on what many recognized as the dividing line between air and space, a region that might precipitate conflict over sovereignty. And he was an idealist because he believed in the power of international law to cut off these developments at the pass. He hoped that the United Nations, to which he submitted his Magna Carta in May, would adopt the document hook, line, and sinker. But its absolutist provisions prevented its serious consideration by international space lawyers. Ban war in space "forever"? Governments must bestow the fruits of resource exploitation in space to "all people"? Such pretention smacked of the defunct League of Nations Covenant or the Kellogg-Briand Pact. Despite receiving a Papal Medal from Pope Paul VI for his efforts, Hyman struggled to capture the audiences necessary to legitimate his charter.[39]

Although the Magna Carta of Space gained little traction, one could observe the circulation of more substantive precedents for negative law. Two days before *Sputnik*, Polish Foreign Minister Adam Rapacki submitted to the UNGA a plan to demilitarize Central and Eastern Europe by banning nuclear weapons from the region. Though the "Rapacki Plan" was rejected by the United States and NATO, it became the foundation for the "nuclear free zone" treaties negotiated from the 1960s through the mid-2000s; eventually all of Latin America and the Caribbean, the South Pacific, Southeast Asia, Africa, and Central Asia would be forbidden soil for the bomb.[40] In December 1959 twelve nations with resource and territory interests in Antarctica signed a landmark agreement that closed off the continent from claims of national sovereignty, nuclear weapons testing, and "any measure of a military nature." The accord declared that Antarctica was to be used for peaceful purposes only and established freedom of scientific exploration for all nations. Four years later, the Soviet Union, the United States, and dozens of other countries signed onto the LTBT, which prohibited signatories from conducting nuclear weapons tests underwater, in the atmosphere, and in outer space.

Negative legal approaches also manifested themselves in a series of UNGA resolutions on space. On December 20, 1961, the UNGA adopted Resolution 1721 (XVI), which encapsulated many of the principles that would later become binding articles of the OST. It commanded states to apply international law, including the UN Charter, to outer space and other celestial bodies and declared them free for all nations to explore and exploit without traditional claims of national sovereignty. The resolution attempted to position the United Nations at the center of global space cooperation by encouraging states to report launchings to the secretary general and to provide for the free exchange of information relating to space activities. Two years later, the UNGA issued resolutions 1884 and 1962. The former called on states to refrain from orbiting any object carrying nuclear weapons or other weapons of mass destruction, installing such weapons on celestial bodies, or stationing them in space "in any other manner" (see chapter 5). The latter recapitulated a number of positions from the 1961 resolution, but notably added that states would bear international responsibility for damage caused by spacecraft, that such craft would remain under national jurisdiction while outside national borders, and that the exploration and use of outer space was to be carried out "in the interests of all mankind." Astronauts, likewise, were to be considered "envoys of all mankind" and thus to be given international assistance in times of distress and emergency.[41]

By the middle of the 1960s, however, many interested observers grew wary that continued development in space technology would outpace the capacity for mere declarations of policy to regulate behavior in space. "It must be remembered that international law is effective only to the extent that it is accepted by nations through their accession to a treaty or convention or through custom and practice over a period of years," Senator Albert Gore (D-TN) reminded Americans. "A resolution adopted by a substantially divided vote in the UNGA on a controversial cold-war issue will not be accepted as 'law' by those nations which vote 'no,' and it will not have the force and effect of law."[42] Arthur Goldberg, the US ambassador to the United Nations, thus warned the UNGA in December 1965 that within a few years "the need for a treaty governing activities on the moon and other celestial bodies will be real."[43]

This estimate proved conservative. Within six weeks, the Soviet Union's *Luna IX* spacecraft became the world's first vehicle to make a soft landing on the moon and to send photographic data of its surface back to Earth. *Surveyor 1*, the US soft lander, quickly followed suit in May. Legalists in the State Department and in the COPUOS Legal Subcommittee fretted that the timeline for a manned lunar landing might preclude a formal agreement banning sovereign territorial claims to space. Such claims, warned a State Department position paper, "would extend existing conflicts beyond the confines of earth" and restrict national freedom of access, exploitation, and research of celestial bodies. Best to preempt the moon mission through the law.[44]

East and West, North and South

The United States took the initiative regarding a binding agreement on space activities when President Johnson issued a statement to news correspondents at his Texas ranch on May 7, 1966. He proposed a treaty that might encapsulate many of the principles outlined in the Antarctic Treaty and the 1961–1963 resolutions. These included provisions excluding claims of national sovereignty on the moon and other celestial bodies; ensuring freedom of scientific research and facilitating international cooperation; sanctioning studies to ensure states could avoid harmful contamination of space; mandating the aid of astronauts of one country by other parties to the treaty; and the banning of nuclear weapons, weapons tests, and military maneuvers. "I am convinced," Johnson claimed, "that we should do what we can—not only for our generation, but for future generations—to see to it that serious political conflicts do not arise as a result of space activities." Two days later, Arthur Goldberg submitted a letter to COPUOS chair Kurt Waldheim requesting an early convening of the group's Legal Subcommittee to discuss the feasibility of negotiating such a treaty. Waldheim agreed, and Goldberg quickly submitted a draft treaty to the Soviet Union's representative to the United Nations, Nikolai Fedorenko. By the end of the month, Gromyko requested that negotiation of an outer space treaty be added to the agenda for the twenty-first session of the UNGA. Soviet and US delegates submitted their respective drafts on June 16.[45]

The two drafts were remarkably similar. Each contained provisions forbidding the stationing of nuclear weapons on celestial bodies and in space; recognized the role of international organizations in governing the development of space activities; outlawed claims of sovereignty and national appropriation on celestial bodies; declared that space activities were to be only "peaceful" in nature; provided for the freedom of exploration and use of space on the basis of equality; and proclaimed activities in space subject to international law. As negotiations began in Geneva on July 12, US and Soviet delegates were in 80 percent agreement, Goldberg later recalled. This fact derived in large measure from the transplantation—sometimes word for word—of principles and provisions expressed in earlier agreements on space and Antarctica that the United Nations had ironed out over the course of the past half-decade.[46]

Similarities between the draft treaties, as well as later testimony that the negotiations had been "swift," "courteous," and "business-like," masked significant disagreements that hampered and eventually stalled the talks in August.[47] For starters, the US treaty applied only to celestial bodies, whereas the Soviet draft extended to space generally. The US draft provided for the reporting of all space activities to the secretary general (read: the public), whereas the Soviet draft was silent on the issue. Over the course of more than three weeks, delegates clashed on liability for damage caused by spacecraft, the safe return of astronauts to their countries of national origin, and options to escape the treaty, among other differences. Some of these problems proved uncomplicated: for example, US delegates quickly agreed to the notion that the treaty should be extended to all of space; the implications of such an extension did nothing to interfere with American plans for space. The Soviets, in turn, quickly withdrew their reservations about US "barracks" on the moon, understanding that this expression did not imply a "military" installation.[48]

Two issues proved far thornier. The first was the right to inspect foreign installations on the moon and other celestial bodies to verify that parties to the treaty were abiding by its principles. US negotiators insisted on "full access at all times," a more liberal and literal provision than the Soviets' insistence that "visits" be conducted "on a basis of reciprocity," an arrangement whereby parties to the treaty would have to acquire the permission from the host country to inspect a given facility. The Soviets fairly pitched this language as necessary to protect a base's operations and personnel from unsafe interruptions, but US negotiators were quick to label their position simply another iteration of the old Soviet refusal to allow inspections of any kind. Indeed, the US delegation feared that the Soviet Union might use the "reciprocity" clause to argue that if the USSR did not inspect the facilities of other nations, those nations, in turn, had no right to inspect Soviet installations.[49]

Equally contentious was the issue of whether parties should be able to erect satellite-tracking facilities on foreign soil, regardless of the host country's wishes. The Soviet delegation insisted that any signatory that might provide access to tracking stations to another signatory must, by law, open access to *all* parties on a basis of equality. Gromyko was afraid of "discrimination against the Soviet Union" if "relatives" of the United States extended privileges to their allies but shuttered their doors to Soviet engineers. In contrast, the US delegation insisted that signatories should negotiate access to tracking facilities on a bilateral basis. Any sovereign nation had the right, per the UN Charter, to maintain control over access to its territory on a case-by-case basis. In a meeting between US and Soviet diplomats on September 22, Secretary of State

Dean Rusk insisted that while the United States was happy to open access to tracking facilities for the Soviet Union on US territory, it could not compel other countries to do the same. Rusk reasoned that if Soviet negotiators clung to total equality of access, the two sides "would have to face the very real possibility that the US and the USSR would be the only two countries signing the outer space agreement."[50]

State Department officials considered the impasse over tracking stations an intentional ploy by the Soviet Union to stall talks until political conditions made a space treaty more favorable.[51] The US delegation received help from influential delegates from Brazil, Mexico, Japan, and Australia (each of these countries had negotiated bilateral agreements with the United States on tracking facilities), who vocalized their objections to the Soviets' equal access provision.[52] Goldberg remarked that the idea of leasing land to whatever foreign spacefaring nations were so inclined set off a "hailstorm" of criticism in the Legal Subcommittee.[53]

With two deadlocks in place, a nervous energy pervaded the State Department's communications with its UN delegation as the talks in Geneva ended on August 4. Rusk emphasized to Goldberg that agreement on the outlying issues in the treaty should be ironed out as soon as possible (the negotiations would come to Manhattan in mid-September), for a quickly negotiated treaty would show that the superpowers could cooperate "even in the face of the Vietnam conflict." Early agreement on a space treaty would also help exclude matters of space law that might complicate the treaty, namely the issue of communications satellites. In attempting to negotiate a treaty as quickly as possible, however, the US delegation should not forfeit its position on the most important differences, such as access to foreign installations on the moon and other celestial bodies.[54]

Following his orders on September 22, Goldberg agreed, in a token concession, to include a statement in the operative clauses of the treaty that parties should "consider" granting access to tracking facilities to all signatories but fell short of Soviet demands. Gromyko replied that this phrasing did nothing to settle the issue and that US negotiators needed to "show more imagination."[55] The Soviets softened nevertheless, for they not only recognized that their position was politically untenable but recoiled at the idea that the United States, with the sanction of international law, might establish tracking stations on Soviet soil! In exchange for agreement that visits to foreign installations on the moon and other celestial bodies be carried out on a basis of reciprocity and with "reasonable advance notice," Fedorenko surrendered equal access to tracking stations. Construction of such facilities would continue to develop bilaterally.⁵⁶

Major differences between Soviet and American versions of the space treaty confirmed the widely held view that negotiations were a matter of superpower politics, a tired East-West confrontation. Yet from the very beginning of the negotiations, and indeed beforehand, it was clear that agreement on any international space treaty would be as much a matter of North-South tensions. Surely, the emergence of this dynamic did not come as a surprise: the talks unfolded amid an explosion of "developing world" participation in UN activities during the 1960s as European colonialism gasped its final breaths in the Global South. From 1958 to 1967, when the General Assembly voted to approve the OST, UN membership grew from 82 to 123, with most of the new additions having newly won their independence. Adding fuel to the fire was the fact that many of these countries—Algeria, Guyana, and Kenya, for example—were then contributing territory and facilities to Western space agencies for the purposes of either launching or tracking satellites. The construction of some of these facilities had occurred under colonial rule. US and Soviet officials well understood these circumstances and attempted to "head off the snow-balling of support for unsound or disadvantageous international measures" forwarded by the "have-not" countries, mainly by including references in their respective draft treaties about the need to explore space for the benefit of all.⁵⁷

Mollifying the developmental and technological aspirations of the former colonies proved easier in theory than in practice. Leaders in Egypt and India spearheaded a growing frustration among developing nations over neglect by the UN's economic agencies. This animosity generated a transformation within the United Nations in which security matters, which had once preoccupied UN energies, gave way to problems of wealth disparity and development, a trend highlighted by the creation of the G-77 in mid-1964. "Put crudely," observes the historian Paul Kennedy, "the 'have-nots' (the South), encouraged by the socialist bloc and First World radicals, were challenging the 'haves' (the North and its institutions) about the existing balance of economic power. Distribution, not growth, was back on the agenda."⁵⁸

These attitudes quickly permeated the COPUOS's considerations of the treaty drafts. The Brazilian delegation to the Legal Subcommittee insisted that the language in the Soviet draft regarding the necessity of carrying out space activities "for the benefit of all peoples irrespective of their degree of economic or scientific development" be moved from the preamble to Article I, where the provision would become operable.[59] Similarly, the United Arab Republic (UAR), a sovereign political union consisting of Egypt and Syria, motioned that a new paragraph be added to the first article ensuring that "States engaged in the exploration of outer space undertake to accord facilities and to provide possibilities to the non-space powers."[60]

The donation of these facilities would help developing countries participate in the modern economy built by space technology. If outer space were to become "the province of all mankind," Global South representatives agreed, the fruits of exploration should be shared with all, regardless of whether one country or another had contributed labor, technology, or capital to the effort. As J. W. Fulbright, chair of the Senate Foreign Relations Committee, would speculate about the developing world's intentions in senatorial review of the OST, "it leaves the impression that they were expecting the use of . . . Comsat [the Communications Satellite Corporation, created by the US government] be given to them free of charge, so to speak."[61]

The COPUOS's developing world representatives were occupied with broadening the treaty's ability to bar the spacefaring states from advancing their national power through space technology. The UAR, for example, proposed an amendment to the treaty whereby broadcast satellites would be forbidden from disseminating "hostile propaganda," a phrase that was ultimately expunged in favor of broadcasting that would promote "friendly relations" between nations.[62] India's delegation, moreover, considered it "disastrous" that the United States had not expanded its articles on nuclear weapons in space to include conventional weapons as well. Mongolia agreed, arguing that the superpowers' provisions legitimizing the use of military equipment had created a breach through which future weaponization would unfurl.[63] That satellites for weapons targeting and orbiting military bases would still be permitted was anathema for emerging space powers reading the drafts' soaring language about the exploration of space for peace.

Representatives from the developing world feared that outer space would become merely the latest pie to be divided among hungry, neocolonial competitors. It was the "self-interest of nations," thought many African and Asian members of the COPUOS, that prevented the realization of a global commons in space.[64] In language resembling the legal absolutists of previous generations, some urged halting manned exploration altogether until humanity could prove itself mature enough to extend civilization to other planets. The most eloquent proponent of restraint was Liberia's longtime president William V. S. Tubman, who insisted that the space programs of the United States, the Soviet Union, and Europe represented the "unbridled, unthinking, hollow quest for knowledge" rather than a "purposeful, thoughtful, creative search for understanding." Tubman placed space technology at the end of a long line of scientific and technological achievements that had offered salvation and liberation but delivered, for many, social calamity. He connected the cotton gin and the pyramids to slavery, the spinning jenny to child labor, and the discovery of quantum mechanics to the atomic bomb. "History," he reminded his colleagues at the United Nations, "demonstrates how easily, how often and how unhappily man's knowledge outstrips his understanding." The developing nations of the world were watching "in awe and apprehension" as humanity reached into the heavens "with the same imperfect hands and inadequate understanding [that] have marked his progress through the ages." He concluded that man was as yet too immature, too undisciplined, too obsessed with knowledge (rather than wisdom) to pry open the secrets of the universe. "We can only believe that man owe[s] it to himself to *pause and reflect* before plunging willfully on the path he has chosen."[65]

Tubman called for an international moratorium on space exploration to last up to ten years and urged his "sister African States" to spearhead negotiations for the ban. The global consequences of exploration justified the Global South's active and equal participation in political decisions about the cosmos. Liberia and her neighbors were not participants in the space race, "but no matter how small, no matter how struggling, no matter how poor we may be, we share the same heavens with the greatest Powers." By connecting the history of European imperialism to the future of the developing world, Tubman forcefully argued for an equal stake in global space politics: "Their catastrophes are usually ours; their failures of understanding affect our lives as intimately as their own; their concentration of money, imagination, scientific endeavor and national ambition on a headlong, impatient and wasteful race for knowledge which they cannot even take the time to study affects our lives, our hopes, our future just as it does their own."[66]

Appeals such as Tubman's did not fall on deaf ears. Though Global South negotiators failed to achieve more sweeping rules on the developmental and military elements of the space treaty, the final version of the document reflected a measure of success. The preamble's promise that space exploration would be carried out "for the benefit and in the interest of all countries" irrespective of wealth or technology was transplanted to the treaty's first article. The Soviet Union agreed that access to foreign soil for tracking stations should be negotiated on a bilateral basis. Per Article II, neither the United States nor the USSR could lay claim to the moon or any other celestial body, a stipulation that seemed to signify the coming end of colonialism. Most important, perhaps, representatives from developing countries ensured that they had a continued voice in space politics, for the treaty mandated that member states conduct space activities in accordance with the UN Charter.

Having secured a measure of compromise between negotiators East and West, North and South, the OST received a unanimous endorsement from the Political Committee of the General Assembly on December 17, 1966. Two days later, the UNGA voted—again unanimously—to approve the accord.

Doors Left Ajar

Reaction to the OST was mixed. With few exceptions, US and Soviet officials, particularly those who had negotiated the agreement, were ebullient. Johnson, Rusk, Goldberg, and others waxed triumphantly about the treaty having marked a "historic year for all humanity."[67] Johnson predicted that upon landing on the moon, US astronauts and Soviet cosmonauts could now meet each other as "brothers" rather than as national rivals. Dobrynin held out hope that the agreement would contribute to the settlement of international conflict back on Earth; at the very least, the treaty was a good omen for the possibility of agreement on the NPT then being negotiated in the UN's Eighteen-Nation Committee on Disarmament. The OST seemed to prove, noted the *Wall Street Journal*, that the Soviets "don't feel constrained to shun any agreement with the US until the war in Vietnam ends."[68] Editors across the country congratulated the diplomats who participated for exhibiting "vision" that had been absent at the dawn of the atomic revolution. Negotiators had profited from previous experience: it was easier to secure agreements before technological developments made such concords impossible. The negotiators could be praised for anticipating the pressing international implications of spaceflight and setting parameters by which competition would be minimized, cooperation maximized, and nuclear weapons barred from yet another shared commons. "The world surely will have occasion to look back to this day," the *Washington Post* opined, "as one that set the nations on the right path and on which the great powers made a wise decision to shun the

FIGURE 18. In the wake of the OST, two competing visions of space politics permeated public discourse. One was Johnson's and others' idea that the OST might provide momentum for the superpowers to find agreement on earthly political issues. The other interpretation, exhibited in this 1966 Herbert Herblock cartoon, was that the treaty was highly ironic—and limited—in the face of racial tensions in the United States, the arms race with the Soviet Union, and the war raging in Vietnam. A 1966 Herblock Cartoon, © The Herb Block Foundation.

military exploitation of celestial bodies and of outer space for narrow, nationalistic purposes."[69]

There were healthy doses of skepticism, too (figure 18). The most common criticism levied against the treaty negotiators was that they had worked to forbid military activities that neither the Soviet Union nor the United States sought to perform in the first place. "There are no 'wars of liberation' brewing on Jupiter or Canopus," chided the *Chicago Tribune*, and "nothing to gain ... from a military installation on the moon."[70] The *New York Times* agreed that the treaty's benefits were "more psychological than practical." The editors argued that far from stemming the arms race on Earth, the treaty was signed amid an escalation in superpower tensions.[71]

CHAPTER 6

The triumphalism of ratification was further undermined by persistent confusion—stemming from the treaty's "fuzzy" language—about what the OST had prescribed. What did it mean, exactly, that space was henceforth to be "the province of all mankind"? Did the treaty's provision about exploring the cosmos "for the benefit of all countries" connote technology transfer, resource allocation, or merely data sharing? To be sure, the treaty included unambiguous passages banning military bases, weapons testing, and maneuvers, but even these terms were poorly defined. Many observers asked why a military communications station did not count as a "base," or why spying on other countries with satellites constituted a "peaceful" activity. Others were frustrated that "weapons testing" applied only to celestial bodies—research on laser, kinetic energy, and fractional orbit weapons for near-Earth space could continue. It was possible, some observers argued, that the treaty's opaque language would precipitate, not prevent, conflict. Joshua Lederberg, breaking with his more laudatory colleagues at the *Post*, predicted that within a few years' time, the Soviet Union and developing countries would wield the opaque language in the treaty as "a propaganda club" against any US space activity suspected of having military applications.[72] Provisions for access to space facilities "on the basis of reciprocity," seconded one lawyer, had opened a window for the Kremlin to deny the United States access to future Soviet space stations and moon bases. Soviet leaders could reasonably claim that if the USSR refrained from inspecting US facilities, it could fairly deny access to American or UN inspectors. "The treaty is a step forward," he concluded, "but those who would hail it as a solution to all the problems of war in space are overoptimistic."[73]

Some critics went as far as to say there should be no treaty at all, that human beings were better off aggressively militarizing space so that nation-states could eventually transition international conflict to—and contain it in—the void. Air Force leaders had throughout the 1960s envisioned a future for war in which, conflict having moved to space, civilian populations might be spared. The moon, especially, might prove a battleground on which political outcomes could be decided, a recess playground where power hierarchies could be created and managed.[74] If the OST reflected a willingness on the part of the Soviet Union to make concessions, insisted the *Tribune*, "it would be more sensible to outlaw war on earth, where people can get killed, and to do our fighting in outer space or, better still, on the moon."[75] By the time the US Senate ratified the treaty, no less than Senator Eugene McCarthy (D-MN), soon-to-be

Democratic presidential candidate, endorsed the idea that it was "preferable to have [war] in space rather than where people lived." Instead, McCarthy scolded, the United States and the Soviet Union should be working on a treaty pacify the Earth and to wage war *only* in space.[76]

The breadth and depth of criticism levied against the OST must have unsettled its proponents, for the treaty had not yet been ratified by the US Senate. When the dust cleared, Goldberg and others had little to fear, for as with the UNGA vote, the Senate ratified the treaty unanimously. And yet, just as with the UNGA vote, the support the OST received is misleading. When the Senate Foreign Relations Committee conducted hearings on the treaty beginning in March 1967, it was far from clear that members of Congress, particularly those concerned about how the OST would affect national defense, would agree on the advisability of a treaty that impinged on US military space activities. A litany of security-related questions permeated the Senate over the course of three meetings: Would the Soviets now have access to US reconnaissance satellites? (No.) Would the United States be able to inspect a Soviet moon base? (Yes.) Would the treaty preclude observation from space? No. If at some point in the future the United Nations established a space police, would *it* be permitted to orbit nuclear weapons? (No.) Could the United States inspect other satellites to ensure that there were no nuclear weapons on board? (No.) Despite having been briefed by the director of defense research and engineering about questions that could possibly arise during the hearings, Rusk, Goldberg, and State Department counsel Leonard Meeker labored to convince the nation's leaders that the OST would benefit US security.[77]

To help reassure the Senate, Rusk brought in Earle "Bus" Wheeler, chair of the Joint Chiefs of Staff, then busy trying to stabilize the war in Southeast Asia. Wheeler assured the assembled politicians that the treaty would not quash American military space activities; in fact, he predicted, it would likely result in a doubling of efforts. A ramped-up satellite reconnaissance program, for example, would be needed to ensure that the Soviet Union was abiding by the rules of the treaty, and a renewed commitment to research and development for ASAT ordinance was needed to counteract the conventional space weapons that were still permitted.[78] In a world increasingly armed to the teeth with weaponry, Wheeler added (quoting a report the Foreign Relations Committee had itself composed seven years earlier), "arms control ought to be viewed as *part* of our military and national security policy and not as an alternative to military policy." "Strength alone will not suffice," he reminded the Senate, and "threats to the peace can be reduced and perhaps eliminated by stabilizing expectation and demands in the frontier areas of modern technology." The OST would supply that stabilization.[79]

Endorsement from the one of the Pentagon's highest-ranking officers was welcome, but doubt persisted. Many congressional leaders argued that the OST would fail to enhance national security given the Soviets' propensity for duplicity, furtiveness, and disregard for international law. The treaty was "nothing more than a scrap of paper" to the Soviet Union, insisted Strom Thurmond (R-SC).[80] Because the treaty did not permit inspection of space vehicles in outer space—as distinct from the moon and other celestial bodies—the United States had to rely on its own reconnaissance and strategic analysis capabilities to determine whether the USSR was complying with the treaty's provisions against nuclear weapons in space. Deputy Secretary of State Cyrus Vance assured skeptics that while the Soviet Union could indeed orbit a small number of nuclear weapons without the US government's knowledge, such a small contingent of weapons would not constitute a significant threat. As had been elucidated in the NSC, the Congress, and the press over the past ten years, bombs launched from orbit were technically and economically unfavorable compared with traditional ground- or submarine-launched ICBMs. In any case, the US capacity to detect such weapons was growing by the day. Between 1963 and 1965, the United States had launched twelve Vela satellites, which were designed to detect nuclear detonations in the wake of the LTBT. NASA's MOL (which would be canceled in 1969), was capable of the same detection.[81]

When it came time for a vote on ratification on April 25, the Senate swallowed its earlier skepticism with a 88-0 "yes" vote.[82] Here, the considerable propaganda value of the treaty was manifest: simply put, no senator wanted to vote down an agreement that barred nuclear weapons from space, forbade military installations and maneuvers on the moon, and declared space free to explore by all—especially on one's own.

Within months, the treaty confronted its first real test, and again it appeared to fall short of its promises. The widely touted notion that the agreement was the foundation of an apolitical and pacific utopia in space immediately bumped up against Soviet tests of a terrifying new weapon in near-Earth orbit. Just two days before US officials shook hands with Dobrynin, the Soviet Union attained its first fully successful launch of the fractional orbital bombardment system that had so frightened the Kennedy administration earlier in the decade (figure 19).

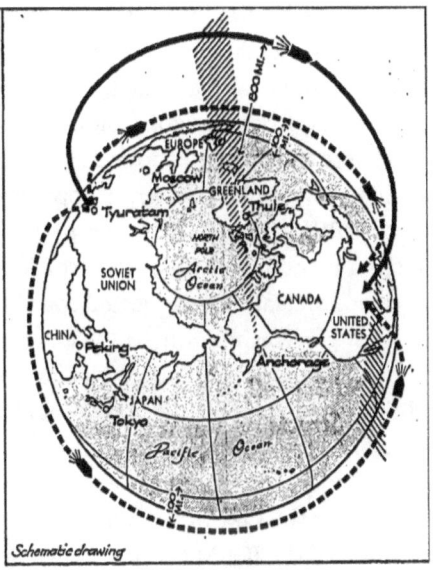

FIGURE 19. Accompanying the text of McNamara's statement on FOBS in the *New York Times*, this drawing depicts the orbital trajectory of a FOBS missile compared with a traditional land-launched ICBM. The shaded areas emanating from Anchorage, Alaska, and Thule, Greenland, represent the US Ballistic Missile Early Warning System (BMEWS). FOBS, the theory went, could orbit a warhead over Antarctica and penetrate US airspace from the south, thus bypassing the BMEWS. *New York Times*, November 4, 1967, 2.

This modified super-heavy ICBM was engineered to enter a partial orbit of the Earth, where retro-fire rockets would slow the missile enough to lower it out of orbit and into a predetermined trajectory. Tests of the weapons had been going on since at least December 1965, but the Kremlin had given them no public designation. The CIA would soon listen in on Soviet boasts of the weapon during a parade in Moscow: these missiles could "carry the most powerful nuclear warheads and deliver them to any point on Earth. These are ballistic rockets; they can also be used for orbital flights—a very heavy type of rocket."[83]

In a press conference assembled so hastily that State Department officials—concerned that FOBS was a violation of the treaty—were unable to attend, Defense Secretary Robert McNamara briefed the nation about the Soviets' new weapon and tried in vain to reassure Americans that the balance of power had not been turned upside down. To assuage concerns that FOBS would "confound our defense and come in through the back door," McNamara reminded his audience that while FOBS missiles were indeed harder to detect by radar and faster to their targets, orbital bombardment in general was far less accurate and less destructive than systems using conventional land- or submarine-launched ICBMs. In any case, he added, the United States would soon operationalize an "over-the-horizon" radar capable of bouncing signals off the ionosphere to detect weapons using orbital trajectories. This effort represented a level of instrumental sophistication in missile technology, whereas FOBS remained "a system in which the disadvantages far outweigh the advantages as far as the attacker is concerned."[84] Rather ingenuously, too, McNamara suggested that FOBS missiles did not violate Article IV of the freshly minted OST because they did not make a "complete" orbit of the Earth. Although some officials, particularly at NASA and State, considered FOBS to be a clear violation, the Johnson administration was careful not to rattle cages over a comparatively disadvantageous weapon already in the testing phase.

The NSC agreed that for the sake of the space treaty FOBS should be considered "an extension of the ICBM problem."[85] Article IV of the OST was not intended to cover ICBMs and weapons such as FOBS that presumably would carry warheads only in times of war. "We must be careful to avoid vague charges which cannot be substantiated that the Treaty has been violated," wrote the NSC's Spurgeon Keeny. "Such hasty actions can lead to counter charges that we are interested in employing the Treaty for a tactical, political advantage when it so serves our purposes. This can only serve to degrade the Treaty in the eyes of the world."[86]

To most observers outside the White House, however, it appeared as though the Soviets were thumbing their noses at the new agreement. To ordinary Americans, FOBS was just another "unholy device" of a destabilizing, costly, and unpredictable arms race.[87] Stewart Alsop wrote to his *Saturday Evening Post* readership that because FOBS was likely a compensation for the Soviets' lag in bombers and conceived as a second-strike weapon, the orbiting missile was essentially "an anti-people weapon," a city-buster.[88]

Accordingly, many feared that the United States would react to FOBS by initiating its own orbital bombardment program. Others, in turn, demanded such a course of action. Representative William L. Dickinson (R-AL) went so far as to call for McNamara's resignation considering the secretary's nonchalant response to what the latter considered to be merely the Soviet's latest toy.[89] "It is plain that there are gaping loopholes in [the OST]," the *New York Times* presciently observed two days after McNamara's statement, "loopholes which military technologists have no intention of leaving unexploited."[90]

Four years later, in 1971, the Soviet Union would cease testing of fractional orbit missiles. McNamara's assertion that the FOBS weapons tested from 1965 did not carry nuclear weapons was true: the warheads were kept in separate storage facilities until 1972, after testing had stopped. The extent to which the OST entered into Soviet calculations about whether the FOBS program should continue is a matter for debate, although a report filed by the Russian Strategic Military Forces in 1991 suggested that the United States' lack of an efficient ABM system in the 1970s—not international law—accounted for the program's slowdown.[91] Though the treaty survived the FOBS scare, many were left wondering if the peaceful cooperation in space during the decade, reflected in the 1975 joint US-Soviet *Apollo-Soyuz* Test Project, could be attributed to the ethical standards set by the treaty or, more cynically, the mutual restraint of the superpowers amidst fiscal, technical, and strategic limitations.

The Outer Space Treaty in History

How are we to judge the Outer Space Treaty in history? Was the agreement, as Dolman has argued, merely "the jumbled consolidation . . . of conflicting precedents"?[92] Did it accomplish nothing more, as the *Chicago Tribune* accused late in 1966, than the "sale [of] our peace-loving consciousness"?[93] There are reasonable arguments to be made for the affirmative. No nation has transgressed the treaty's article on nuclear weapons, but the provisions in question do nothing to prevent the conventional weaponization of space.[94] Nor does the treaty addresses the "real need" of space law: environmental protection, particularly debris mitigation.[95] Although the OST includes a provision requiring any space activity with the potential to contaminate outer space be subject to international consultation, it does little to hold member states accountable for pollution. Observers looked through the treaty in vain when in 2007 the Chinese government shot down its own weather satellite using a projectile ASAT weapon, or when a US *Iridium* satellite collided head-on with a Russian *Kosmos* vehicle two years later, a crash that spread vast clouds of debris along the satellites' former orbital trajectories. Fifty years on, the OST remains a flawed document.[96]

Inflicting further harm to the OST's legacy is the swiftness with which the COPUOS's Legal Subcommittee negotiated the agreement, as well as the unanimity it enjoyed in both the UNGA and US Senate. This apparent ease led some observers, both at the time and since, to speculate that nothing worthwhile had been accomplished. How could it have been so easy, after all? Both superpowers had significant interests in space exploration—particularly military applications—and each had imbued its space activities with the power to convey its government's social, economic, technological, political, and moral qualities. Over the previous decade, US and Soviet officials had clashed on a legal definition for space, the legitimacy of satellite reconnaissance, the creation of the COPUOS, and an agreement to ban nuclear weapons from space and celestial bodies. Why, many critics asked, should provisions as comprehensive and ambitious as those in the draft treaties require any less combativeness, mistrust, and obstinacy?

But neither the speed with which the OST was negotiated nor the unanimous support it enjoyed in the United Nations should distract from the opposition levied against it from both cold warriors who thought it would hamper US military space programs and voices on the left, like McCarthy, who thought its provisions did not go far enough. During the negotiations, Soviet obstinance regarding access to foreign soil for tracking stations, international access to future moon facilities, and the definition of "military" space activities threatened to torpedo the accord, as did US stubbornness about the necessity of reporting space activities to the secretary general. Global South representatives in the Legal Subcommittee advanced a formidable agenda that ensured space exploration would not continue without the interests of the developing world at hand. Whatever the vote tallies may at face value indicate to historians, then, agreement on such comprehensive rules for space was no cakewalk.

Jaundiced views of the OST that appeared in senatorial debate, contemporary editorials, and in some recent scholarly analysis stem in part from a one-dimensional view of the treaty as an instrument of international law. Because the treaty demilitarized space in only the spirit—not the letter—of the law, it was, and remains, easy to label the OST merely an empty box "wrapped in many silken flags and tied with much gold braid."[97] But, as its deeper history suggests, the treaty was more than the sum of its legal parts. It was a reaffirmation of political ideas about outer space that had been circulating in Europe and the United States well before the space age had begun: that exploration should be advanced for peaceful purposes only, that it should be cooperative, and that its benefits should be shared with all. Space exploration, boosters convinced the public, would be a fixture of humanity's future—best to preemptively bar from this future the "viruses" of war, colonialism, nationalism, and historic animosity. In this way, it is just as easy to view the OST as the apotheosis of H. G. Well's socialism, Konstantin Tsiolkovsky's utopianism, and John McConnell's pacifism as it is to view it as an image-making tool of the United States and the Soviet Union. Indeed, the treaty received the unanimous support it did not as a juicy piece of Cold War propaganda but rather a genuine argument that the Cold War should stop at the edge of space. This was an argument easily adopted not only by two large, spacefaring countries trapped in an arms race but also to the numberless people for whom spaceflight was still a distant possibility.

The OST also offered much of practical use. It served as the most important link in a chain of international legal arrangements for space dating from the late 1950s to the mid-1980s. The treaty built on and codified important precedents laid out in the 1959 Antarctic Treaty, the "bombs-in-orbit" resolution of 1963, and the UN Declaration of Legal Principles, adopted that same year. It also established basic legal tenets from which subsequent agreements were built, a role that has made the OST "quasi-constitutional."[98] From 1968 to 1984, the OST birthed four additional agreements establishing rules for activities on the moon; liability for damages caused by spacecraft; the safe return of fallen astronauts; and the registration of space vehicles.

It is important to remember, too, what Earle Wheeler reminded the Foreign Relations Committee during Senate ratification: *with or without a treaty*, it was possible for the United States or the Soviet Union to enact policies or begin weapons programs in space that harmed the interests of peace. With a treaty on the table, Wheeler observed, "there are psychological and world opinion factors which would ... exert a restraining influence upon a state contemplating violation of the treaty."[99] Indeed, in global debates over Soviet tests of FOBS, space-based antiballistic systems such as SDI, kinetic energy weapons, and ASAT weapons, the OST served as a platform upon which the case against militarization and weaponization was laid. It is reasonable to speculate that if basic principles on the exploration and use of outer space had not been the subject of agreement among OST signatories in the late 1960s, then international rivalry in space might have compelled national governments to pursue the vision of space propounded by the USAF earlier in the decade.

More immediately, the OST had seemingly arrived in the nick of time, for the superpowers were accelerating their headlong sprint into space. Less than a month later after the signing ceremony *Venera 3* impacted the Venusian surface, becoming the first spacecraft to land on another planet. *Luna 10* orbited the moon on April 3, 1966. Over the course of the year NASA's Gemini program demonstrated the feasibility of extravehicular activity, practical work capability in space, direct-ascent rendezvous, and spacecraft docking. On January 27, 1967—the very same day that US, Soviet, and British leaders met in the White House to celebrate the OST— a launch simulation failure led to a fire that killed *Apollo 1* astronauts Virgil "Gus" Grissom, Ed White, and Roger Chaffee; the lessons learned from this horrifying accident (no crewed missions flew for twenty months) increased flight control accountability and supplied future spacecraft with a slew of additional safeguards. Despite blows like *Apollo 1*, it was clear that the treaty—and by extension the entire "sanctuary" paradigm—would soon meet its greatest test: men were headed for the moon.[100]

Chapter 7

Stairway to Heaven?

Like the Outer Space Treaty before it, the first moon landing appeared a triumph for promoters of a space sanctuary. When on July 20, 1969, Neil Armstrong and Edwin "Buzz" Aldrin stepped out of their Lunar Module (LM) onto the moon, they did so prepared to commemorate their historic mission not only as evidence of American bravery and technical mastery but also the world's unity and commitment to peace. The astronauts carried with them ceremonial tokens intended to signify *Apollo 11* as the culmination of peaceful, scientific enterprise and a symbolic act representing the unification of humanity. Though Armstrong and Aldrin planted an American flag in the lunar soil, their spacecraft also brought miniature flags from every UN member nation to be presented to heads of state upon the mission's return. Before departing the moon, moreover, they left a tiny silicone disc inscribed with well wishes from leaders of more than seventy countries; a golden olive branch; and, in an act defying the space race, two medals commemorating Soviet cosmonauts Yuri Gagarin and Vladimir Komarov, who had recently died in tragic crashes.[1] Perhaps most famously, the astronauts also left a small plaque recording, for any future passersby, that when "men from the planet Earth" had first set foot upon the moon, they had come "in peace for all mankind."

Having extended human activity to another celestial body, *Apollo 11* sparked hopes that the process of social, ethical, and political transformation among human societies that interplanetary philosophers had promised could now begin. On the day of the landing Pope Paul VI anticipated that Apollo's "sublime victory" would be a victory over violence and hatred, that further exploration in space would mark "true progress toward the temporal and moral good of humanity."[2] Armstrong himself speculated that in the next century visitors to the moon would read his declaration of peace left at Tranquility Base to find "that this was the age" when world friendship "became a fact." If the "spirit of *Apollo*" was heeded, the *Los Angeles Times* editorialized, "we need not wait another three decades for a world that can live at peace with itself."[3] Humans had broken the film that separated their ignorant past from a more enlightened future. They now had room to grow up.

Such was the version of Apollo's history that US leaders, NASA representatives, and other influencers of popular opinion had already begun to write. Over the eight days that the mission unfolded, and in the weeks and months after the astronauts splashed down, three narratives with special significance to the idea of a "sanctuary" in space began to emerge. The first was that the moon landing, though cloaked in the garb of American superiority, was an achievement of the entire human race and evidence that space exploration could unify the species in the quest for new knowledge. US engineers, while taking pride in their accomplishment, were quick to point out the contributions of thousands of foreign contractors working in tracking facilities from Lima to Zanzibar to Guam.[4] Of the six scientific experiments that Armstrong and Aldrin conducted in their brief time on the moon, newspapers noted that the first—a collection of solar wind samples—had been designed by Swiss scientist Johannes Geiss.[5]

Apollo 11's globalist patina led many to conclude that the mission would open the door for diverse societies to rediscover their common humanity. It was easy in mid-1969 to believe that if countries could cooperate in space science, they could live in harmony too. Upon returning from a global tour, President Nixon reminded Americans that the "spirit of *Apollo*" could "bring the people of the world together in peace."[6] The view of Earth from space, especially, would break down old tribal loyalties and make room for a commitment to shared interests. Carl Sagan echoed many in his assessment that the orbital perspective had begun to soften the human heart. The "unexpected final gift of *Apollo*," he wrote later, was the visual transcendence of nationalism:

"I'm struck again by the irony that spaceflight—conceived in the cauldron of nationalist rivalries and hatreds—brings with it a stunning transnational vision. You spend even a little time contemplating the Earth from orbit and the most deeply ingrained nationalisms begin to erode. They seem the squabbles of mites on a plum."[7]

A second narrative permeating popular discourse was that the *Apollo 11* mission reflected the eminently peaceful nature of the US space program and of American science generally. Although the astronauts had been culled from the military services, they conduced none other than peaceful experiments and, as the metal plaque they left behind on the moon indicated, "came in peace for all mankind." Here, the first moon landing was the natural culmination of civilian control of space exploration, of NACA's 1958 victory over the Pentagon. *Apollo* had succeeded in exorcising the ghosts of the atomic bomb.[8]

Last, Americans argued that the moon landing marked the death knell of empire. Although the *Apollo* 11 crew had stepped onto a vast, open frontier teeming with valuable mineral resources (and no native population), it had no intention of claiming the moon for the motherland. NASA issued a legal declaration that the astronauts' flag raising on the moon was "not to be construed as a declaration of national appropriation."[9] In one of the first histories of the Apollo mission, the journalist Richard Lewis wrote of the radical departure of the first moon landing from the Western tradition of exploration. Whereas earlier eras of discovery were motivated by the desire for land, trade, resources, or slaves, the space age represented a new age of exploration motivated purely by scientific inquiry.[10] Nixon had foreshadowed the argument six months before *Apollo 11* in his first inaugural: "As we explore the reaches of space, let us go to the new worlds together—not as new worlds to be conquered, but as new adventures to be shared."[11]

In the years that followed, events back on the ground belied all three narratives. How could it be that *Apollo* had united humanity when the nation that had organized the mission was more divided politically than at any time since its civil war? Throughout the buildup to the moon landing, racial, political, and other forms of violence, rooted in one way or another in difference, wracked the country. Following the assassination of Martin Luther King, Jr. in April 1968, black Americans in cities from Los Angeles to Washington erupted in anger, setting city blocks ablaze, looting local businesses, and rioting. That same August, at the Democratic National Convention for president in Chicago, police officers and national guardsmen clashed with antiwar protesters and other activists on live television. In late July 1969, just as the moon landing was happening, Pennsylvania governor Raymond Shafer declared a state of emergency over racial gang violence in York.

The ongoing war in Southeast Asia debunked the pretense that *Apollo* either heralded a coming reconciliation among nations or represented the peaceful character of American science. When the *Saturn V* rocket carrying the astronauts launched from Cape Canaveral on July 16, 1969, US troops were still reeling from the Tet Offensive launched by the North Vietnamese Army at the end of that January, the lunar new year. By then the war had reached its zenith, with US deployment having reached its highest point—more than half a million men—since the beginning of the conflict. In June, a month before the moon landing, *Life* magazine published its now-infamous cover story about the scores of young soldiers who had lost their lives in a "average" week in Vietnam. Only a few months after the astronauts' safe return, Americans learned, also from *Life*, about the horrific massacre committed at My Lai.[12]

In short, the cosmopolitan light with which officials bathed the moon landing illuminated the darker corners of national behavior at home and abroad and cast doubts on the peaceful trajectory of US space policy.[13] Irony became the reigning method of expressing the meaning of *Apollo 11*.[14] "We find ourselves rich in goods," Nixon remarked in his first inaugural address, "but ragged in spirit; reaching with magnificent precision for the moon, but falling into raucous discord on earth."[15] "The man jus' upped my rent las' night/'cause Whitey's on the moon," poet Gil Scott-Heron wrote famously in 1970. "No hot water, no toilets, no lights/ but Whitey's on the moon."[16] Rockets found themselves juxtaposed to oxcarts and straw-thatched roofs in Vietnam, handsome astronauts to starving urban children in Detroit and Philadelphia. Everywhere was the question asked: "If we can fly people to the Moon, why can't we . . . ?"

At one turn, *Apollo* marked the triumph of sanctuary politics and displayed evidence of its impending demise. Even as Nixon and other senior officials marked *Apollo* a victory for all humanity, other public voices, particularly in Congress, heralded the landing as the culmination of purely American courage and know-how, a position reflected in the government's choice to fly only its own flag during the landing ceremony. And though NASA and the State Department did much to dispel notions of US sovereign interests in the moon, American aerospace firms and mining interests fought hard to maintain an open door to its natural resources. When the space race finally concluded, it was difficult to say whether *Apollo* marked a victory for "space for peace" or the beginning of something else entirely.

"Never Closer Together Before"

For more than a decade after *Sputnik* NASA administrators and congressional politicians of every stripe had propagated a vision of space exploration as a project capable of uniting the entire world, and in at least one sense the first moon landing fulfilled their promise. When Armstrong and Aldrin took their first steps on the moon, more than six hundred million people from around the world—one-fifth of its population—watched them in unison on television. For nearly forty continuous hours all three major US television networks provided full coverage of the astronauts' descent, landing, and departure. Together CBS, NBC, and ABC spent nearly $6 million to broadcast live images from Tranquility Base and to collect reactions from capitals across the globe. City officials turned London's Trafalgar Square, Manhattan's Central Park, and Tokyo's tallest skyscraper into makeshift auditoriums from which the public could view the historic event. After the flag planting, Armstrong and Aldrin received a scheduled phone call from Nixon, who remarked that "for one priceless moment in the whole history of man, all the people on this earth are truly one." The pair of astronauts acknowledged the sentiment, replying that it was a great honor to represent "not only the United States, but men of peace of all nations." Indeed, Aldrin would later recall that he and his fellow crew members had perceived an "almost mystical identification of all the people in the world at that instant."[17] Here, embodied, was what historian Alexander Geppert has called the "synchronicity" of the moon landing. It was an event that all, in at least the mythology of the moment, seemed to experience together.[18]

Achieving human oneness had been at the heart of the cultural and political narrative of space exploration since the middle of the nineteenth century, but by the time of the Apollo program—indeed in large measure *because* of the moon mission—the notion that spaceflight would spark the transcendence of human difference had become axiomatic.

The *Apollo 8* mission in 1968 suggested that the prophecy of unification would soon be fulfilled. While orbiting the moon on Christmas Eve astronauts Frank Borman, James Lovell, and Bill Anders witnessed the Earth rise, as if the sun, in the small window of their capsule. Gazing on the lambent blue sphere floating in the void, all three men perceived, with the force of an epiphany, that the world was fundamentally different from the home they saw from the ground. Earthly differences and conflicts now seemed mere trivialities, personal experiences infinitesimal next to the grandiosity of their new, interplanetary viewpoint.[19]

FIGURE 20. *Earthrise.* Courtesy of NASA.

"Raging nationalistic interests, famines, wars, pestilences don't show from that distance," reported Borman. "From out there it really is 'one world.'"[20] That view would force people to wonder: "Why the hell can't we learn to live together like decent people?"[21]

Though commentators the world over could not fully understand the crew's emotions from the ground, they propounded similar ideas once they had seen the photographs *Apollo 8* had beamed back to NASA, particularly Anders's famous *Earthrise* picture (see figure 20). The *Boston Globe* rhapsodized on "how infinitely tawdry are the differences that separate its races, its nations, its men one from another—these infinitesimal grains of sand on an earth which is but a grain of sand itself, exaggerating their nonexistent differences, forgetting their oneness in the brotherhood they do not want to recognize."[22] The *New York Times* thought the mission "a sobering perspective on man's puny earthly works and rivalries, reminding all humanity that nature is the basic antagonist, not other men."[23] Arthur C. Clarke, the subject of even greater fame after Stanly Kubrick's film adaptation of his novel *2001: A Space Odyssey*, predicted that Christmas 1968 would mark a dividing line in history, one separating humanity's slumber from its awakening. "The second Copernican revolution is upon us," he wrote in *Look* magazine shortly after the flight, "and with it, the second Renaissance." Children born after this inflection point would become something more than Americans or Russians, perhaps "citizens of the United Planets."[24]

The first moon landing reignited these ideas. "The greatest thrill I can imagine for myself," wrote astronomer Walter Orr Roberts shortly after the mission, "is to stand on the moon's surface and to look back from the harshness of the lunar landscape to the luminous hospitable earth. From that vantage point, I believe, I could view the earth in its oneness." As interplanetary writers had predicted from the 1890s through the interwar period, the world now perceived that the view of Earth from space possessed a mind-altering quality that might strip human beings of their chauvinism, insularity, and aggression. Space writer Frank White would later call it the "Overview Effect," astronaut Ron Garan the "Orbital Perspective." Whatever the name, and whether through photographs or the experience of space travel itself, few denied that a certain "magic" existed in viewing one's home planet from the lonely isolation of the cosmos.[25]

Michael Collins, who had been left behind in the *Apollo 11* command module, provided powerful testament to the transformative power of seeing the whole Earth and its potential consequences for international relations. He hoped to escort political leaders to a distance of 100,000 miles to "see how there are no borders and how small the differences between nations really are."[26] From such a distance, Collins wrote, the "tiny globe would continue to turn, serenely ignoring its subdivisions, presenting a united façade that could cry out for unified treatment." The cumulative effect of erased borders, diminished racial differences, and the fragility of Earth could be none other than transcendental. "By causing [people] to realize that the planet we share unites us in a way far more basic and far more important than differences in skin color or religion or economic system," Collins would write in his widely sold memoir, *Carrying the Fire*, the orbital vantage point would be "invaluable in getting people together to work out joint solutions."[27]

It was a message that public officials, pundits, and social commentators reinforced upon the crew's return to Earth. Nixon, greeting the astronauts aboard the *USS Hornet* shortly after their splashdown on July 24, thought the mission to have been "the greatest week in the history of the world since Creation," for the world had become not only "bigger, infinitely," but also "never closer together before."[28] UN Secretary General U Thant echoed the sentiment in a celebration of the mission he hosted in front of the General Assembly building in New York three weeks later. Thant considered the moon landing to have been experienced "vicariously" by all who had witnessed the event through television and radio. He remarked that the plaque the astronauts had left behind on the moon, a copy of which the astronauts presented to Thant, highlighted "the common identity of all the inhabitants of this planet and our never-ending search for peace."[29] Coming from the leader of the world's greatest international body, the idea that the entire world was "privileged to share in [*Apollo*'s] achievement," became all the more cogent.

Aside from the moon landing, nothing did more to promote the image of US space exploration as a world-uniting experience than the global tour undertaken by the *Apollo 11* crew two months after the mission. At the behest of Nixon, who lent the astronauts one of his jets, Armstrong, Aldrin, and Collins departed from Houston with their wives on September 29 to spread the "spirit of *Apollo*" around the globe. Over thirty-eight days, the three couples, accompanied by an entourage of NASA and State Department officials, visited twenty-nine cities in twenty-four countries, crossing the equator six times in the process. In motorcades, press conferences, and other venues, more than 100 million people came out to see the astronauts, who would greet and shake hands with no less than 25,000 of them.[30]

"Giantstep," as the *Apollo* world tour was known, was among the greatest public diplomacy opportunities of the Cold War. The tour's itinerary reflected the interests of the State Department in healing tense relationships abroad and strengthening old alliances. In addition to a robust list of stops in Europe (which included London, Paris, Rome, Madrid, and Brussels), the astronauts visited a number of cities in the Global South (Mexico City, Buenos Aires, Bogota, Rio de Janeiro, Kinshasa, Tehran, Bombay, and Dacca), not only to promote a vision of American technological supremacy but create the impression that the United States "shared" the experience of *Apollo* with the entire world, regardless of whether one nation or another could then launch rockets, orbit satellites, or land a spacecraft on the moon.[31]

At each stop, the astronauts were to present gifts to foreign dignitaries: a replica of the plaque left on the moon mounted on a walnut backing (presented to the leading official in each city); a replica of the Goodwill Message disc left on the moon, an eight-power magnifying glass, and a framed photograph (for the signers of each of the individual messages that were contained in the original disc); as well as autographed color photos of the *Apollo 11* mission (for other high-level officials).[32] As ambassadors for democratic capitalism, US-led science and technology, and the "American Way," no public representative could match the appeal or aura of the men who had gone to the moon.

In both the symbolic acts they performed on the moon and in their tour appearances, the astronauts, Norman Mailer observed, "exhibited as much sensitivity to an audience as any bride on her way down the aisle."[33]

Space exploration had been a fixture of US public diplomacy since the late 1950s. In the wake of the United States' first space feats, NASA cooperated with the State Department and the USIA to create countless films, pamphlets, books, exhibits, and radio broadcasts on the space program to be disseminated to countries all over the world. When Alan Shepherd became the first American to inhabit outer space in May 1961, NASA, USIA, and DoD sent packets about the flight to the governments of more than eighty nations and lent Shepherd's historic space capsule to the International Air Show in Paris and the International Science Fair in Rome. A year later, after astronaut John Glenn became the first American to make a full orbit of the Earth, NASA sent Glenn's capsule, *Friendship 7*, on an even more peripatetic tour, "a Fourth Orbit" for the capsule. This public space diplomacy was aimed at increasing the perception of the US space program as not only superior to that of the Soviet Union (a hard case to make given Soviet accomplishments) but more open as well.[34]

But something was qualitatively different about the *Apollo* world tour. Whereas NASA's global outreach had, under Kennedy, emphasized mastery of technology, masculine competence in spaceflight, and the benefits of democratic capitalist management over Soviet technocracy, Giantstep seemed to subordinate notions of superiority to *Apollo*'s "human" element. The tour's philosophical and political objective was to convey to every person, in every country, that to share in the transcendent experience of the moon landing, one needed only to be a human being. Everywhere onlookers reached out to touch the astronauts' bodies, as if through physical contact each person could somehow acquire some moon magic. "Half the kids in town chased us through the police lines and in and out of the cars, with everyone wanting to touch, touch and to embrace," reported one tour official as the *Apollo 11* motorcade made its way through Mexico City.[35]

The Giantstep tour, the crew's visit to the United Nations, and television coverage of the moon landing did much to legitimate *Apollo*'s status as an omen for the unification of the world's people. But as both contemporary observers and historians since 1969 have pointed out, perceptions of the landing's political and moral consequences often derived from where one stood. From Huntsville, Houston, or Cape Canaveral, it was easy to be sanguine. From Watts, Baltimore, or the Cambodian villages then being bombed under Operation Menu, things appeared quite different.

That *Apollo* had united the world was a particularly difficult notion to accept given divisions over the moon mission itself. A yawning gap opened between enthusiasts who considered the moon landing the "single greatest technological achievement of all time" and critics who thought it a wasteful "Moondoggle."[36] In the countdown to *Saturn V*'s launch, during the mission, and well after the astronauts returned, it was impossible to disentangle the miraculous events in space with sober realities and political infighting back on the ground. It was a sign of the times when Archibald MacLeish, who had written the most famous commemoration of the *Apollo 8* mission (see chapter 1), refused Nixon's offer to memorialize the *Apollo 11* ceremonies, having "thought twice about doing anything with Nixon connected with it."[37] The world at once "celebrated and recoiled" from *Apollo*, as historian Roger Launius has put it.[38]

Most divisive was the program's cost. As widespread as enthusiasm for the moon landing was in July 1969, NASA found itself on the defensive from a formidable subset of the country who felt that the bottomless sums spent on the Apollo program (roughly $25 billion) had been wasted and would have been better spent elsewhere. On a basic level, this charge emanated from profound racial and class inequalities at home and abroad. Emerging nations and civil rights leaders in the United States had decried the moon project throughout the 1960s as wasteful and ignorant of more pressing matters affecting the world's poor. The sentiment came through most powerfully when Southern Christian Leadership Conference president Ralph Abernathy led a Poor People's Campaign protest on the lawn just outside the Kennedy Space Center in Cape Canaveral on the day of the *Apollo 11* launch. With donkeys and wagons juxtaposed to the towering *Saturn V*, Abernathy called on NASA administrator Thomas O. Paine to divert the enormous resources then dedicated to the Apollo project to improving the lives of African American citizens in the nation's crumbling cities and rural backwaters.[39]

The argument extended beyond the impoverished. "For that kind of money," remarked Kurt Vonnegut on the *Evening News*, "the least [NASA] can do is discover God."[40] Invited by CBS to provide the devil's advocate perspective on the moon mission, Vonnegut spoke for millions when he argued that humanity "should be humbled by his own waste and stupidity." Noted historian Arnold J. Toynbee agreed, suggesting that expenditures were a matter of basic morality. "In a sense," he offered, "going to the moon is like building the pyramids or Louis XIV's palace at Versailles. It's rather scandalous, when human beings are going short of necessities, to do this."[41]

Whether poor or elite, fiscally grounded criticisms of *Apollo* did not stray far from public opinion. A 1967 Gallup poll found that more than half of Americans—fully 60 percent—did not think it was important to send a man to the moon before the Soviets. Similarly, when asked in the lead-up to *Apollo* whether the United States should set aside money for a manned Mars expedition, a majority replied "No."[42]

Apollo's substantial cost contributed to a sober ambivalence that crept through the pages of editorials across the world in the weeks and months after *Apollo*, particularly the feeling that the mission ultimately did as much to bring human character into doubt as it did to elevate it. Indeed, as a philosophical exercise *Apollo* was a study in contrasts. It was impossible to escape the notion, as Dean Rusk put it, that from their vantage point on the moon the astronauts had "seen us as we really are." When they returned home Aldrin, Armstrong, and Collins would find both "brotherhood and armed conflict, tolerance and blind prejudice, sacrificial service and narrow selfishness, generosity and greed, luxury alongside of demeaning poverty, soaring hope and abject despair."[43] For columnist Flora Lewis, the central meaning of *Apollo* was the gap between "the capacity of man and our deficiencies." As "It is heaven and earth," she wrote in the *Los Angeles Times*, "inspiration and meanness. It is the challenge of hope and the degradation of fear. It is ingenious courage and niggling suspicion."[44] The German-born sociologist Amitai Etzioni echoed many when he railed against the disconnect between the space program's "investment in things" and the need for "investment in people"; the "glory of rocket-powered jumps" and the shortage of "critical self-examination."[45]

Similar views could be heard in countries all over the globe. British journalist George Lichtheim reported that the rhetoric of *Apollo* "fell on deaf and uncompromising ears" in his native country. The ravages of the Second World War, the uncertainty of Cold War defenses, and an increasingly unstable economy had made the British "temporarily disposed to believe that glowing promises of a better future are never likely to materialize."[46] The *Times of India* argued with resentment that humanity had strove to reach the moon simply because it was "there."[47] Commentators from the Eastern bloc drew a contrast between what US leaders promised through space exploration and the actions of their government. Surely, one East German newspaper remarked, the half-million American troops then stationed in South Vietnam could not claim to have come "in peace for all mankind." The indiscriminate violence exacted on the Vietnamese, civilians especially, marked the true nature of the United States in the world.[48]

"In Peace for All Mankind"

Indeed it was the United States' bloody, protracted war in Southeast Asia that most undermined the second major narrative NASA and US officials told about the first moon landing—that it represented the peaceful nature of the American people and the US space program. The coincidence of the war's major events with the unfolding of the *Apollo 11* mission provided a stark reminder that the promises of space travel remained ill-equipped to solve earthly problems. In the weeks and months surrounding the moon landing, US troop deployment had reached its peak at fully 540,000, and more than 33,000 had been killed, surpassing the Korean War's deadly toll. In April 1969 five-hundred Harvard students seized University Hall, threw out the deans, then locked themselves inside to protest the escalation. In May, forty-six men of the 101st Airborne Division died (around four hundred others were wounded) fighting for "Hamburger Hill" in the A Shau Valley near Hue, a strategically useless scrap of land that the army quickly abandoned once the firefight was over. That same month the *New York Times* broke the news of Nixon's secret bombing campaign in Cambodia. And, perhaps most damaging for the image of *Apollo 11*, the US Army charged Lieutenant William Calley, just over a month after the completion of the mission, with murder for his role at My Lai.

It was for the Vietnamese people, in particular, that the irenicism of the first moon landing was a bald lie. In 1968 Lyndon Johnson had seen fit to include Ho Chi Minh in his list of recipients for a special print of NASA's *Earthrise* photograph, an act that the North Vietnamese leader had appreciated; now, only a year later, even that scrap of goodwill had evaporated. For many Vietnamese, the savage war being fought in their villages and towns—and the often-indiscriminate violence with which that war was being waged—directly contradicted the notion that Americans visited foreign lands either "in peace" or "for all mankind." For one Hanoi-based journalist, *Apollo* and Nixon's global tour were merely a ploy to "cover up" the United States' "aggressive . . . designs" and to sell the president's "fake peace." His readership knew better. The moon landing was not some "magic wand" the United States could wave to achieve a reversal of its shattered image aboard. NASA had failed to divert "the attention of the world's people, who are constantly vigilant against US imperialists' criminal actions and plots on earth."[49]

It was easy to dismiss such sentiments as the propaganda of an embattled enemy. But the peacefulness of *Apollo* and the US space program generally began to come under wider scrutiny when journalists revealed that NASA had been aiding the war effort since at least 1965. Late that year, USAF General James Ferguson visited NASA to brief its Office of Defense Affairs on the operational difficulties the US military was then experiencing in Southeast Asia. The enemy's use of guerilla tactics and the nature of a limited war fought in thick jungles had made it almost impossible to find combatants, to land aircraft, and to achieve the element of surprise. Fighting an unfamiliar battle, they needed new ideas. NASA administrator James Webb enthusiastically offered his agency's "full technical support." His engineers would make every effort "to uncover those NASA solutions to problems, devices, or techniques, that might be of assistance to our forces in Southeast Asia." Within a month the space agency had established a "Limited Warfare Committee" to pool resources, manage personnel, and lobby laboratories all over the country for assistance.[50]

By the end of 1966 Webb had acquired the cooperation of every NASA center and assigned thirty-five of the agency's engineers to research for the war. The committee's budget (funneled through a NASA account labeled "Special Support Projects") swelled to $4 million annually by 1967 and its personnel had grown to one hundred scientists working in labs from MIT to Berkeley. Together NASA and the Pentagon collaborated on more than eighty-nine different projects for the war: an acoustical net to pinpoint the source of mortar fire; improved aircraft target markers; a beacon for locating downed fighter pilots; mountain-top and airborne radio relays; miniaturized circuitry for backpack radio sets (reducing the weight of equipment); steerable parachutes to drop men and supplies into otherwise unreachable forest clearings; and noiseless aircraft engines for approaching enemy troops without prior warning.

The most pertinent technologies attempted to improve US reconnaissance capabilities, particularly the ability of US troops to find and destroy enemy forces in the jungle after sundown. These included the retooling of NASA's Applications Technology Satellites (ATS) to provide real-time weather reports to US pilots stationed in the region and a proposal for a giant aluminized Mylar mirror—2,000 feet in diameter—that would reflect the sun's light onto jet-black areas of National Liberation Front–controlled territory.[51]

Notable among these projects was NASA's adaptation of its lunar seismometers for use in Operation Igloo White, a covert mission involving the use of seismology to create "an electronic battlefield" along the Ho Chi Minh trail in Laos and Cambodia between 1968 and 1973. NASA's engineers developed thousands of such Air-Delivered Seismic Intrusion Detectors, or ADSIDs, which, disguised as tropical vegetation, detected vibrations from passing military convoys and picked up voices from microphones. The sensors would immediately transmit these "hits" to USAF planes circling continuously overhead, which in turn relayed the data to the Infiltration Surveillance Center, then the largest building in Southeast Asia. There, two IBM 360-65 computers—identical to those being used in Houston for the Apollo program—quickly analyzed the data and, within five minutes, conveyed target coordinates to the closest available bomber.[52]

Though enthusiastic about the aid NASA might provide to the war effort, the agency's leaders were keen to keep it secret. To begin with, Webb and company relied on the territory, labor, and facilities of dozens of foreign nations, particularly in the Global South, to launch, track, and retrieve spacecraft, conduct experiments, relay communications, train new personnel, and of course, spread the gospel of US-led spaceflight around the world. If the rhetoric of the Space Act was contradicted by NASA's deeds, perhaps foreign governments would temper, or withdraw outright, their support. Moreover, agency authorization to engage in DoD projects was, as the *Washington Post* pointed out, "fuzzy at best." The Space Act was contradictory on the matter. On the one hand, it barred the agency from engaging in activities "peculiar to, or primarily associated with, with development of weapons systems, military operations, or the defense of the United States." That was the Pentagon's sole jurisdiction. On the other hand, just a few lines below this declaration, the legislation directed the agency to make available to the services any "discoveries that have military value or significance." NASA officials came to argue that, as long as the agency refrained from designing or developing weapons, military research could continue. "We are not developing anything that shoots a bullet or a missile at somebody," insisted one NASA representative. "I don't think anybody is so naïve that he might feel an agency spending $4 billion a year on technology shouldn't spend some of it trying to win a war we're fighting."[53]

Despite its best efforts, NASA could not extricate the space program either from its contemporary association with militarism or its roots in previous conflicts, a problem that journalist Rudy Abramson termed the "Apollo Paradox." All of the most significant engineering feats that had made it possible for *Eagle* to land on the moon, he pointed out, "were forced to happen before their time by the pressure of war." The army had stolen the United States' first space rockets after World War II from Germany, who had designed them to rain terror on European civilians. *Apollo* had trained its astronauts in flight tests of military aircraft. Thus, it was not a surprise when George Wald, a Harvard biochemist, reported that his students saw the moon mission as "an exercise in great wealth and power, heavy with political and military overtones." Many of his students, rather than inspired or proud, felt "a little more trapped, a little more disillusioned, a little more desperate."[54]

The increasing association between Project Apollo and violence was evidenced by a renewed interest in the culpability of Wernher von Braun, the German émigré with whom Americans most associated the success of the US space program. For Norman Mailer, von Braun embodied the price that modern societies had paid to achieve technologies such as the space rocket. His Nazi past, including the use of slave labor at Peenemünde, constituted the historical "essence" of American space exploration. Nazism and "NASAism," Mailer wrote, were quite similar: both were grand projects rooted in technological achievement, and both entailed the transcendence of vast new spaces for the revitalization of the nation.[55] When the Italian journalist Oriana Fallaci visited Marshall Space Flight Center in Huntsville, Alabama, to interview von Braun, she felt compelled to leave early upon smelling his lemon-scented detergent, the same used by German soldiers who had ransacked her home as a child during the war. Human spaceflight, Fallaci argued, could not be disentangled from von Braun and the Nazis. Theirs "was the story we would have to tell the Martians and Venusians when, filled with admiration, they watched us coming down in our spacecraft and asked us 'But how did you do it? How did it happen?'"[56]

Americans' identification of space exploration with war extended even to literature. In *Slaughterhouse Five* (1969), Kurt Vonnegut likened the desolate scenes of bombed-out Dresden to the moon no less than a dozen times.[57] The protagonist of Saul Bellow's *Mr. Sammler's Planet* (1970), a Holocaust survivor who rails against grand, sixties-era plans, is most skeptical about escapist fantasies of deep-sea and space exploration. "New worlds? Fresh beginnings?" he asks with scorn at the novel's outset. Not likely. Experience had led him to believe that wherever human beings traveled, they would bring their violent and selfish nature with them.[58] Three years later, Thomas Pynchon, Jr. won a National Book Award for *Gravity's Rainbow* (1973), a roving, complex novel centering loosely on the construction of the Germans' "vengeance weapon" (the title refers to the parabolic trajectory of the missile). Pynchon, whose authorship of the book had been coterminous with the Apollo missions, pitched the rocket not as a vehicle with which to reach the moon but rather a celebrated implement of war and corporatism.[59] As in Bellow's story, the violence of which human beings are capable cannot be extinguished, only transformed as they move to new places. As one literary scholar pointed out, *Gravity's Rainbow* drove home the fact that the United States' technological and economic supremacy derived from the "dark negotiations" that had brought Nazi scientists to America.[60]

A Lunar Empire?

If the Vietnam War and its reverberations in the wider culture sewed doubt in NASA's claims about the peacefulness of US space exploration, neither were critics inclined to believe Nixon's third assertion: that the first moon landing had inaugurated a postimperial epoch. The optics of the *Apollo 11* ceremonies presented one problem. The national flag raising on the moon closely resembled past images of conquering armies, including Theodore Roosevelt's rough riders on San Juan Hill in the Spanish-American War and the Marine Corps' victory against Japan at Iwo Jima toward the end of the Pacific War. Although the OST, signed just two years prior, forbade national sovereign claims to the moon, many commentators found it difficult to forget earlier calls to "Claim the Moon!" and the Air Force's warning that whoever controlled space would control the world.[61] As the *Apollo* astronauts brought the first moon rocks back to Earth, many speculated about the opening of new resource vistas to be exploited.

NASA's greatest defense against charges of imperialism was the fact that although Armstrong and Aldrin had erected their national flag on the moon, the act did not represent, as it had throughout the age of colonial exploration, a sovereign claim to new territory. The astronauts' erection of Old Glory merely represented national pride and "legally means nothing," USAF jurist Martin Menter explained.[62] Lunar rituals simply filled "an emotional need" and had "absolutely no practical, territorial—or lunatorial—significance," one space lawyer agreed. They were no more important than the Soviet Union's deposit of pennants bearing the Hammer and Sickle on the moon in 1959, when *Luna II* had impacted the lunar surface.[63] Dean Rusk, newly retired as secretary of state, considered the moon landing only the most recent iteration of American restraint exhibited despite awesome power. The United States, he wrote, though it had emerged from World War II as the world's most powerful country, "did not . . . exploit [its] unmatched power in imperialist adventure." The nation had demobilized rapidly, funneled millions of dollars to Europe and Japan for reconstruction, and rehabilitated its former enemies rather than exacting vengeance. A quarter-century later, notwithstanding the economic and technological supremacy that *Apollo* represented, the United States remained completely uninterested in "reaching out for more territory in the name of a doctrine of 'lebensraum.'"[64]

Mindful of *Apollo*'s symbolic significance, NASA had been cautious and deliberate in its planning for the first moon landing. Five months before the scheduled launch, the agency's acting administrator, Thomas O. Paine, appointed a special Committee on Symbolic Activities for the First Lunar Landing to be headed by Willis Shapley, then associate deputy administrator. Paine charged the committee with two principal goals: first, develop a set of symbolic acts that might "signalize" the moon landing as both a US accomplishment and one of the entire human race, and second, ensure that other nations did not perceive the United States to be "taking possession of the moon" in violation of the OST.[65] Shapley led deliberations on articles to be taken to the moon and returned to Earth, such as photographs, flags, and tokens; articles to be attached to the descent stage of the LM, such as plaques, documents, and microfilm; and, most sensitively, objects to be left on the moon.

As expected, the weightiest decision for the committee to make was which flag the astronauts would hoist once on the surface of the moon. Several options presented themselves. The most obvious choice was the US banner, which was already an accessory to the *Apollo* spacecraft, the *Saturn V*, and the spacesuits worn by the astronauts. NASA's leadership also discussed planting the American flag alongside the United Nations standard and placing decal flags of UN-member nations on the LM descent module. In its internal deliberations NASA attempted to toe a line between acknowledging the preeminent role the United States played in the mission while also preserving the cosmopolitanism with which it wrapped the *Apollo* program.⁶⁶ Undersecretary of State U. Alexis Johnson wrote to Paine urging his agency to privilege options that would "enhance our posture abroad and to encourage other countries to further identify their interests in the exploration of space with our own."⁶⁷ Because the moon flight "is being done in the name of all humanity," NASA's head agreed little more than a month before take-off, any ceremony should "reflect the philosophy that, while it is an American spaceship, it also is an undertaking of all mankind."⁶⁸

But Congress made its voice heard, rejecting the UN flag out of hand. Speaking for millions of Americans, Representative Robert Michel (R-IL) argued that planting any standard other than the national one would be "a travesty and a double cross" to the men NASA would fly to the moon. It would give the Soviet Union, China, and other adversaries "a full share in what has been solely a United States project." The United States had taken the risks, developed the necessary technology, and footed the bill; why shouldn't it take the credit? The United States maintained the United Nations financially, charged Michel's colleague Joe Evins (D-TN). "It is American technological skill and it is the daring of Americans that made the moon trip possible. We earned the right to proudly plant our flag on the moon not as conquest but as a peaceful venture. By what right has the U.N. flag earned such an honor![?]" The United States should "make no apology" for its triumphs in space, agreed Wallace F. Bennett (R-UT). The American flag did not suggest a claim for ownership over the moon but rather that "this great accomplishment came from a free republic and was supported by a people who believed in liberty, freedom, and justice."⁶⁹

These convictions turned into threats that a UN flag raising would put continued NASA funding in jeopardy. On June 10, Representative Richard Roudeboush (R-IN) proposed an amendment to the appropriations bill for fiscal year 1970 prohibiting NASA from deploying any other flag than the national one on missions paid for solely by the US government.⁷⁰ As a consolation, miniature flags of all the UN-member countries, the fifty US states, the District of Columbia, and US territories would be stowed in the LM and returned to the leaders of those governments upon the astronauts' return. In lieu of national flags, *Apollo 11* would leave behind on the moon a half-dollar-size silicon disc inscribed with the miniatured goodwill messages of more than seventy countries.

The decision by Congress to sidestep an international representation for the moon landing did not go unnoticed. On the contrary, it proved explosively controversial. Citizens across the country wrote to local and national news outlets demanding that *Apollo 11* project a more cosmopolitan face. One likened Congress's verdict on the flag to Nikita Khrushchev's "childlike glee" at having impacted the moon with the Soviet flag a decade earlier. "Americans ought to be more sophisticated," many agreed.[71] The most popular alternatives recommended were to fly, as some NASA engineers had originally proposed, the United Nations flag, perhaps the flags of every nation in the world, or even the banner of some international scientific group such as the National Geographic Society. *New York Times* correspondent Max Frankel was particularly rueful about the political implications of striking the Stars and Stripes into the moon. It proved that "the moon will be no more exempt from mundane strife than the New World has been these last 500 years. The rivalries of nations will spill into the ocean of space just as surely as they spilled into the Atlantic in Columbus's wake." Frankel concluded, "If there were an earth flag flying on the moon today, the men up there would obviously be tempted to label it the banner of a race of lunatics."[72]

Sure enough, Michel, Evins, and the other senators had defenders as well. Many Americans considered the lunar ceremonies an ideal opportunity to contrast the open, scientific US space program with the secret Soviet effort hidden away in a vast military bureaucracy. In his pledge to send men to the moon Kennedy had predicted that the world would witness outer space governed "not . . . by a hostile flag of conquest, but by a banner of freedom and peace," and it was widely understood in the lead-up to *Apollo 11* that only the American flag could justifiably represent that freedom and peace. In any case, many agreed with the idea that because *Apollo* was "an extension of the exploring and enterprising spirit that built America," none other than the US flag was a fair choice: "No apologies necessary." Aldrin recalled in his memoirs the "sudden rush of pride for the country" after he and Armstrong completed the flag-raising ceremony. In October singer Judy Carne appeared on the *Ed Sullivan Show* to perform a new ballad called "American Moon," in which she insisted "Apollo Eleven delivered our heavenly right to say, 'The man in the moon is a citizen of the USA. Stand up and brag for your grand old flag waving on the moon tonight.'"[73]

Because the United States had forsaken any claim to the moon through its ratification of the OST, this kind of jingoism did not translate to accusations of an American desire to conquer the Earth's natural satellite; after 1967, that prospect simply failed to hold much water. But political and legal conversations about US imperial designs on the moon refused to die out. For one, the language of exploration—"conquest" of gravity and space, expansion into vast new "frontiers"—alluded to an imperial age that, while coming to an end, still lingered in many corners of the globe. Space boosters continued to speak in colonial tones about the moon in particular, an "Africa-sized world" teeming with rich mineral resources.[74] Representatives at NASA and leaders in government emphasized that the material wealth of the lunar soil could be dug up without negative implications for the native population—there was none.

The moon's intrinsic value was undisputed. From the colonial period, US scientists had pointed to it as an ideal astrological observatory; operating outside the Earth's dense atmosphere, telescopes could penetrate deeper into the void.[75] As we've seen, in the wake of *Sputnik* the Air Force developed plans and theories about the military utility of a lunar base. Homer Boushey, Thomas Power, and other generals were convinced that such bases would prove "impregnable" from Earth-based military capabilities and thus serve as a powerful deterrent against Soviet aggression. Beginning in the mid-1960s, moreover, science writers wrote with increasing frequency about the moon's rich deposits of silicon, magnesium, titanium, aluminum, calcium, cobalt, nickel, and iron.[76] It also possessed stores of carbon, nitrogen, and helium-3. In April 1964 Hugh Dryden announced that "Geologically, we have no reason to doubt that the moon and the nearby planets, being solid bodies, may be rich in rare mineral resources, possibly offering economic returns far outweighing the costs of exploration."[77] By the time of *Apollo 11*, *New York Times* correspondent C. L. Sulzberger saw fit to dub the moon an "eighth continent," one replete with the resources that would usher the world into a new generation of prosperity.[78]

As one might expect, the development of US lunar mining interests was concomitant with the moon mission itself. As NASA shifted its manned spaceflight program from orbital to lunar operations, the United States and many of the nation's largest technology firms devoted increasing attention to how the mineral wealth of the moon might be extracted and processed. Beginning in 1962 NASA annually convened a Working Group on Extraterrestrial Resources (WGER) to "evaluate the feasibility and usefulness" of extraterrestrial minerals, a goal aimed at reducing the dependence of space exploration on Earth-based materials. The group consisted not only of NASA officials but representatives from the JPL, the Rand Corporation, the Air Force, the Office of Engineers of the US Army, and the US Bureau of Mines (USBM). WGER's special Committee on Lunar Mining and Processing was charged with establishing the quantity and quality of mineral deposits, syncing mining operations with available logistical capabilities and the lunar environment, and identifying areas of technology in which further research was necessary.[79]

Toward those ends, through the 1960s and early 1970s WGER's members submitted proposals for lunar construction, remote detection, life support, food synthesis, vacuum detonations, lunar mapping, transportation, luminescence techniques, power generation, and countless other topics relevant to prying commodities from the lunar soil. As historian Megan Black has pointed out, the USBM was an especially important player in the WGER's constellation of mining interests. Beginning in 1965, the Bureau's scientists, working at seven locations from Reno to College Park, began developing methods to extract water and air from moon rocks, melt and weld lunar materials, build underground shelters for lunar miners, and operate conveyor belts and drills in extreme conditions.[80] Some USBM scientists believed that rocket fuel could be found on the moon in the form of acetylene, trapped in the rocks of ancient volcanoes. The possibilities, the *Washington Post* wrote of the Bureau's work, appeared "limitless."[81]

Of particular importance for extracting and processing lunar resources was drilling, an expertise that USBM's corporate partners were happy to supply. Westinghouse's Defense and Space Division, under contract with NASA's Marshall Space Flight Center, developed a diamond rotary drill equipped with a retrievable wireline innertube assembly "patterned after the system in common use for Earth exploration." Northrop Space Laboratories proposed a down-the-hole, gas-operated percussive drill. Martin-Marietta gave NASA plans for its Apollo Lunar Surface Drill (ALSD) scheduled for use during the *Apollo 13* mission in 1970. When full-scale excavation became a reality, USBM scientist R. L. Schmidt reported to the WGER, the technology acquired from exploratory drills would become the foundation of "production drills . . . that will serve the same function on the Moon as they do on Earth."[82]

In painting their "cornucopian fantasies" about the moon's mineral riches, NASA and WGER maintained that lunar resource cultivation was intended exclusively for further operations in space, for the cost of lugging minerals back to Earth would more than counteract the value of products retrieved from the mining operation. "Even if we found pure platinum on the moon," said one USBM research director, "it would cost too much to bring it back." Another said that even the discovery of "pure diamonds" on the moon would fail to justify the prodigious costs of cis-lunar transportation. Implied in these observations, of course, was the assertion that the United States had little intention of mining resources on the moon for profit, trade, or use in earthly industries. US utilization of lunar resources would, according to NASA, resemble the economical means by which native Plains Indians had lived off the North American bison: future astronauts would recycle every mineral, gas, and water supply to further the quest of discovery on which *Apollo* had set humankind. Mining would be practical, prudent, and advanced in the interests of world scientific knowledge.[83]

But broader ambitions revealed themselves when, early in 1971, the Soviet Union proposed a new international treaty to the United Nations COPUOS, one pertaining specifically to the moon (Argentina had submitted its own proposal the year before, but it went largely unnoticed). "Desiring to prevent the Moon from becoming a scene of international conflict," the draft restated many of the OST's provisions and applied them with renewed legal force to the moon. It barred any state from claiming sovereignty over any part of the moon; banned the testing of "any type of weapon" as well as the construction of military bases, installations, and fortifications (with new language prohibiting the threat or use of "any . . . hostile action"); outlawed the orbiting of nuclear weapons around the moon and their installation on or below the moon's surface; declared the universal freedom to explore the moon; and aligned the conduct of states on the moon and in cis-lunar space with the UN Charter.

Other provisions were decidedly more ambitious and hence controversial: Article 8 of the draft (its text would become Article 11 of the final treaty) declared that neither states, international organizations, nongovernmental entities, nor individuals could claim "the surface or subsoil of the Moon as their property." Any objects constructed on the moon, it added, "shall not create a right of ownership" over any part of the moon. Crucially, it also decreed that "portions of the surface or subsoil of the Moon may not be the object of concession, exchange, transfer, sale or purchase, lease, hire, gift, or any other arrangements or transactions with or without competition."[84]

US officials exhibited a measure of support, proposing additional provisions, amendments, and eventually its own draft.[85] The State Department was certainly interested in measures that would reinforce and build on the stipulations outlined in the OST. Yet as the draft treaty was debated and amended over the course of the 1970s, it became increasingly clear that the United States was unprepared to accept the more prohibitory provisions regarding the extraction and use of moon resources. By the time the United Nations formally adopted the Moon Treaty in December 1979, US officials and business interests had rallied in sufficient numbers to invalidate the agreement and keep lunar resource vistas open.[86]

Foremost among the Moon Treaty's opponents was the L-5 Society, a coalition of spaceflight enthusiasts, science-fiction authors, and aerospace corporations that sought to make the colonization of space a reality. The organization drew its name from the Lagrangian or "L" points (see chapter 4) that allowed for large objects to remain stable between the gravitational pulls of two celestial bodies. The Society's founders, Carolyn Meinel and her husband Keith Henson, were inspired by Princeton physicist Gerard K. O'Neill, who published an influential article on space habitation in September 1974. After O'Neill invited them to a conference at Princeton the following year, the couple founded the Society to raise money for and spread awareness of the physicist's ideas.[87]

L-5 members argued that the treaty's provisions regarding the equitable sharing of moon resources, particularly along lines that would benefit "the developing world," would squash any incentive to invest in space research among the industrialized nations. "No one from Rockwell International or Boeing is going to manufacture moon-mining equipment when they know that control and profit from such technology will be shared with countries such as Sri Lanka," implored one space booster. Capitalism was the only force capable of enticing rich nations like the United States to invest the resources necessary to colonize the harsh space environment. If such nations perceived that returns on their investments would be redistributed by "an international socialist regime," efforts to fulfill O'Neill's dreams would never get off the ground. To many technology firms especially, the Moon Treaty was a thinly veiled attempt to close off space from corporate development, to fundamentally reallocate the industrial world's wealth to developing nations and to ensure that the power to explore and develop space remained vested only in national governments.[88]

The truth was more complicated. Whereas many in the West interpreted the principle of common heritage as establishing the moon as *res nullius*, a no-man's land capable of being expropriated if not claimed, representatives from developing nations considered the clause to mean that the moon was *res communis humanitatis*, the common property of all people. Such an interpretation had played out in the 1970 UN

Convention on the Law of the Sea (UNCLOS), in which the convened nations agreed that the natural resources of the seabed were the "common heritage of mankind" and therefore ineligible for extraction. "The commonness of the 'common heritage'" clause, insisted Ambassador M. C. W. Pinto of Sri Lanka, "is a commonness of ownership and benefit. The minerals are owned in common by your country, and by all the rest as well. In their original location, these resources belong in undivided and indivisible share to your country and to mine, and all to the rest to mankind, whether organized States or not. If you touch the nodules at the bottom of the sea, you touch my property. If you take them away, you take away my property."[89]

Leaders of the developing world applied the same principle to the moon. Until nations could agree on an arrangement whereby all countries would be able to "directly participate or benefit" from the utilization of space," they argued, the United Nations should issue a strict moratorium "on any development whatsoever."[90] As with the negotiation of the OST, proponents of the Moon Treaty felt a sense of urgency in attempting to codify "sound rules" for lunar exploitation; it would be "far easier now," the *New York Times* noted, when the moon still seemed distant, than "when the journey is easier and when powerful nations will see feasible means of exploiting the moon economically and perhaps even militarily."[91]

The L-5 Society leveraged the sentiments of poorer nations in their assault on the Moon Treaty as it was debated in the Senate over the summer of 1980. The Society retained Leigh Ratiner to make its case before the Senate Foreign Relations Committee. The United States, the attorney argued, should not "buy a pig in a poke" by rushing to ratify the agreement. First, the government should seek to make clear exactly what "common heritage" implied in space. For Ratiner the term seemed to connote the redistribution of resources culled by the industrial West for the benefit of the developing world. Exploiting space for the benefit of all implied that "certain parts of mankind should be severely restricted in their access to the resources—particularly the industrialized countries, and more particularly the United States of America." What company, Ratiner asked, would invest the billions of dollars necessary to establish mining operations on the moon or on an asteroid before agreement was reached as to the rules and regulations governing the extraction of resources? At this early stage, the United States could still withdraw from the treaty without being the "bad boys" at the United Nations; the government should assemble an international vote for a protocol on the "common heritage" principle.[92]

Of all the L-5 Society's allies—which included personalities ranging from Timothy Leary to Barry Goldwater—the most important in its campaign against the Moon Treaty were US tech firms, who reasonably concluded that the agreement threatened to sink their efforts at commercial development on the moon. Alexander Haig, who Reagan would soon nominate as secretary of state, lobbied the government as president of United Technologies Corporation. He wrote to the State Department's legal adviser that the common heritage concept was clearly designed to achieve a redistribution of global wealth. In forwarding *res communis humanitatis*, Haig argued, the developing countries had "indicated they intend to gain control of critical raw materials and to gain access as a matter of right to the technology needed to exploit them."[93] Many of these technologies, other firms noted, were identical to important military hardware—transfer of such technologies to hostile or nonaligned nations in the Global South would be extremely dangerous.

With these considerations flowing freely through congressional hearings and the press, criticism of the treaty quickly became vociferous. Charles Sheffield, president of the American Astronautical Society predicted that ratification would be "the single worst mistake in United States diplomatic history." Future generations would regard it as more important than the Louisiana Purchase or the acquisition of Alaska, yet in this case "the US will be the loser, not the gainer."[94] Permitting the developing world to dictate the pace and quality of space exploration, one editorial emphasized, made "about as much sense as fish setting conditions under which amphibians could colonize the land." The editorial continued, hyperbolically: "If space resources are developed under this Treaty, the inhabitants of space may have no place to work to save the 'Regime,' no place to spend their wages except the Company Store, and no place to live but the Barracks."[95]

This offensive generated significant results. Although State Department officials attempted to shirk the perceived threat to free enterprise—"You can still make a buck off the moon, if there's a buck to be made," one aide said—ratification of the treaty was in serious jeopardy. Letters from agitated spaceflight enthusiasts poured into legislative offices. L-5 lobbyists were seen "prowling" the halls of Congress. In the House, several representatives began circulating a resolution against the treaty.[96] Pressure from aerospace firms compelled Senators Jacob Javits (R-NY) and Frank Church (D-ID) to write a letter to Secretary of State Cyrus Vance urging restraint. The treaty's provisions on resource extraction, they insisted, would benefit the Soviet Union by erecting barriers to economic development. Because the United States relied on industrial and business interests to engage in exploration, the USSR was free to exploit space under the guise of scientific investigation. Its space program had been state led from the very beginning. "Seen from a long-term geopolitical perspective," wrote Javits and Church, "we believe this outcome could be damaging to fundamental American security requirements."[97] By December a third of the Foreign Relations Committee had committed in writing to oppose the treaty.[98]

The agreement was finally trounced for good when Reagan ascended to the presidency in January 1981. The new administration made clear its opposition to the treaty on the same grounds outlined by Ratiner, Haig, and scores of other treaty opponents.[99] Even though the Moon Treaty had been passed unanimously in the UNGA, without a vote, it was never sent to the Congress for ratification. Only eighteen countries have ever signed the agreement; only seven have ratified. The Soviet Union, China, and Europe (with the exception of Austria and France) have neglected to acknowledge it as well.

The defeat of the Moon Treaty convinced many observers that although US aspirations on the moon had little to do with territory, they nevertheless contained imperial dimensions. United Technologies, the Martin Company, Boeing, and other aerospace firms could applaud the political insignificance of the moon landing, for they perceived the window for lunar development to have remained open. After all, had the United States needed to annex Guatemala, Costa Rica, or El Salvador for the United Fruit Company to establish banana-growing operations? Did ARAMCO require a territorial empire to extract Saudi oil? In the late 1970s and early 1980s, US leaders surmised that the Moon Treaty would spell disaster for US corporatism in space. According to their critics, they had defeated the agreement to protect not a territorial empire but a corporate empire, a technological empire, and the rest of what historian William Appleman Williams called "informal" imperialism. Amid the wave of decolonization washing the Global South since the 1950s, territorial control had become politically dubious, retrograde, even unnecessary.[100] To those calling for Armstrong and Aldrin to claim the moon, there came the easy question: What for?

The Shape of Things to Come

As a symbolic act, the first moon landing represented the fulfillment of the interplanetary dream that science-fiction writers, philosophers, engineers, and political leaders had sustained since before the publication of

The War of the Worlds in 1898. The *Apollo 11* mission created a scene markedly different from Wells's dystopia. Outer space seemed to have become a source of human unification, cooperation, and peace rather than cataclysmic war based in difference and a scarcity of resources. Although the astronauts had erected an American flag in the lunar surface, the act did not represent a claim to it. To "conquer" space, as the language of the time expressed it, meant nothing more than scientific discovery and a test of humanity's limits. For decades figures like Arthur C. Clarke had promised that space travel would facilitate the coming together of nations on Earth. For a brief moment, many believed those prophecies to have come true.

The reality, as others pointed out before the mission had even begun, was more complicated. Political and racial divisions at home—over the war in Southeast Asia, poverty, and social injustice—belied the astronauts' observations about the oneness of the Earth's inhabitants. Minority communities in the United States and non-white peoples across the Global South were particularly critical of and suspicious about the prevailing discourse. For all the high talk about leveraging NASA technologies to revive crumbling inner cities and poor countries around the world, to "spin off" the astronauts tools' to the benefit of those most in need, *Apollo* produced only embarrassing contrasts between glittering rockets and asbestos-infested tenements, liquid-hydrogen fuel and animal-driven ploughs. Nor did Project Apollo convince observers that US space exploration was a peaceful enterprise. The rising death toll in Vietnam, particularly among civilians, negated Nixon's claim about *Apollo* having brought the world's people together in peace. NASA's development of defense technologies, which at times derived directly from the Apollo program, further undermined the notion that the moon mission would usher in an era of international goodwill.

Not even the moon proved immune from conflict. For those whom the Interplanetary School had won over since *Sputnik*, this was among the most disappointing developments. *Apollo* had failed to convince people that the United States had made the moon safe from political rivalries back on Earth. "The diplomats have thus far thought of the Moon as a harmless wasteland that can be domesticated as easily as Antarctica has been," wrote one columnist five months after the first moon landing, "but that is a rash assumption."[101] It seemed fitting when during the lead-up to the launch, a Jordanian rebel asked the astronauts to paint the moon black so that it would not illuminate raids on Israel. By the mid-1960s US aerospace firms and mining interests had made the seemingly benign question of "what is the Moon made of?" a profoundly political one. As the astronauts returned to Earth with the first lunar samples, the prospect of bringing extraterrestrial resources home for profit or for building enterprises in space had become realistic and controversial possibilities. Commentators again began to speak of the moon as "a strategic way station" to yet more celestial riches, a steppingstone to a twenty-first-century China Market.[102] Despite the precedents set by the Antarctic Treaty, the ongoing UNCLOS conventions, a procession of UN resolutions, and the absolutist language of the space treaty, the question of whether one nation or another could claim lunar rights refused to go away. As they had since the late 1950s, most space lawyers brooked no contrarians on the matter. Situated deeply in the moral and political arguments against national sovereignty, they continued, long after the ratification of the OST, to produce books, articles, and pamphlets about why the moon *should* be exempted from historic claims. Yet whether it *could* was a subject of interest as long as the legal minutia of space exploration were up for debate and for as long as lunar possessions seduced governments with new economic horizons.

As the negotiation of the OST had been, the moon landing was an awakening, from dreams and nightmares alike. On the one hand, *Apollo 11* demonstrated that US officials responsible for national space policy had roused to the immediate dangers entailed in transplanting Cold War rivalries to the cosmos. The ceremonies at Tranquility Base constituted an important milestone in the political and cultural construction of outer space as a zone apart from earthborn conflicts. It is notable that Kennedy's goal of landing men on the moon before the end of the 1960s superseded every military space project of the decade, not only in the public eye but in terms of government expenditure and political capital. Whereas at the end of the 1950s the United States was studying the possibility of testing nuclear weapons on the moon and establishing a lunar military base, by the end of the following decade *Apollo* had reinforced the idea that the nation would exempt the moon, as well as subsequent cosmic bodies on which it might land, from interstate conflict, sovereign claims to new colonies, and the accompanying forces that had driven European powers to Asia and the New World in previous thrusts of "exploration."

On the other hand, zealous promoters of a sanctuary in space awoke as well. They discovered that the realities of Cold War competition were more immune to the metamorphic influence of space exploration than they had initially imagined; that differences in wealth and power between spacefaring nations and those stuck on the ground had rendered the Interplanetary Project grossly incomplete; and that despite the triumph of *Apollo*, there remained formidable limitations for exploration to reform human hearts and minds.

This tension between symbol and politics, between myth and reality, and between dreams and nightmares was the shape of things to come. Certainly no one expected that the *Apollo 11* crew would, overnight, inspire the world to halt the Cold War at the edge of space or inaugurate a new, warless era in international affairs. But looking into the 1970s and beyond, it seemed that the cosmos would continue to be subject to the whims of world history. Of the many ironies that characterized the first moon landing, perhaps the most consequential was that at this zenith of the space race, it remained unclear whether the Interplanetary Project had succeeded in achieving its basic aims. Indeed, it was an uncomfortable possibility that because *Apollo 11* was a pinnacle, there was nowhere left to go but down.

Conclusion

Among the millions who witnessed the first moon landing on television, there sat one especially interested observer. In July 1969 James Mangan still reigned as ruler of Celestia. He was elated. Initially wary of trespassers, he had issued a twenty-year lease to the US government for space exploration and accorded special privileges to the lunar mission. Two decades after he claimed the cosmos in the name of peace, Mangan issued passports to the crew of *Apollo 8* and, with *Earthrise* beamed back home, felt that "my job is done." The transcendent symbolism of Tranquility Base seven months later seemed to prove his judgment correct.[1]

Mangan was lucky to behold the moment. He died less than a year later in Oak Lawn, Illinois, where he had lived his entire life.[2] Celestia passed to his family. His son and daughter became prince and princess of the nation. His grandchildren inherited galaxies. One of them, Dean Stump (Duke of the Milky Way), later wrote that Celestia's birth after World War II had not been a coincidence. The totalitarian movements of the 1930s had "left much of the world in [a] state of exhaustion, despair, and exploitation." To overcome these afflictions of the spirit, Mangan had founded Celestia on one philosophical principle: "Magnanimity." He believed, Stump remembered, "that the promotion of peace is essential," and that perhaps the world needed a "State" of magnanimity to serve as a model.[3] "By achieving bigness of mind," Mangan had insisted, "the people of the world can find lasting peace."[4] Reaching out into space was merely the clearest path to achieving those new mental horizons.

James Mangan's life spans precisely the years covered in this book. His birth in 1896 preceded by mere months the popularization of interplanetary fiction that *The War of the Worlds* and *Auf Zwei Planeten* brought about. His Chicagoan adolescence paralleled the maturation of rocketry and, with it, interplanetary thought in Europe, Russia, and the United States. The Western Hemisphere's first planetarium, Adler, opened in 1930 just miles from Mangan's home. Celestia's founding in 1949 occurred just as the cosmic philosophies of Arthur Clarke and Olaf Stapledon found purchase in postwar Britain. Throughout the 1950s and 1960s, as he scurried about protesting the incursions of Soviet cosmonauts and satellites, US officials set about creating the legal and political architecture that made up what James Clay Moltz calls "strategic restraint": an emphasis on diplomacy and international law; pursuit of arms control; self-discipline in the development of military space technology; civilian control of the space program; and a public commitment to use space "exclusively for peaceful purposes."[5] In other words, a space sanctuary.

For all the successes that posture brought about, Mangan's dream of a "bulwark of international peace" in space, one that might transcend the Cold War, died with him. As policy, strategic restraint lasted just a dozen years. Ambivalence toward *Apollo 11* was but one of many omens signaling that the idea of a space sanctuary—and its power to transform international relations back on Earth—had peaked too soon. Indeed, if the moon landing was a climax for the space-for-peace consensus, it was also, as the term "climax" implies, the beginning of a swift and inexorable decline.

Consider the strange circumstances by which notions of a space sanctuary entered the vernacular. In the decade after *Sputnik* very few people had actually used the word "sanctuary" to describe their hopes for space. Instead, they used synonyms. Lyndon Johnson saw space as an "ocean of peace." Arthur Schlesinger, Jr. called it a "zone of peace." The *Times of India* alluded to a "Warless World." And so on. Though commentators chose disparate terms and phrases, the idea pulsed with vitality. There was the feeling, as late as 1969, that "the cold war has been left behind as man reaches toward the moon and, eventually, the planets." Without using the magic word, Johnson had captured the essential element of a sanctuary when he signed the Outer Space Treaty:

the goal was "to enlarge the perimeters of peace by shrinking the arenas of potential conflict."[6]

Ironically, the term "space sanctuary" came into popular use only during the 1970s, when Americans began to call its very existence into doubt. As satellites became part of the warp and woof of military operations on both sides of the Iron Curtain, journalists, pundits, and military leaders confidently announced the death of the sanctuary era. In 1976 a special panel commissioned by Gerald Ford concluded that "treating space as a sanctuary [was] neither enforceable nor verifiable." In his administration's final defense authorization, Ford's Defense Secretary Donald Rumsfeld argued that space could not remain a "relative sanctuary" because space technology had become powerful enough to decide the outcome of future wars. In fact, wrote an engineer at Lockheed, to treat space as a sanctuary was a "genocidal hoax."[7]

Why did the space-for-peace regime self-destruct in the 1970s, and why so abruptly? The answers lay in the nooks of conservatism's revival in US politics; the collapse of détente; structural shifts in national security policy during the late Cold War; and, of course, the changing nature of space technology itself. To relay all the sordid details would require another book entirely, but the reader has slogged this far and deserves at least an orientation to the variables at play.[8]

Perhaps most consequential was the decline of manned spaceflight itself, to which notions of human rebirth in the cosmos were so obviously and intimately attached. From the very beginning, human space travel had been linked to the transference of civilization from war, overpopulation, and ignorance to peace, material abundance, and moral clarity. For nearly one hundred years the narrative power of exploration to effect change dictated that the realm be sheltered from violence and political competition. But beginning in the 1970s, space travel did not seem the urgent national, philosophical, or metaphysical project it did only a decade before.[9] The Soviets had been beaten in the prestige race to the moon—NASA struggled to answer what would come next. The agency forwarded the Space Shuttle as a reusable transportation system to boost heavy loads into orbit at low cost, but operational expenses far outstripped initial expectations, undermining public trust and casting the overall US space mission into doubt.[10]

Suddenly, the human future no longer seemed to be taking place in space. The growing environmental movement argued throughout the post-*Apollo* period that "there is no Planet B"; all that remained was stewardship of Spaceship Earth. The overwrought national budget, the war in Vietnam, and racial unrest at home forced Nixon to slash NASA funding. A 1969 Harris poll showed that most Americans thought the Apollo program had been too costly to begin with.[11] In any case, the development of robotics had outstripped crewed spaceflight in the 1960s and appeared after *Apollo* to make human missions both unnecessary and irrationally expensive by comparison. Over the course of the 1970s, despite severe budget shortfalls, NASA's *Mariner*, *Viking*, and *Voyager* spacecraft succeeded in flying by, imaging, mapping, and in some cases landing on nearby planets. Humans, by contrast, made their final voyage to the moon in December 1972—and have yet to return.[12]

Drifting further and further into the human future, space increasingly came to be understood in a narrower and more immediate sense—that is, as a platform from which to gain advantage on Earth. Because interplanetary theorists had predicted that human beings would evolve morally, socially, and politically as they moved through and built new civilizations in space, it was a logical corollary that because space travel was slowing down, the promised revolution in human psychology and spirit would have to be postponed as well. All that was left to do was hunker down and begin reconciling the new technologies with old political forms.

Arming for the Space Age

In the wake of the OST, it appeared for a time that military competition in space was winding down. In June 1969 the DoD cancelled its massive MOL intended to ascertain what military role humans would play in space. In all, the Pentagon scaled back its military space operations by more than 35 percent between 1969 and 1975, from $6 billion to $3.8 billion. In 1971 the Soviet Union, too, suspended its testing of the FOBS weapons that had so terrified US defense planners in the previous decade. Richard Nixon and Leonid Brezhnev achieved further momentum the following year with SALT I, in which both sides agreed to prohibit interference with satellites that helped to monitor arms control accords. The future seemed bright.[13]

But as with détente generally, these developments proved illusory. The superpowers' increased dependence on satellites for military operations dictated that they become legitimate objects of conflict. The first sign of trouble came late in 1975, when the Soviets resumed ASAT testing. Over the next six years the Soviet Union would go on to conduct thirteen total ASAT tests. Accordingly, US spending on satellite survivability ballooned from $19 million in 1978 to more than $33 million three years later.[14]

The reinitiation of Soviet ASAT tests was jarring enough to wake the dormant US effort. At the end of the Ford administration, ASAT capabilities, satellite surveillance, and satellite survivability were grouped together under one umbrella: the Space Defense program, for which US spending increased from a measly $100,000 in 1974 to more than $41 million four years later.[15] In 1977, the Air Force awarded a $58.7 million contract to Vought to begin development of the Pentagon's most robust new ASAT system, the miniature homing vehicle (MHV). Added to this program were development projects using modified ABM weapons and ICBMs, particle beams and lasers, and electronic jamming.[16]

In conjunction with his authorization of ASAT weapons, Jimmy Carter, as part of a two-track foreign policy, initiated bilateral talks with the Soviets to reign in the new arms race. Three rounds of ASAT talks began in June 1978, but they were doomed from the start. Negotiators faced all the old disagreements about verification, as well as continued fears about technical asymmetry. US officials could not countenance negotiating in an area of weapons technology in which they were already behind. The thorniest issue was that arms controls for ASATs implied massive oversight of dozens of military systems with only ancillary relevance to satellites, for seemingly every weapon was deemed capable of adaptation to the ASAT role. So-called residual satellite-killers lurked in the ABM programs of both countries; in their SLBMs, IRBMs, and ICBMs; in air-launched capabilities, directed-energy research, radio transmission, and rapidly accelerating satellite technology itself. On the surface, it appeared that no combination of measures possessed the capacity to curb all the relevant projects, at least none that either side was remotely willing to accept.[17]

Failure to reach agreement on ASAT weapons was, perhaps, attributable less to any specific issue than the general deterioration of diplomatic relations between the United States and the Soviet Union in the late 1970s. The relative stability that had earlier prevailed in bilateral space politics succumbed to the collapse of détente late in 1979. In December the Soviet Union began its ill-fated invasion of Afghanistan, an act that prompted an exhausted Carter not only to boycott the Moscow Olympics in 1980 but request the Senate to postpone ratification of the SALT II agreement, which contained articles bearing on ASAT ordnance. Ronald Reagan's triumph in that year's presidential election signaled the dawn of a more hardline approach to the Soviet Union. It could only follow that space policy trace this reorientation.

To the High Ground

Indeed, the Reagan administration thought of outer space differently from its predecessors. Space, not merely a frontier from which to display US competence in technology, organization, and science, was a platform to project power. It was now incumbent on the nation's leaders to maintain capabilities that would ensure free access to space, deny the enemy use of space during wartime, and "apply military force from space if that becomes necessary," as one defense planning document phrased it.[18]

Not content to pursue agreements for image-making, the United States would now take a hard-headed, realist approach toward arms control in space. If it determined that limits to existing or future systems were not in the interests of national security, the new administration would simply avoid talks aimed at those limits. Eugene Rostow, whom Reagan plucked from Yale Law School to head the ACDA, explained the matter in Congress: though it was a "noble aspiration" to preserve outer space as a "sanctuary from the conflicts that beset us here on earth," it did not reflect facts on the ground. "Outer space," Rostow testified, "is not a distant place way out there." Missiles passed through it, and the orbits of satellites enabled them to pass over nearly any point on the globe. "We cannot regard space as a place totally apart—it is an inextricable element of our national security concerns."[19]

These determinations quickly manifested themselves in policy. In October 1981, Reagan announced the start of what would be massive support for the air-launched ASAT system begun under Carter, dealing Vought and Boeing contracts worth nearly $419 million for R&D. A year later the Air Force formed a distinct Space Command to consolidate all space launch and missile warning operations under one roof and leverage space for military operations on Earth. Reagan's National Security Decision Directive 42 (NSDD-42), issued that July, established a senior interagency group on space to be chaired by the national security adviser. At an Independence Day address celebrating the return of the fourth Space Shuttle launch, Reagan expressed confidence that aside from establishing "a more permanent presence in space," his administration would begin "strengthening our own security by exploring new methods of using space as a means of maintaining the peace."[20]

This posture found its purest expression in the Strategic Defense Initiative (SDI), Reagan's proposal for a space-based ballistic missile defense (BMD) system. Reagan's goal for SDI was to replace mutually assured destruction (what the president referred to as "a suicide pact") with "assured survival," to eliminate US vulnerability to nuclear weapons once and for all. By constructing a vast network of sensors, guidance systems, and high-energy lasers, the United States could retake the initiative in the Cold War and perhaps render nuclear weapons "impotent and obsolete." Drawing from his career in acting, Reagan pitched Star Wars as if in a television commercial. "What if free people could live secure in the knowledge that their security did not rest upon the threat of instant U.S. retaliation to deter a Soviet attack, that we could intercept and destroy strategic ballistic missiles before they reached our own soil or that of our allies?" he asked in a now-(in)famous March 1983 speech on national defense. After consulting with technologists and military leaders in his administration, he reported, "I believe there is a way."[21]

SDI turned the logic of a space sanctuary on its head. Whereas Eisenhower, Kennedy, and Johnson had placed their faith in mutual restraint in space weapons, Reagan placed his own confidence in space weapons themselves. Whereas as SDI's opponents considered outer space a pristine new realm in which to transcend the Cold War, its supporters viewed it as a battlefield upon which the United States could finally win it. Cooperation in space technology, a hallmark of the sanctuary vision, was incompatible with Reagan's notion of pitching "our great industrial base" head on against communism. In 1958, at NASA's founding, the partition of civilian space exploration from military programs had been designed to put the Soviets at ease. Now, almost thirty years later, US officials intended for Star Wars to frighten the Kremlin toward the negotiating table or else propel it into an arms race that it could neither afford nor engineer.

The cost, of course, was to bring an escalatory research program into an environment that US and Soviet leaders had at least feigned to be a weapons-free zone for the past three decades. The numerous proposals submitted to and by the Strategic Defense Initiative Organization (SDIO), the DoD office responsible for development of Star Wars projects, conceived of space in the same way that space-power theorists had done at the dawn of the space age: as "the high ground" from which to enforce deterrence and dominate the twentieth-century battlefield.[22] Project Excalibur, a brainchild of Edward Teller at Livermore, envisioned packing large numbers of X-ray lasers around a nuclear device, which, once detonated, would propel the lasers toward targeted missiles. Such a weapon could conceivably take down several missiles in a single shot. Brilliant Pebbles, another Livermore project, entailed orbiting a space-based "architecture" of thousands of interceptors that would detect the rocket motors of enemy ICBMs during the boost phase and destroy them. Hundreds of such interceptors would be orbiting over the Soviet Union at any given time.[23] Under George H. W. Bush, the SDIO would approve deployment of a program designed to defend against missile attacks from the developing world—the Global Protection Against Limited Strikes, or GPALS. The system would utilize more low-orbit satellites (called Brilliant Eyes) as well as a series of ground-based systems to destroy missiles that did not reach the altitudes necessary for Pebbles interceptors to detect.

Advocates of these weapons argued that the scope and cost of SDI would force the Soviets to negotiate; if space-based BMD could make nuclear weapons ineffective, perhaps Mikhail Gorbachev would agree to substantial, if not complete, nuclear disarmament. It was an optimism that ignored completely the history of Soviet behavior since the beginning of the Cold War, for although the Kremlin was indeed interested in limiting Star Wars to the laboratory and showed interest in political trades for arms control, Soviet leaders were not content, as they never had been, to let the imperialists set the terms of the strategic debate.

In the late 1970s and early 1980s, the Soviet Union funded two massive R&D studies to explore countermeasures for the United States' imagined BMD programs: an orbital platform that would shoot down SDI satellites with a carbon-dioxide laser, the *Polyus* or *Skif* spacecraft; and Kaskad, which would destroy satellites with missiles from a craft stationed in orbit.[24] When Ronald Reagan walked away from the Reykjavik disarmament summit in October 1986, it became an unappreciated wrinkle of the late Cold War that at the moment of his decision to preserve SDI at the expense of arms control, the Soviet Union was closer to orbiting space weapons than was the United States. Just months after the summit, the Soviet Union launched a demonstration model of its *Skif* weapon. The cumbersome platform suffered a software malfunction upon separating from the rocket and disintegrated in the atmosphere as it fell toward the Pacific. Though the launch failed, the project's ambition—and the power of the *Energia* rocket that had propelled the weapon—foreboded a costly arms race in space.

The Ghosts of Space Supremacy

Such a race would surely have been in the offing had the Soviet Union not precipitously collapsed over the following three years. On their face the astonishing political developments of 1989–1991 appeared to augur a return to the ideas that had defined outer space as a special realm to be exempted from earthly conflict. With the Cold War ended, what need was there for a robust US program to militarize the cosmos? The Japanese and European space programs were years behind and, in any case, unthreatening. China and India had made enormous strides in space technology during the 1970s and 1980s, but were still years away from an ASAT capability. Yet the end of the Cold War, rather than precipitating the decline of the space supremacy paradigm, hastened its absorption into American political culture. The notion that space technology provided an avenue for international reconciliation, though it never expired, gradually gave way to the competing idea that these tools were levers the United States could pull to maintain military, economic, and political hegemony. Space itself, viewed less as an oasis of peace after 1991, came increasingly to be portrayed as a high ground from which to enforce a Pax Americana.

This shift, already well underway because of the Reagan-era policies, accelerated during the First Gulf War, a conflict that many would later dub "the first space war."[25] The moniker was not without justification: DoD's Global Positioning System (GPS)—known as NAVSTAR, or Navigation System Using Timing and Ranging—though it only featured sixteen satellites at the beginning of 1991, was critical to the Coalition's victory against Iraq. Operation Desert Storm required that Coalition forces fight and navigate in inhospitable, featureless deserts. It required nimble artillery fire, effective communications, and reactive troop movement based on weather and the enemy's location. Satellites helped secure all these needs.[26] Over the one-hundred hours that the lopsided ground war transpired, NAVSTAR provided three-dimensional coverage of the battlefield for nineteen hours at a time. Whereas earlier, land-based GPS provided coordinates with an expected error margin of up to eight miles, the system employed in Desert Storm plotted points to within sixty feet. GPS receivers allowed infantry divisions to link phase lines and thus maintain effective command and control. Artillery divisions successfully leveraged satellite positioning to direct fire on dense concentrations of Iraqi armor. Before long, commanders on the ground replaced the artillery surveyor's compass, aiming circle, slide rule, and other trade tools with GPS.[27]

The swiftness of victory and the outsize role that space technology had played in achieving it convinced observers that the world stood on the threshold of a Long Peace backed by American techno-military supremacy. William J. Perry, who had served Carter and Reagan as undersecretary of defense for research and engineering, thought GPS had proven itself a revolutionary force multiplier. To engage an enemy without such technology, he wrote in *Foreign Affairs*, would resemble a cavalry equipped with tanks meeting one of horses.[28]

Desert Storm's success exhumed the corpse of "space war" from its shallow grave. The 1990s witnessed an avalanche of military analysis regarding the role space technologies would play in future wars and the necessity of controlling the space environment to deny opponents an opportunity to "shape" the battlefield, both on Earth and in space.[29] The new wave of literature, offering little in the way of updated interpretation, regurgitated the predictions of the 1960s analysts. Generously quoting Bernard Schriever, Thomas White, and Homer Boushey (see chapter 4), military officers and defense analysts once again crowned outer space the new high ground of modern warfare. Few writers neglected to apply old Mahanian ideas about critical choke points to space; defense scholar Colin Gray's publication of "The Influence of *Space* Power upon History" in 1996 drove the point home.[30]

By 1998 Lieutenant Colonel David Lupton, dean of the new space-power theorists, declared sanctuary policy "a fallen star." Dreams of a space sanctuary, he wrote, had been conjured by "reasonable, peace-loving men" who appreciated the potential for space technology to strengthen nuclear deterrence through treaty verification, test detection, and photoreconnaissance. It was a viable approach for the early Cold War, when trust was in short supply and crises abounded. But continued technological revolution had made the idea obsolete by the middle of the 1970s. The military value of space for observation; the increasing sophistication of satellites for mapping, communications, and weather prediction; and the superpowers' budding ASAT programs had belied "rose-colored" space rhetoric. Although some dreamed of a return to the politics of peace, Lupton concluded, "like lost virginity, the ideal sanctuary is irretrievable."[31]

More than a half-century has passed since the beginning of the space age, yet we are inhabiting a similar political world. When the G-20 Summit convened in Bali, Indonesia, in November 2022, Joe Biden reassured reporters that the United States could avoid "a new Cold War" with China. Despite escalating tensions over trade policy, Taiwan, and the Russian invasion of Ukraine, the president held out hope that the two countries could responsibly manage competition and avoid conflict. Yet the very utterance of the phrase belied the truth. Growing bipolarity and protracted rivalry between two nuclear-armed superpowers?

Yes. Geopolitical flashpoints in Europe, East Asia, and the Middle East? Check. Intense technological competition in areas deemed central to military and economic power? Most assuredly. And because of the Chinese Communist Party's total control over government, even communism is a topic of discussion again. Try as politicians might to deny it, and fret as we might about the onset of another "twilight struggle" between democratic capitalism and communist autocracy, the Second Cold War is already well underway.[32]

Space technology and "astropolitics" generally will play an even greater role in this redux than they did in the years after *Sputnik*. As part of its now decades-long drive for military modernization, the People's Republic of China has achieved rapid progress in counterspace weapons, launch capabilities, and the integration of space-based assets with everyday military operations. China proved a direct-ascent ASAT capability when it shot down an aging weather satellite in January 2007, an experiment that generated more than 3,000 individual pieces of trackable debris, a full sixth of the total then orbiting Earth. Recent Chinese research into robotic arms in space—to clean up the mess!—also reflects a latent co-orbital weapon. Long-forgotten fears of orbital bombardment have reawakened, too: twice in 2021 China successfully tested a nuclear-capable hypersonic glide vehicle that, after riding a Long March rocket into orbit, descended to Earth at speeds exceeding Mach 5, confirming USAF fears of China's "potential for global strikes . . . from space."[33]

Not least, the United States has replayed the First Cold War by all but proclaiming a new space race in which the American Artemis Program will compete with China's manned lunar program, Chang'e, to return humans to the moon by the end of the 2020s. Nor does this latest sprint appear a benign display of technological prestige or scientific curiosity. "We better watch out that they don't get to a place on the moon under the guise of scientific research," NASA administrator Bill Nelson explained to interviewers in January 2023, for "it is not beyond the realm of possibility that they say, 'Keep out, we're here, this is our territory.'"[34] His counterpart Ye Peijian, director of China's Lunar Exploration Program, harbors the same fears. "If we don't go [to the moon] now even though we are capable of doing so, then we will be blamed by our descendants. If others go there, then they will take over, and you won't be able to go even if you want to. This is reason enough."[35]

For its part, Russia—now the junior partner vis-à-vis Beijing—will also play an adversarial role in space. Putin's war in Ukraine is a case in point: Russia has spoofed the locations of US and British warships, jammed GPS-guided weapons supplied by NATO, and threatened to destroy civilian Starlink satellites, which have provided internet services to Ukrainian forces. Moreover, for nearly a decade Russia has conducted numerous tests of its Nudol missile system, a satellite-killer. On November 15, 2021, one of those tests directly intercepted a Soviet-era intelligence satellite, creating at least half as much debris as the earlier Chinese test. Two weeks later Russian state television warned that the military could destroy all thirty-two GPS satellites in geostationary orbit and "blind [NATO's] missiles, planes and ships, not to mention the ground forces." A sign of the times, Russia has even announced that it will withdraw from the International Space Station.

Far from a "sanctuary," space has assumed in the Second Cold War new, more ominous monikers. Official US space strategy defines the medium as fundamentally "congested, contested, and competitive."[36] When Donald Trump announced that he would create a distinct branch of the military, the US Space Force (USSF) in 2019, he dubbed the cosmos "a warfighting domain," a classification NATO affirmed that same year when it officially declared space a fifth operational medium alongside air, sea, land, and cyber.[37] The USSF's inaugural doctrine was clear-eyed in its assessment of the situation. "Military spacepower [is] a crucial manifestation of the high ground in modern warfare," it began. As a war-fighting branch, the Space Force must "steadfastly prepare to prosecute the appropriate amount of violence against an opponent subject to strategic objectives, legal, and policy restraints."[38] For years now, US politicians, military leaders, and journalists have assumed the inevitability, as CNN put it in a lurid documentary, of "War in Space: The Next Battlefield."[39]

Two options are open to the United States and the wider world in the Second Cold War. The first is to accept the certainty of space conflict and reconcile with it. As they have always done, the thinking goes, states will fight to protect their national interests. Toward this end they will build space weapons and counterspace weapons. They will lay claim to orbits and moons. Man, replaying history, will bring his fallen nature to each new domain he inhabits. These harsh realities dictate that we huddle in the command center, acquaint ourselves with space-war doctrine, and welcome the coming dark.

Political scientists have well prepared us for this first option and have fed the fatalism at its core. "The militarization and weaponization of space is not only a historical fact, it is an ongoing process," wrote Everett Dolman more than two decades ago. Astropolitics, both in the past and future, "is not pretty or uplifting or a joyous sermon for the masses."[40] Nor has the narrative changed. Violence, we've been told more recently, is innate to space technology, its "original sin" and therefore an immortal one. Space war—and human expansion into the cosmos generally—is accelerating the closure of Earth by machine civilization and hence propelling us to cataclysm.[41] What's left to do but throw up our hands?

This brings us to the second option. What if, instead of resignation, we adopt hope? What if we learn from the example set by the generation of writers, engineers, and politicians who first faced outer space as a profound new arena of international relations? After all, though that besieged cohort faced challenges every bit as daunting as our own, it achieved stunning success. In the span of a single decade, it ushered the United Nations to the center of space politics; suspended nuclear weapons tests in space and subsequently banned nations from stationing warheads there; and stifled, for a time, the futuristic ambitions of the Air Force, a victory that helped cool Soviet military activity in space. Most notably, it succeeded in crafting a binding international treaty to govern the use of space for all nations. That US and Soviet officials were able to stave off an arms race in space at the height of the Cold War should serve as ample instruction for those seeking to curb the weaponization of space in our own time. We needn't draw up a fresh astropolitical recipe for the Second Cold War, for one is already available, tucked away at the bottom of the cupboard—that is to say, in history.

In the end it may very well be that the idea of a space sanctuary was just a flight of fancy. Few areas of politics, indeed, have been as vulnerable to mythmaking and false prophecy as space. The cosmos served as legend, parable, and origin story for ancient peoples. Medieval civilization affixed its angels, demons, and spirits to the stars. And in the nineteenth and twentieth centuries, space fueled fantasy, imagination and, in no small measure, propaganda. "I do not subscribe to some 99% of what is written about this subject—space exploration—as having any validity," James Van Allen told Congress in 1957.[42] This remains sage advice.

But political imagination begets political *innovation*. We may ground space policy in optimism and nonetheless find ourselves better off. Consider the story just told. It was difficult, no doubt, for cold warriors like Eisenhower or Kennedy to imagine that space could be protected from conflict. Why should it be so? Yet for all their pragmatism, they imbued space with enormous symbolic power. They assigned to the cosmos all the transformative qualities that utopians like H. G. Wells, Arthur C.

Clarke, and even Mangan had dreamt up. Their record was not half bad. Placebos can be very effective.

Here is the truth: space is inherently neither a high ground nor a sanctuary. It is what human beings will make of it, the sum product of an innumerable quantity of small decisions, rational and fanciful alike. And careful must we be: like aiming a telescope—if the astronomy analogy is permitted—the slightest change in amplitude or altitude may deliver us a galaxy away from our intended destination. Let us calibrate the lenses. Let us aim straight and true, and through the eyepiece find a world named Peace.

Notes

Abbreviations Used in the Notes

APP   *American Presidency Project*
AW&ST   *Aviation Week and Space Technology*
AWIST   *Aviation Week Including Space Technology*
BAS   *Bulletin of Atomic Scientists*
BG   *The Boston Globe*
BSP   Bernard Schriever Papers (Library of Congress)
CBNH   Celestial Bodies Negotiating History (Lyndon Johnson Library)
CLP   Curtis LeMay Papers (Library of Congress)
COPUOS   Committee on the Peaceful Uses of Outer Space
CSM   *Christian Science Monitor*
CDT   *Chicago Daily Tribune*
CT   *Chicago Tribune*
DDE   Dwight D. Eisenhower
DDEL   Dwight D. Eisenhower Presidential Library
DDRS   Declassified Documents Reference System
DoD   Department of Defense
DSB   *Department of State Bulletin*
DNSA   Digital National Security Archive
FBIS   Foreign Broadcast Information Service
FCJ   Files of Charles E. Johnson (Lyndon Johnson Library)
FJC   Files of Joseph Califano (Lyndon Johnson Library)
FRUS   *Foreign Relations of the United States*
GF   General File
JBIS   *Journal of the British Interplanetary Society*
JFKL   John F. Kennedy Presidential Library
JMP   John McConnell Papers (Swarthmore College Peace Collection)
LAT   *Los Angeles Times*
LBJL   Lyndon B. Johnson Presidential Library
LBP   Lincoln Bloomfield Papers (MIT Special Collections)
LLIU   Lilly Library, Indiana University
LS   Legal Subcommittee (UN COPUOS)
LOC   Library of Congress
MD   Manuscripts Division (Library of Congress)
MIT   Massachusetts Institute of Technology, Special Collections
MOSC   Militarization of Outer Space Collection (National Security Archive)

NSAM National Security Action Memorandum
NSF National Security File
NYHT *New York Herald Tribune*
NYT *New York Times*
RG Record Group (US National Archives)
SCPC Swarthmore College Peace Collection
TWP Thomas White Papers (Library of Congress)
UNA United Nations Archives, New York City
USNA United States National Archives
WP *Washington Post*
WSJ *Wall Street Journal*
WvB Werner von Braun Papers

Introduction

1. "Chicago Man Calls Space His Property," *The Sun* (Chicago, IL), January 7, 1949; "Chicago Man Stakes Claim to Outer Space," *Science Illustrated* (May 1949): 42–43; Virgiliu Pop, "The Nation of Celestial Space," *Space Policy* 22 (2006): 205–13. Mangan's grandson, Dean Stump, whom Mangan dubbed "Duke of the Moon," would later claim that Celestia had more than 100,000 "members": see the preface to James T. Mangan, *The Secret of Perfect Living* (Conshohocken, PA, 2006).

2. "Claims South Half of Space," *NYHT*, February 12, 1950; "'Hey! That's My Space!' Yank's Protest to Reds," *CDT*, October 6, 1957.

3. Virgiliu Pop, *Unreal Estate: The Men Who Sold the Moon* (Cardiff, UK, 2006), 31–32.

4. James T. Mangan, "Sky Merchandisers Notified They Are Out of Order," press release, Nation of Celestial Space, March 5, 1963.

5. Mangan, "Sky Merchandisers"; "Trespassers Beware! Designer Got Deed to All Outer Space," *LAT*, January 19, 1949.

6. "Chicago Man Stakes Claim to Outer Space"; "'Ruler' of All Space Would Like to Bring Ike Up to Date," *BG*, May 7, 1958.

7. Charter of Celestia, Nation of Celestian Space, accessed May 15, 2017, http://nationofcelestialspace.com/history/.

8. David W. Ziegler, "Safe Heavens: Military Strategy and Space Sanctuary Thought" (master's thesis, School of Advanced Airpower Studies, 1998); Joseph E. Justin, *Space: A Sanctuary, The High Ground, Or a Military Mission?* (RAND Report P-6758, Santa Monica, CA, 1982); Brian Weeden, "The End of Sanctuary in Space," *War Is Boring*, January 7, 2015, https://medium.com/war-is-boring/the-end-of-sanctuary-in-space-2d58fba741a; Dale Armstrong, "American National Security and the Death of Space Sanctuary," *Astropolitics* 12, no. 1 (March 2014): 69–81; Bruce M. DeBlois, "Space Sanctuary: A Viable National Strategy," *Airpower Journal* 12, no. 4 (Winter 1998): 41–57; Michael E. O'Hanlon, *Neither Star Wars nor Sanctuary: Constraining the Military Uses of Space* (Washington, DC, 2004); Elbridge Colby, *From Sanctuary to Battlefield: A Framework for a US Defense and Deterrence Strategy for Space* (Washington, DC, 2016); Karl P. Mueller, "Totem and Taboo: Depolarizing the Space Weaponization Debate," *Astropolitics* 1, no. 1

(June 2003): 4–28; Tommy C. Brown, *Violating the Sanctuary: The Decision to Arm Space* (Fort Leavenworth, KS, 1994); Stephen Buono, "Sanctuary or Battlefield? Fighting for the Soul of American Space Policy," *Perspectives on History*, July 15, 2020, https://www.historians.org/publications-and-directories/perspectives-on-history/summer-2020/sanctuary-or-battlefield-fighting-for-the-soul-of-american-space-policy; Robin Dickey, "Space Has Not Been a Sanctuary for Decades," War on the Rocks, September 16, 2020, https://warontherocks.com/2020/09/space-has-not-been-a-sanctuary-for-decades/.

9. "Sanctuary" (n.), *Online Etymology Dictionary*, accessed January 24, 2022, https://www.etymonline.com/word/sanctuary.

10. Kendrick Oliver, *To Touch the Face of God: The Sacred, the Profane, and the American Space Program, 1957–1975* (Baltimore, MD, 2013).

11. Alton Frey, "The Military Danger," *Atlantic Monthly* 212, no. 2 (August 1963): 46–50. Lyndon B. Johnson address before meeting of CBS affiliates, January 14, 1958, "Space Notebook [removed from binder]," box 359, CMTE on Aeronautical and Space Sciences, United States Senate, 1949–61, Papers of LBJ, LBJL; Edward R. Finch, "Outer Space Can Help the Peace," *International Lawyer* 7, no. 4 (October 1973): 898.

12. Alexander C. T. Geppert, ed. *Imagining Outer Space: European Astroculture in the Twentieth Century* (London, 2018); Alexander C. T. Geppert, "Space Personae: Cosmopolitan Networks of Peripheral Knowledge, 1927–1957," *Journal of Modern European History* 6, no. 2 (September 2008): 262–85.

13. David Lasser, *The Conquest of Space* (Burlington, ON, 1931), 15, 114, 137, 181. Also see the appendix item, in the same book, "Annual Report to the American Interplanetary Society," April 13, 1931; De Witt Douglas Kilgore, *Astrofuturism: Science, Race, and Visions of Utopia in Space* (Philadelphia, PA, 2003), 38–39.

14. Walter McDougall, . . . *the Heavens and the Earth: A Political History of the Space Age* (New York, 1985); William E. Burrows, *This New Ocean: The Story of the First Space Age* (New York, 1998); Matthew Brzezinski, *Red Moon Rising: Sputnik and the Hidden Rivalries that Ignited the Space Age* (New York, 2008). The works in footnote 23 are also representative.

15. Kenneth Osgood, *Total Cold War: Eisenhower's Secret Propaganda Battle at Home and Abroad* (Lawrence, KS, 2006), chap. 10; Yanek Mieczkowski, *Eisenhower's Sputnik Moment: The Race for Space and World Prestige* (Ithaca, NY, 2013); Teasel Muir-Harmony, *Operation Moonglow: A Political History of Project Apollo* (New York, 2020); David Meerman Scott and Richard Jurek, *Marketing the Moon: The Selling of the Apollo Lunar Program* (Cambridge, MA, 2014); Michael Allen, *Live from the Moon: Film, Television and the Space Race* (London, 2009), chap. 8.

16. David H. DeVorkin, *Science with a Vengeance: How the Military Created the US Space Sciences after World War II* (New York, 1992); Paul B. Stares, *The Militarization of Space: US Policy, 1945–1984* (Ithaca, NY, 1985); Matthew Mowthorpe, *The Militarization and Weaponization of Space* (New York, 2003); Jack Manno, *Arming the Heavens: The Hidden Military Agenda for Space, 1945–1995* (New York, 1984); Sean Kalic, *US Presidents and the Militarization of Outer Space, 1946–1967* (College Station, TX, 2012); Steven J. Zaloga, *The Kremlin's Nuclear Sword: The Rise and Fall of Russia's Strategic Nuclear Forces, 1945–2000* (Washington, DC, 2002); Nicholas M. Sambaluk, *The Other Space Race: Eisenhower and the Quest*

for Aerospace Security (Annapolis, MD, 2015); Gerald L. Borrowman, "Sovet Military Activities in Space," *Journal of the British Interplanetary Society* 35, no. 2 (1982): 86–92; Nicholas L. Johnson, *Soviet Military Strategy in Space* (Coulsdon, UK, 1987); Curtis Peebles, *High Frontier: The United States Air Force and the Military Space Program* (Washington, DC, 1997); David N. Spires, *Beyond Horizons: A Half Century of Air Force Space Leadership* (Peterson AFB, CO, 1997); R. Cargill Hall and Jacob Neufeld, *The US Air Force in Space: 1945 to the Twenty-First Century* (Washington, DC, 1998). Useful overviews of the military space historiography include Peter L. Hays, "Space and the Military," in *Space Politics and Policy, an Evolutionary Perspective*, ed. Eligar Sadeh (Dordrecht, Netherlands, 2002); Dwayne A. Day, "The State of Historical Research on Military Space," *Journal of the British Interplanetary Society* 50 (1997): 203–6; Roger D. Launius, "The Military in Space: Policy-Making and Operations in a New Environment," in *A Guide to the Sources of United States Military History: Supplement IV*, ed. Robin Higham and Donald J. Mrozek (North Haven, CT, 1998); and Stephen B. Johnson, "The History and Historiography of National Security Space," in *Critical Issues in the History of Spaceflight*, ed. Steven J. Dick and Roger D. Launius (Washington, DC, 2006).

17. John Logsdon, *John F. Kennedy and the Race to the Moon* (London, 2010); Deborah Cadbury, *Space Race: The Epic Battle Between America and the Soviet Union for Dominion of Space* (New York, 2007); Douglas Brinkley, *American Moonshot: John F. Kennedy and the Great Space Race* (New York, 2019); Asif Siddiqi, *Challenge to Apollo: The Soviet Union and the Space Race, 1945–1974* (Washington, DC, 2000); Asif Siddiqi, *Sputnik and the Soviet Space Challenge* (Tallahassee, FL, 2003); Alan J. Levine, *The Missile and Space Race* (Westport, CT, 1994); Michael D'Antonio, *A Ball, a Dog, and a Monkey: 1957—The Space Race Begins* (New York, 2008); Matt Bille and Erika Lishock, *The First Space Race* (College Station, TX, 2004).

18. McDougall, *Heavens and the Earth*, 9 (emphasis added). The book won the Pulitzer Prize for history in 1986.

19. Hal Brands and John Lewis Gaddis, "The New Cold War," *Foreign Affairs* 100, no. 6 (November–December 2021), 10–21.

20. Trump quoted in David Montgomery, "Trump's Excellent Space Force Adventure," *Washington Post Magazine*, December 3, 2019, https://www.washingtonpost.com/magazine/2019/12/03/trumps-proposal-space-force-was-widely-mocked-could-it-be-stroke-stable-genius-that-makes-america-safe-again/.

21. David E. Sanger and William J. Broad, "China, Testing New Weapon, Jolts Pentagon," *NYT*, October 27, 2021.

22. Daniel Deudney, *Dark Skies: Space Expansionism, Planetary Geopolitics, and the Ends of Humanity* (New York, 2020); Joan Johnson-Freese, *Heavenly Ambitions: America's Quest to Dominate Space* (Philadelphia, PA, 2009); James Clay Moltz, *Crowded Orbits: Conflict and Cooperation in Space* (New York, 2014); Namrata Goswami and Peter A. Garretson, *Scramble for the Skies: The Great Power Competition to Control the Resources of Outer Space* (Lanham, MD, 2020); Everett C. Dolman, *Astropolitik: Classical Geopolitics in the Space Age* (London, 2002); Bleddyn E. Bowen, *Original Sin: Power, Technology and War in Outer Space* (New York, 2022). For a representative sample of other works that reflect the

political science debate over international space security, see Colin S. Gray, "Space and Arms Control: A Skeptical View," in *America Plans for Space: A Reader Based on the National Defense University Space Symposium* (Washington, DC, 1986); Joan Johnson-Freese, *Space as a Strategic Asset* (New York, 2007); Joan Johnson-Freese, *Space Warfare in the 21st Century: Arming the Heavens* (London, 2016); James Clay Moltz, *The Politics of Space Security: Strategic Restraint and the Pursuit of National Interests*, 3rd ed. (Stanford, CA, 2019). All agree that in the near-term, space will remain "congested, contested, and competitive," as the US National Security Space Strategy phrased it in 2011. See Department of Defense, *National Security Space Strategy: Unclassified Summary* (Washington, DC, 2011), i.

23. Donald E. Kash, *The Politics of Space Cooperation* (West Lafayette, IN, 1967); Arnold Frutkin, *International Cooperation in Space* (New York, 1965); Dodd L. Harvey and Linda C. Ciccoritti, *US-Soviet Cooperation in Space* (Miami, FL, 1974); Matthew J. Von Benke, *The Politics of Space: A History of US Soviet/Russian Competition and Cooperation in Space* (Boulder, CO, 1997).

24. See Stares, *Militarization of Outer Space*, appendix 2.

25. CIA, "The Soviet Space Program: Expenditure Implications of Soviet Space Programs," April 1969, 8. In 1984, Colin Gray and B. R. Schneider, estimated that 70 to 80 percent of Soviet launches were military, and another 15 percent were "partly" military. See Colin Gray and B. R. Schnieder, "The Soviet Military Space Program," *Signal* (December 1984): 69.

26. National Aeronautics and Space Act of 1958, Pub. L. No. 85-568, 72 Stat. 426-438 (July 29, 1959), Sec. 102(a). For the transition of this utopian language to the UN resolutions, see P. J. Blount, "Peaceful Purposes for the Benefit of All Mankind: The Ethical Foundations of Space Security," in *War and Peace in Outer Space: Law, Policy and Ethics*, ed. Cassandra Steer and Matthew Hersch (New York, 2021), 115-16; P. J. Blount and David Miguel Molina, "Bringing Mankind to the Moon: The Humans Rights Narrative in the Space Age," in *NASA and the Long Civil Rights Movement*, ed. Stephen P. Waring and Brian C. Odom (Gainesville, FL, 2019).

27. John Lewis Gaddis, *The Long Peace: Inquiries into the History of the Cold War* (New York, 1987), chap. 7.

28. Treaty on Principles Governing the Activities of States in the Exploration and Use of Outer Space, Including the Moon and Other Celestial Bodies, Signed at Washington, London, and Moscow, January 27, 1967, in *Cornerstones of Security: Arms Control Treaties in the Nuclear Era*, ed. Thomas Graham, Jr. and Damien J. LaVera (Seattle, WA, 2003), 35.

29. Summary of Speech of Professor Lincoln Bloomfield before New England Regional Conference on Space, Cambridge, November 13, 1962, Outer Space Conference 1962, box 4, Lincoln P. Bloomfield Papers (LBP), MIT Archives.

30. Rhodes, the arch-imperialist prime minister of the Cape Colony in South Africa, once regretted that the heavens were so distant and inaccessible, for "I would annex the planets if I could."

31. John F. Kennedy, "Inaugural Address," January 20, 1961, American Presidency Project, https://www.presidency.ucsb.edu/documents/inaugural-address-2.

1. Imagination

1. Frank Borman, *Countdown: An Autobiography* (New York, 1988), 212.
2. Archibald MacLeish, "A Reflection: Riders on the Earth, Brothers in the Eternal Cold," *NYT*, December 25, 1969.
3. Robert Poole, *Earthrise: How Man First Saw the Earth* (New Haven, CT, 2008), 36–55; Benjamin Lazier, "Earthrise; or, The Globalization of the World Picture," *American Historical Review* 16, no. 3 (2011): 602–30; Kelly Oliver, *Earth and World: Philosophy after the Apollo Mission* (New York, 2015); Denis Cosgrove, "Contested Global Visions: *One-World, Whole-Earth*, and the Apollo Space Photographs," *Annals of the Association of American Geographers* 84, no. 2 (1994): 270–94; Joshua Yates, "Mapping the Good World: The New Cosmopolitans and Our Changing World Picture," *Hedgehog Review* 11, no. 3 (Fall 2009): 7–27; Sheila Jasanoff, "Image and Imagination: The Formation of Global Environmental Consciousness," in *Changing the Atmosphere: Expert Knowledge and Environmental Governance*, ed. Clark A. Miller and Paul N. Edwards (Cambridge, MA, 1996); Robin Kelsey, "Reverse Shot: Earthrise and Blue Marble in the American Imagination," in *New Geographies 4: Scales of the Earth*, ed. El Hadi Jazairy and Mellisa Vaughn. (Cambridge, MA, 2011). For a complete list works linking space exploration to the emergence political and environmental globalism, see Neil H. Maher, *Apollo in the Age of Aquarius* (Cambridge, MA, 2017), 241, n. 4.
4. Poole, *Earthrise*, 37–55; Denis Cosgrove, *Apollo's Eye: A Cartographic Genealogy of the Earth in the Western Imagination* (Baltimore, MD, 2001).
5. David Lasser, *The Conquest of Space* (Burlington, ON, 1931), 138, 164.
6. Frank White, *The Overview Effect: Space Exploration and Human Evolution* (Boston, 1987).
7. H. G. Wells, *The First Men in the Moon*, ed. Simon J. James (1901; repr., New York, 2017), 126–27.
8. Alexander Bogdanov, *Red Star: The First Bolshevik Utopia*, ed. Loren R. Graham and Richard Stites, trans. Charles Rougle (1908; repr., Bloomington, IN, 1984), 47.
9. Frank Winter, *Prelude to Space: The Rocket Societies, 1924–1940* (Washington, DC, 1983). Beryl Williams and Samuel Epstein, *The Rocket Pioneers: On the Road to Space* (New York, 1955); Chris Gainor, *To a Distant Day: The Rocket Pioneers* (Lincoln, NE, 2008); Wyn Wachhorst, *The Dream of Spaceflight: Essays on the Near Edge of Infinity* (New York, 2000).
10. Bradley G. Shreve, "The US, the USSR, and Space Exploration, 1957–1963," *International Journal on World Peace* 20, no. 2 (June 2003): 78; Karsten Werth, "A Surrogate for War—The US Space Program in the 1960s," *Amerikastudien/American Studies* 49, no. 4 (2004): 563–87.
11. Arthur C. Clarke, "The Challenge of the Spaceship: Astronautics and Its Impact upon Human Society," *JBIS* 6, no. 3 (December 1946): 42.
12. White, *Overview Effect*; Konstantin Tsiolkovsky, "Planets Are Inhabited by Living Creatures," Cosmic Philosophy, accessed January 12, 2019, https://tsiolkovsky.org/en/the-cosmic-philosophy/planets-are-inhabited-by-living-creatures-1933/.

13. Lionel Kochan, *Russia in Revolution 1890–1918* (New York, 1966); Mark D. Steinberg, *The Russian Revolution, 1905–1921* (New York, 2017).

14. Sidney Harcave, *First Blood: The Russian Revolution of 1905* (London, 1964), 19–26.

15. Richard Stites, "Fantasy and Revolution: Alexander Bogdanov and the Origins of Bolshevik Science Fiction," in *Red Star*, 1–16, esp. 7; Loren R. Graham, "Bogdanov's Inner Message," in *Red Star*, 241–54; Mark B. Adams, "'Red Star': Another Look at Aleksandr Bogdanov," *Slavic Review* 48, no. 1 (Spring 1989): 1–15; Nikolai Krementsov, *A Martian Stranded on Earth: Alexander Bogdanov, Blood Transfusions, and Proletarian Science* (Chicago, IL, 2011).

16. Michael G. Smith, *Rockets and Revolution: A Cultural History of Early Spaceflight* (Lincoln, NE, 2014), 42; A. N. Shuspanov, "Alternative Social Ideals in Russian Utopian Novels and Science Fiction at the Beginning of the 20th Century," in *Aleksandr Bogdanov Revisited*, ed. Vesa Oittinen (Helsinki, Finland, 2009), 262–63.

17. Stites, "Fantasy and Revolution," 6–7.

18. Howard P. Segal, *Technological Utopianism in American Culture* (Syracuse, NY, 1985), 1, 7.

19. H. G. Wells, *Experiment in Autobiography: Discoveries and Conclusions of a Very Ordinary Brain (Since 1866)* (1934; repr., Boston, 1962); Lovat Dickson, *H. G. Wells: His Turbulent Life and Times* (New York, 1969); Norman Mackenzie and Jeanne Mackenzie, *H. G. Wells: A Biography* (New York, 1973); Anthony West, *H. G. Wells: Aspects of a Life* (New York, 1984); David C. Smith, *H. G. Wells: Desperately Mortal* (New Haven, CT, 1986); Adam Roberts, *H. G. Wells: A Literary Life* (New York, 2019); Andrea Lynn, *Shadow Lovers: The Last Affairs of H. G. Wells* (London, 2001).

20. Frank McConnell, *The Science Fiction of H. G. Wells* (New York, 1981); Justin E. A. Busch, *The Utopian Vision of H. G. Wells* (Jefferson, NC, 2009); Peter Furshow, *Modern Utopian Fictions from H. G. Wells to Iris Murdoch* (Washington, DC, 2007).

21. West, *H. G. Wells*, 232.

22. David Y. Hughes and Harry M. Geduld, eds., *A Critical Edition of* The War of the Worlds: *H. G. Well's Scientific Romance* (Bloomington, IN, 1993), 184.

23. Bernard Bergonzi, *The Early H. G. Wells: A Study of the Scientific Romances* (Manchester, UK, 1961), 134; Peter Fitting, "Estranged Invaders: *The War of the Worlds*," in *Learning from Other Worlds: Estrangement, Cognition, and the Politics of Science Fiction and Utopia*, ed. Patrick Parrinder (Durham, NC, 2001), 143, n.13; Mark Rose, *Alien Encounters: Anatomy of Science Fiction* (Cambridge, MA, 1981), 76.

24. Hughes and Geduld, *Critical Edition of* The War of the Worlds, 52. Wells's complicated attitudes toward race and eugenics are outside the scope of this study, but they are ably summarized in Duncan Bell, "Pragmatic Utopianism and Race: H. G. Wells as Social Scientist," *Modern Intellectual History* 16, no. 3 (November 2019): 863–95; and John S. Partington, "The Death of the Static: H. G. Wells and the Kinetic Utopia," *Utopian Studies* 11, no. 2 (2000): 96–111.

25. Mahesh Rangarajan, "Environment and Ecology Under British Rule," in *India and the British Empire*, ed. Douglas M. Peers and Nandini Gooptu (New York, 2012).

26. Wells quoted in David Seed, "The Course of Empire: A Survey of the Imperial Theme in Early Anglophone Science Fiction," *Science Fiction Studies* 37 (2010): 233; Aaron Worth, "Imperial Transmissions: H. G. Wells, 1897-1901," *Victorian Studies* 53, no. 1 (Autumn 2010): 65-89.

27. Emelie Jonsson, "The Human Species and the Good Gripping Dreams of H. G. Wells," *Style* 47, no. 3 (Fall 2013): 296-315; Peter Kemp, *H. G. Wells and the Culminating Ape: Biological Imperatives and Imaginative Obsessions* (1982; repr., New York, 1996); J. P. Vernier, "Evolution as a Literary Theme in H.G. Wells's Science Fiction," in *H. G. Wells and Modern Science Fiction*, ed. Darko Savin and Robert M. Philmus (London, 1977).

28. Norman and Jeanne Mackenzie, *H. G. Wells*, 56.

29. H. G. Wells, "Zoological Retrogression," in *H. G. Wells: Early Writings in Science and Science Fiction*, ed. Robert M. Philmus and David Y. Hughes (Berkeley, CA, 1975), 158, 168.

30. H. G. Wells, "Human Evolution, an Artificial Process, *Fortnightly Review* 60, no. 358 (October 1896): 590.

31. H. G. Wells, "On Extinction," *Chambers Journal* 10 (Sept. 30, 1893): 623-24.

32. H. G. Wells, *The Island of Dr. Moreau* (1896; repr., Garden City, NY, 1929), 130.

33. Mackenzie and Mackenzie, *H. G. Wells: A Biography*, 128-29.

34. Mackenzie and Mackenzie, *H. G. Wells: A Biography*, 129.

35. Hughes and Geduld, *Critical Edition of The War of the Worlds*, 192-193.

36. Wells, *Experiment in Autobiography*, 106.

37. David Lake, Introduction to H. G. Wells, *The First Men in the Moon*, ed. David Lake (New York, 1995), xiv; Karen ní Mheallaigh, *The Moon in the Greek and Roman Imagination* (Cambridge, UK, 2020), chap. 5.

38. Philmus and Hughes, *H. G. Wells*, 182; Patrick Parrinder, *H. G. Wells* (Edinburgh, UK, 1970), 21, 28.

39. H. G. Wells, *The First Men in the Moon* (Leipzig, 1902), 257, 280, 291.

40. Wells, *First Men* (1902), 271-78.

41. Wells, *First Men* (1902), 281-83.

42. Wells, *First Men* (1902), 286-87.

43. John S. Partington, *Building Cosmopolis: The Political Thought of H. G. Wells* (Burlington, VT, 2003), 4.

44. H. G. Wells, *The War and the Future: Italy, France and Britain at War* (London, 1917), 8, 143.

45. W. Warren Wagar, *H. G. Wells and the World State* (New Haven, CT, 1961); Or Rosenboim, *The Emergence of Globalism: Visions of World Order in Britain and the United States, 1939-1950*, 212-216.

46. Wells, *First Men in the Moon* ([1901] 2017), 112-13.

47. H.G. Wells, *In the Days of the Comet* (1906; repr., New York, 1924).

48. Wells, *Days of the Comet*, 207, 215-16.

49. Wells, *Days of the Comet*, 232-33.

50. Wells, *Days of the Comet*, 215-16, 267-70.

51. Wells, *Days of the Comet*, 273-74.
52. Wells, *Experiment in Autobiography*, 558-59.
53. H. G. Wells, letter to the editor, *The New Age*, mid-October 1907, in *The Correspondence of H. G. Wells*, vol. 2, *1904-1918*, ed. David C. Smith (London, 1998), 163.
54. Wells, *Experiment in Autobiography*, 549.
55. Within two generations *The War of the Worlds* had been adapted to radio, film, television, comics, graphic novels, and even a stage play. For the extent of its influence see tables 1.1 and 10.1 in Peter J. Beck, The War of the Worlds: *From H. G. Wells to Orson Welles, Jeff Wayne, Steven Spielberg and Beyond* (New York, 2016), 6, 185. The novel has never been out of print.
56. Patrick Parrinder and John S. Partington, *The Reception of H. G. Wells in Europe* (London: Thoemmes Continuum, 2005), 4, 6, 74, 106; Mackenzie and Mackenzie, *H. G. Wells*, 209; William J. Scheick, ed., *The Critical Response to H. G. Wells* (Westport, CT, 1995).
57. George Orwell, "Wells, Hitler and the World State," *Horizon* 4, no. 20 (August 1941): 133-38.
58. David T. Hughes, "*The War of the Worlds* in the Yellow Press," *Journalism Quarterly* 43, no. 4 (Winter 1966): 639-46.
59. Garrett P. Serviss, *Edition's Conquest of Mars* (1898; repr., Los Angeles, CA, 1947), 16-19; Paul K. Alkon, *Science Fiction before 1900* (London, 2002), 108-9; David Seed, "The Course of Empire: A Survey of the Imperial Theme in Early Anglophone Science Fiction," *Science Fiction Studies* 37, no. 2 (July 2010): 235-36.
60. Franz Rottensteiner, "German SF," in *Anatomy of Wonder: A Critical Guide to Science Fiction*, 3rd ed., ed. Neil Barron (New York, 1987), 379-404.
61. Kurd Lasswitz, *Auf zwei Planeten* (Weimar, 1897).
62. Ingo Cornils, "The Martians Are Coming! War, Peace, Love, and Scientific Progress in H. G. Well's 'The War of the Worlds' and Kurd Lasswitz's 'Auf zwei Planeten,'" *Comparative Literature* 55, no. 1 (Winter 2003): 28.
63. Cornils, "The Martians Are Coming!," 37.
64. For a full summary of the novel, see William B. Fisher, *The Empire Strikes Out: Kurd Lasswitz, Hans Dominik, and the Development of German Science Fiction* (Bowling Green, OH, 1984), 127-30.
65. Epigraph by von Braun in Kurd Lasswitz, *Two Planets* (1897; repr., Carbondale, IL, 1971), vii.
66. Cornils, "The Martians Are Coming!," 31; Fisher, *Empire Strikes Out*, 126.
67. Quoted in Fisher, *Empire Strikes Out*, 129; Karl S. Guthke, *The Last Frontier: Imagining Other Worlds, from the Copernican Revolution to Modern Science Fiction* (Bern, 1993), chap. 4.
68. Fisher, *Empire Strikes Out*, 31.
69. H. Bruce Franklin, *War Stars: The Superweapon and the American Imagination* (New York, 1988), chap. 2; I. F. Clarke, "Trigger-Happy: An Evolutionary Study of the Origins and Development of Future-War Fiction, 1763-1914," *Journal of Social and Evolutionary Systems* 20, no. 2 (1997): 117-36; I. F. Clarke, *Voices Prophesying War: Future Wars, 1763-3749* (New York, 1993); I. F. Clarke, "Future-War Fiction: The First Main Phase, 1871-1900," *Science Fiction Studies* 24, no. 3 (November 1997): 387-412.

70. Konstantin Tsiolkovsky, "Exploration of the Universe with Reaction Machines," Doc. I-5 in *Exploring the Unknown: Selected Documents in the History of the US Civil Space Program*, vol. 1, *Organizing for Exploration* (Scotts Valley, CA, 1995), 59–83; Robert H. Goddard, "A Method of Reaching Extreme Altitudes," *Smithsonian Miscellaneous Collections* 71, no. 2 (Washington, DC, 1919); Hermann Oberth, *Die Rakete zu den Planetenräumen* (Munich, 1923); William Sims Bainbridge, *Spaceflight Revolution: A Sociological Study* (New York, 1976), 19–20.

71. George M. Young, *The Russian Cosmists: The Esoteric Futurism of Nikolai Fedorov and His Followers* (New York, 2012), 3; Bernice Glatzer Rosenthal, "Political Implications of the Occult Revival," in *The Occult in Russian and Soviet Culture*, ed. Bernice Glatzer Rosenthal (Ithaca, NY, 1997); Asif Siddiqi, "Imagining the Cosmos: Utopians, Mystics, and the Popular Culture of Spaceflight in Revolutionary Russia," *Osiris* 23, no. 1 (2008): 265; Boris Grois, ed., *Russian Cosmism* (Cambridge, MA, 2018).

72. Quoted in Young, *Russian Cosmists*, 47.

73. Siddiqi, "Imagining the Cosmos," 265–66.

74. Siddiqi, "Imagining the Cosmos," 266.

75. James T. Andrews, *Red Cosmos: K. E. Tsiolkovskii, Grandfather of Soviet Rocketry* (College Station, TX, 2009).

76. Daniel Shubin, *Konstantin Eduardovich Tsiolkovsky: The Pioneering Rocket Scientists and His Cosmic Philosophy* (New York, 2016), 15–75.

77. Andrews, *Red Cosmos*, chap. 2.

78. Konstantin Tsiolkovsky, "Citizens of the Universe," August 2, 1933; "Necessity of Cosmic Mindset," 1934; "Planets Are Inhabited by Living Creatures," 1933, Cosmic Philosophy, accessed May 15, 2017, http://tsiolkovsky.org/en/cosmic-philosophy-by-tsiolkovsky/; Shubin, *Konstantin Eduardovich Tsiolkovsky*, 146; Young, *Russian Cosmists*, 145–54.

79. Siddiqi, "Imagining the Cosmos," 260–88; Asif Siddiqi, "Russia's Long Love Affair with Space: It Started with Utopian Dreams and Rocketeers," *Air and Space*, August 2007, accessed August 25, 2017, http://www.airspacemag.com/space/russias-long-love-affair-with-space-19739095/; James T. Andrews, "Storming the Stratosphere: Space Exploration, Soviet Culture, and the Arts from Lenin to Khrushchev's Times," *Russian History* 36, no. 1 (2009): 77–87; Asif Siddiqi, *The Rockets' Red Glare: Spaceflight and the Soviet Imagination, 1857–1957* (Cambridge, UK, 2014); Asif Siddiqi, "Deep Impact: Dr. Robert Goddard and the Soviet Space Fad of the 1920s," *History and Technology* 20, no. 2 (2004): 97–113.

80. Siddiqi, "Russia's Long Love Affair with Space."

81. Frank Winter, "Birth of the VfR: The Start of Modern Astronautics," *Spaceflight* 19 (1977): 243–56.

82. Michael J. Neufeld, "Weimar Culture and Futuristic Technology: The Rocketry and Spaceflight Fad in Germany, 1923–1933," *Technology and Culture* 31, no.4 (October 1990): 725–52; Alexander C. T. Geppert, ed., *Imagining Outer Space: European Astroculture in the Twentieth Century* (London, 2018).

83. Geppert et al., *Militarizing Outer Space: Astroculture, Dystopia and the Cold War* (London, 2021), 7; John MacCormac, "War with Rockets Pictured by Oberth," *NYT*, January 31, 1931.

84. Hermann Oberth, *Man into Space* (New York, 1957), 1, 151.

85. Jared S. Buss, *Willy Ley: Prophet of the Space Age* (Gainesville, FL, 2017), 16, 34, 45, 56–58.

86. Buss, *Willy Ley*, 66.

87. Quoted in Buss, *Willy Ley*, 69.

88. Winter, *Prelude to Space*, 73–86. Beryl Williams and Samuel Epstein, *The Rocket Pioneers: On the Road to Space* (New York, 1955), 171–203; Chris Gainor, *To a Distant Day: The Rocket Pioneers* (Lincoln, NE, 2008), 50–52. Wyn Wachhorst, *The Dream of Spaceflight: Essays on the Near Edge of Infinity* (New York, 2000); Laurence B. Chase, "Space Travel Since 1640," *Princeton University Library Chronicle* 30, no. 1 (Autumn 1968): 1–19; Bainbridge, *Spaceflight Revolution*.

89. Eric Leif Davin, *Pioneers of Wonder: Conversations with the Founders of Science Fiction* (Amherst, NY, 1999), 27–60; De Witt Douglas Kilgore, *Astrofuturism: Science, Race, and Visions of Utopia in Space* (Philadelphia, PA, 2003), chap. 1.

90. David Lasser, "The Rocket and the Next War" (Presidential Address, American Rocket Society, 1931), in *Fighting the Future War: An Anthology of Science Fiction War Stories, 1914–1945*, ed. Frederic Krome (New York, 2012), 176–83.

91. Lasser, "Rocket and Next War," 183.

92. Lasser, "Rocket and Next War," 183.

93. Lasser, *Conquest of Space*, 114, 137; David Lasser, "Annual Report to the American Interplanetary Society," April 13, 1931 (appendix item), 181; Kilgore, *Astrofuturism*, 39–40.

94. Lasser, "Annual Report to the American Interplanetary Society"; Michael L. Ciancone and Amelia "Mimi" Lasser, "David Lasser: An American Spaceflight Pioneer (1902–1996)," *History of Rocketry and Astronautics* 33, no. 22 (San Diego, CA, 2010), 123.

95. Halford Mackinder, "The Geographical Pivot of History," *Geographic Journal* 23, no. 4 (April 1904): 422; Lucian M. Ashworth, "Realism and the Spirit of 1919: Halford Mackinder, Geopolitics and the Reality of the League of Nations," *European Journal of International Relations* 17, no. 2 (2010): 279–301; Lucian M. Ashworth, "Mapping a New World: Geography and the Interwar Study of International Relations," *International Studies Quarterly* 57, no. 1 (March 2013): 138–49.

96. G. Lowes Dickinson, *The European Anarchy* (London, 1916), 9–10; G. Lowes Dickinson, *The International Anarchy, 1904–1914* (London, 1926), 14; James Bryce, *International Relations* (New York, 1922), 3–4.

97. Robert Vitalis, *White World Order, Black Power Politics: The Birth of American International Relations* (Ithaca, NY, 2015), 57, 65; Madison Grant, *The Passing of the Great Race; or The Racial Basis of European History* (1916; repr., New York, 1922), 263; Arthur Herman, *The Idea of Decline in Western History* (New York, 1997), 182–85.

98. Herman, *Idea of Decline*, 207.

99. E. H. Carr, *The Twenty Years' Crisis: 1919–1939* (1939; repr., New York, 1949), 5–8.

100. Hans Morgenthau, *Scientific Man vs. Power Politics* (Chicago, 1946), 9–10; Alison McQueen, *Political Realism in Apocalyptic Times* (Cambridge, UK, 2018), 163.

101. Carl Schmitt, *The Nomos of the Earth in the International Law of the Jus Publicum Europaeum*, trans. G. L. Ulmen (1950; repr., New York, 2003).

102. Stanley Hoffmann, "Raymond Aron and the Theory of International Relations," *International Studies Quarterly* 29, no. 1 (March 1985): 16.

103. Carr, *Twenty Years' Crisis*, 5, 89.

2. Interplanetary Men

1. "The Spirit That Beats V2," *Nottingham Evening Post*, November 11, 1944.

2. Amy Bell, "Landscapes of Fear: Wartime London, 1939-1945," *Journal of British Studies* 48, no. 1 (January 2009): 162-64.

3. Michael J. Neufeld, *Von Braun: Dreamer of Space, Engineer of War* (New York, 2008), 223.

4. Neufeld, *Von Braun*, 186.

5. Daniel Lang, "A Romantic Urge," *New Yorker*, April 21, 1951, 85-86.

6. Neufeld, *Von Braun*, 244.

7. Neufeld, *Von Braun*, 212-13.

8. Lang, "Romantic Urge," 85-86.

9. Thomas Gangale, *How High the Sky? The Definition and Delimitation of Outer Space and Territorial Airspace in International Law* (Leiden, Netherlands: Brill, 2018), 2.

10. Arthur C. Clarke, *Greetings, Carbon-Based Bipeds! Collected Essays, 1934-1998* (New York, 2001), 106.

11. Jay Winter, *Dreams of Peace and Freedom: Utopian Moments in the Twentieth Century* (New Haven, CT, 2008), chap. 4.

12. John Cheng, *Astounding Wonder: Imagining Science and Science Fiction in Interwar America* (Philadelphia, PA, 2012), 3.

13. Asif Siddiqi, "Imagining the Cosmos: Utopians, Mystics, and the Popular Culture of Spaceflight in Revolutionary Russia," *Osiris* 23, no. 1 (2008): 285.

14. Michael Neufeld, "Introduction: Mittelbau-Dora—Secret Weapons and Slave Labor," in Yves Béon, *Planet Dora: A Memoir of the Holocaust and the Birth of the Space Age* (Boulder, CO, 1997), xix.

15. Bell, "Landscapes of Fear," 163.

16. Alexander C. T. Geppert, "Space Personae: Cosmopolitan Networks of Peripheral Knowledge, 1927-1957," *Journal of Modern European History* 6, no. 2 (September 2008): 279.

17. "Bomb Defense 'Impossible,'" *The Sun*, October 5, 1945; Fraser MacDonald, "Space and the Atom: On the Popular Geopolitics of Cold War Rocketry," *Geopolitics* 13 (2008): 613; "Science Nicks Moon Romance," *Indianapolis Star*, January 27, 1946.

18. C. P. Ives, "Mere Devices Aren't Going to Be Enough," *The Sun*, November 18, 1945.

19. "The Future Is Here," *The Enquirer* (Cincinnati, OH), February 20, 1949.

20. Oliver Dunnett, "The British Interplanetary Society and Cultures of Outer Space, 1930-1970" (PhD diss., University of Nottingham, 2011); Oliver Dunnett, "Geopolitical Cultures of Outer Space: The British Interplanetary Society, 1933-1965," *Geopolitics* 22, no. 2 (2017): 452-73.

21. William R. Macauley, "Crafting the Future: Envisioning Space Exploration in Postwar Britain," *History and Technology* 28, no. 3 (2012): 286.

22. A. V. Cleaver as quoted in William Sims Bainbridge, *Spaceflight Revolution*, 154.

23. A. V. Cleaver, "The Interplanetary Project," *JBIS* 7, (1948) 21–39; Andrew Chatwin, *Val Cleaver: A Very English Rocketeer* (London, 2015).

24. Cleaver, "Interplanetary Project," 27.

25. Cleaver, "Interplanetary Project," 27.

26. Cleaver, "Interplanetary Project," 28.

27. Discussion, "Interplanetary Project," 37.

28. De Witt Douglas Kilgore, *Astrofuturism: Science, Race, and Visions of Utopia in Space* (Philadelphia, PA, 2003), 113. Gary Westfahl, *Arthur C. Clarke* (Urbana, IL, 2018), chap. 1; Neil McAleer, *Arthur C. Clarke: The Authorized Biography* (London, 1993).

29. Arthur C. Clarke, "Extraterrestrial Relays: Can Rocket Stations Give Worldwide Radio Coverage?," *Wireless World* 51, no. 10 (October 1945), 305–8.

30. Arthur C. Clarke, Letters to the Editor, "Peacetime Uses for V2," *Wireless World* 51, no. 2 (February 1945): 58.

31. Arthur C. Clarke, "The Rocket and the Future of Warfare" in *Ascent to Orbit* (Ann Arbor, MI, 1984), 70; Clarke, *Greetings*, 32.

32. Arthur C. Clarke, "The Challenge of the Spaceship: Astronautics and Its Impact Upon Human Society," *Journal of the British Interplanetary Society* 6, no. 3 (December 1946): 66–81.

33. Clarke, "Challenge of the Spaceship"; Robert Poole, "The Challenge of the Spaceship: Arthur C. Clarke and the History of the Future, 1930–1970," *History and Technology* 28, no. 3 (September 2012): 255–280; Kilgore, *Astrofuturism*, 116–19.

34. Clarke, "Challenge of the Spaceship," 42.

35. Michael Adas, *Machines as the Measure of Men: Science, Technology, and Ideologies of Western Dominance* (Ithaca, NY, 1990), chap. 6; John Hersey, "Hiroshima," *New Yorker*, August 23, 1946; Lewis Mumford, *Atomic War—The Way Out* (London, 1949); Lewis Mumford, "Gentlemen: You Are Mad!," *Saturday Review*, March 2, 1946, 5–6; Lewis Mumford, "Anticipations and Social Consequences of Atomic Energy," *Proceedings of the American Philosophical Society* 98, no. 2 (1954): 149–52; "Cosmic Power" in Donald L. Miller, *The Lewis Mumford Reader* (New York, 1986), 301; "The Morals of Extermination," *Atlantic Monthly*, October 1959, 38–44; Michael Sherry, *The Rise of American Airpower: The Creation of Armageddon* (New Haven, CT, 1988), 251–55.

36. "National Affairs: World War III?," *Time*, November 3, 1941; "Foreign News: World War III?," *Time*, March 22, 1943; "INTERNATIONAL: or Else," *Time*, February 15, 1943; "GERMANY: For World War III," *Time*, June 5, 1944; "Science: World War III Preview?," *Time*, July 10, 1944; "Medicine: Germs for World War III?," *Time*, March 25, 1946; "The Nations: The Chances for World War III," *Time*, March 15, 1948.

37. Matthew Connelly et al., "'General, I Have Fought Just as Many Nuclear Wars as You Have': Forecasts, Future Scenarios, and the Politics of Armageddon," *AHR* 117, no. 5 (December 2012): 1435.

38. "Preview of the War We Do Not Want," *Collier's*, October 27, 1951; David Alan Rosenberg, "The History of World War III, 1945–1990: A Conceptual

Framework," in *On Cultural Ground: Essays in International History*, ed. Robert David Johnson (Chicago, IL, 1994).

39. James R. Randolph, "Occupation of Mars?," *Army Ordnance* 31 (March–April 1947): 422–23; "Recommended Occupation of Mars as Strategic Necessity," *The Mercury* (Hobart, Australia, March 22, 1947), 1.

40. R. L. Farnsworth, *Rockets: New Trail to Empire*, 2nd ed. (Glen Ellyn, IL, 1945), 16.

41. Arthur C. Clarke, "The Moon and Mr. Farnsworth," in Clarke, *Greetings*, 26–29.

42. Quoted in Clarke, *Greetings*, 29.

43. Arthur C. Clarke, *The Exploration of Outer Space* (1951; repr., Greenwich, CT, 1959), 176; Arthur C. Clarke, "Space Flight and the Spirit of Man," in *Voices from the Sky: Previews of the Coming Space Age* (New York, 1965), 8.

44. Clarke, "Challenge of the Spaceship," 72.

45. Arthur C. Clarke, "On the Morality of Space," *The Saturday Review*, October 5, 1957, 8–10, 35.

46. Clarke, *Exploration of Outer Space*, 175.

47. Clarke, *Exploration of Outer Space*, 177.

48. Poole, "Challenge of the Spaceship," 259.

49. Clarke, "Space Flight and the Spirit of Man"; Clarke, *Exploration of Outer Space*, 181.

50. Arthur C. Clarke, *Astounding Days: A Science-Fictional Autobiography* (New York, 1990), 99, 181; Poole, "Challenge of the Spaceship," 261; Dale Carter, *The Final Frontier: The Rise and Fall of the American Rocket State* (New York, 1988).

51. Clarke, "Space Flight and the Spirit of Man," 4.

52. Clarke, "Space Flight and the Spirit of Man," 8.

53. Arthur C. Clarke, "Social Consequences of the Communications Satellites" (paper delivered to the symposium on Space Law at the Twelfth International Astronautical Congress, Washington, 1961), in *Voices from the Sky*, 129–41.

54. Arthur C. Clarke, *Childhood's End* (London, 1953), 68–72.

55. Clarke, *Childhood's End*, 68–72, 107.

56. Arthur C. Clarke, *Prelude to Space* (London, 1954), 112.

57. Robert Crossey, *Olaf Stapledon: Speaking for the Future* (Syracuse, NY, 1994), 32.

58. Patrick A. McCarthy, *Olaf Stapledon* (Boston, 1982), 18.

59. Stapledon as quoted in McCarthy, *Olaf Stapledon*, 22.

60. Vincent Geoghegan, "Olaf Stapledon: Religious but Not a Christian," in *Socialism and Religion: Roads to Commonwealth* (London, 2011), 85–108.

61. Olaf Stapledon, "Interplanetary Man?," *Journal of the British Interplanetary Society* 7, no. 6 (November 1948): 220–23.

62. William Olaf Stapledon, *Last and First Men: A Story of the Near and Far Future* (1930; repr., London, 2020), 10.

63. Stapledon, "Interplanetary Man," 220–23.

64. "Last and First Men—Olaf Stapledon (1930)," *Weighing a Pig Doesn't Fatten It* (blog), accessed May 20, 2019, https://schicksalgemeinschaft.wordpress.com/2016/05/30/last-and-first-men-olaf-stapledon-1930/.

65. Stapledon, "Interplanetary Man," 220–23.

66. Stapledon, *Last and First Men*, 9.

67. Robert Crossley, *An Olaf Stapledon Reader* (Syracuse, NY, 1997), 365; "Planetary Colonies of Eugenic 'Quasi-Human' Races Forecast," *The Sun*, October 11, 1948; "USS." (United Solar System), *Time*, October 18, 1948, 67–70.

68. Stapledon, "Interplanetary Man," 229.

69. Stapledon, "Interplanetary Man," 217.

70. Stapledon, *Last and First Men*, 33–34.

71. Stapledon, *Last and First Men*, 33–34.

72. David W. Bath, *Assured Destruction: Building the Ballistic Missile Culture of the US Air Force* (Annapolis, MD, 2020); Gregory P. Kennedy, *The Rockets and Missiles of White Sands Proving Ground, 1945–1958* (Schiffer, 2009); Jacob Neufeld, *The Development of Ballistic Missiles in the United States Air Force, 1945–1960* (Washington, DC, 1990); John C. Lonnquest and David F. Winkler, *The Legacy of the United States Cold War Missile Program, USACERL Special Report 97/01* (Washington, DC, 1996); Christopher Gainor, *The Bomb and America's Missile Age* (Baltimore, MD, 2018).

73. "Stapledon Fears War at Any Moment," *BG*, April 6, 1949; Alistair Cooke, "Dr. Stapledon Tells of Britain's 'Passion for Peace,'" *Manchester Guardian*, March 28, 1949.

74. Waldemar Kaempffert, "Science in Review: Rockets and Atomic Power Suggest Conquest of Solar System by New Species," *NYT*, March 6, 1949; "Course of Empire," *WP*, October 23, 1948; "War or Guided-Missile Bases on Moon Held Possible by '98," *WP*, March 5, 1948.

75. Willy Ley, *Rockets Missiles and Space Travel* (1944); *Rockets: The Future of Travel Beyond the Stratosphere* (New York, 1944); with Wernher von Braun, *The Complete Book of Outer Space* (Maco Magazine, 1953); with Wernher von Braun, *The Exploration of Mars* (New York, 1956); the *Adventure in Space* Series: *Space Pilots* (1957), *Space Stations* (1958), and *Space Travel* (1958); *Satellites, Rockets and Outer Space* (1958); *Mars and Beyond: A Tomorrowland Adventure Adapted for School Use* (1959); *Rockets* (1960).

76. James Mangan, "Report to the Universe: The First Seven Years" (white paper, State Department, Celestia, USA, April 30, 1956), personal collection of Virgiliu Pop.

77. Virgiliu Pop, *Unreal Estate: The Men Who Sold the Moon* (Cardiff, UK, 2006), 31–32.

78. "The Nation of Celestial Space, since 1948," accessed May 23, 2022, https://nationofcelestialspace.com/history/.

79. I thank W. Patrick McCray for this insight.

80. Fraser MacDonald, *Escape from Earth: A Secret History of the Space Rocket* (New York, 2019); M. G. Lord, *ASTRO TURF: The Private Life of Rocket Science* (Walker, 2005); W. Patrick McCray, *Making Art Work: How Cold War Engineers and Artists Forged a New Creative Culture* (Cambridge, MA: MIT Press, 2020); George Pendle, *Strange Angel: The Otherwordly Life of Rocket Scientist John Whiteside Parsons* (New York: Harcourt, 2005); Iris Chang, *Thread of the Silkworm* (New York, 1995); Christpher Gainor, *To a Distant Day: The Rocket Pioneers* (Lincoln, NE, 2008).

81. John Bluth, "Malina, Frank Joseph," *American National Biography*, February 2000, accessed May 9, 2022, https://www.anb.org/display/10.1093/anb/9780198606697.001.0001/anb-9780198606697-e-1302215.

82. Gainor, *To a Distant Day*, 127–32.

83. MacDonald, *Escape from Earth*, 63.

84. McCray, *Making Art Work*, 21.

85. C. R. Koppes, *JPL and the American Space Program: a History of the Jet Propulsion Laboratory* (New Haven, CT, 1982).

86. Frank Malina, interview with Mary Terrall, December 14, 1978, Caltech Archives, chrome-extension://efaidnbmnnnibpcajpcglclefindmkaj/https://oralhistories.library.caltech.edu/149/1/Malina.pdf.

87. Frank Malina to Liljan Malina, September 6, 1944, personal files of W. Patrick McCray.

88. MacDonald, *Escape from Earth*, 121.

89. Gainor, *To a Distant Day*, 132; MacDonald, *Escape from Earth*, 122–23.

90. MacDonald, *Escape from Earth*, 128–29; Fraser MacDonald, "High Empire: Rocketry and the Popular Geopolitics of Space Exploration, 1944–62," in *New Spaces of Exploration: Geographies of Discovery in the Twentieth Century*, ed. Simon Naylor and James R. Ryan (London, 2010), 200.

91. MacDonald, *Escape from Earth*, 130.

92. Constitution of the United Nations Educational, Scientific and Cultural Organization, November 4, 1945.

93. Malina interview with Terrall.

94. McCray, *Making Art Work*, 22–23, 29; MacDonald, *Escape from Earth*, 131–32.

95. McCray, *Making Art Work*, 2, 43–46.

96. Lasser, *The Conquest of Space*, 15.

97. Stapledon, "Interplanetary Man?," 223.

98. Olaf Stapledon, "Mankind at the Crossroads," in Crossley, *Olaf Stapledon Reader*, 211.

99. Konstantin Tsiolkovski, "Creatures Higher Than a Man," June 28, 1939; "Creatures from Different Stages of Evolution," n.d., *The Cosmic Philosophy*, https://tsiolkovsky.org/en/the-cosmic-philosophy/creatures-higher-than-a-man-1939/.

100. Clarke, *Greetings*, 35, 107, 134; Poole, "Challenge of the Spaceship," 259.

101. Stapledon, "Interplanetary Man?," 220–23.

102. Clarke, *Exploration of Outer Space*,181.

103. Stapledon, "Interplanetary Man?," 220–21.

104. Olaf Stapledon, "Interplanetary Man?," 215–16.

105. "Toward Outer Space: An Address of His Holiness Pope Pius XII to the Seventh International Congress of Astronautics," September 21, 1956, *The Pope Speaks* 3, no. 3 (December 1956): 305–8.

3. Star of Hope

1. Robert A. Divine, *The Sputnik Challenge* (New York, 1993), xiii–xiv.

2. Robert M. Weir, *Peace, Justice, Care of Earth* (Kalamazoo, MI, 2007); Robert M. Weir, *Star of Hope: The Life and Times of John McConnell* (Pine Plains, NY, 2006);

Darrin J. Rodgers and Nicole Sparks, "Pentecostal Pioneer of Earth Day: John McConnell," in *Blood Cries Out: Pentecostals, Ecology, and the Groans of Creation*, ed. A. J. Swoboda (Eugene, OR, 2014), 19-20; John McConnell, *Earth Day: Vision for Peace, Justice and Earth Care: My Life and Thought at Age 96* (Eugene, OR, 2011).

3. John McConnell, "Make Our Satellite a Symbol of Hope!," *Toe Valley Review*, October 31, 1957, Star of Hope—news and articles, box 33, John McConnell Papers (hereafter JMP), Swarthmore College Peace Collection, Swarthmore, Pennsylvania (SCPC).

4. McConnell, "Make Our Satellite a Symbol of Hope!"; John McConnell, "Share-a-Star," n.d., Star of Hope planning/organizational, box 33, JMP.

5. Divine, *Sputnik Challenge*; David Callahan and Fred I. Greenstein, "The Reluctant Racer: Eisenhower and US Space Policy," in *Spaceflight and the Myth of Presidential Leadership*, ed. Roger D. Launius and Howard E. McCurdy (Urbana, IL, 1997); Yanek Mieczkowski, *Eisenhower's Sputnik Moment: The Race for Space and World Prestige* (Ithaca, NY, 2013); Roger D. Launius, John M. Logsdon, and Robert W. Smith, eds., *Reconsidering Sputnik: Forty Years Since the Soviet Satellite* (New York, 2000); Asif Siddiqi, *Sputnik and the Soviet Space Challenge* (Tallahassee, FL: University Press of Florida, 2003); Rip Bulkeley, *The Sputniks Crisis and Early United States Space Policy: A Critique of the Historiography of Space* (Bloomington, IN, 1991); Barbara Barksdale Clowse, *Brainpower for the Cold War: The Sputnik Crisis and National Education Act of 1958* (New York, 1981); Wayne J. Urban, *More Than Science and Sputnik: The National Defense Education Act of 1958* (Tuscaloosa, AL, 2010).

6. Walter Sullivan, *Assault on the Unknown: The International Geophysical Year* (New York, 1961); Sydney Chapman, *IGY: Year of Discover; the Story of the International Geophysical Year* (Ann Arbor, MI, 1959); McDougall, *Heavens and the Earth*, chap. 7; Western Working Paper Submitted to the Disarmament Subcommittee: Proposals for Partial Measures of Disarmament, August 29, 1957, *Documents on Disarmament, 1945-1959*, vol. 2 (Washington, DC, 1960), 871.

7. James R. Killian to DDE, "Memorandum on Organizational Alternatives for Space Research and Development," December 30, 1957, accessed June 30, 2017, https://history.nasa.gov/monograph10/doc2.pdf.

8. United States Memorandum Submitted to the First Committee of the General Assembly, January 12, 1957, *Documents on Disarmament, 1945-1959*, vol. 2 (Washington, DC, 1960), 733.

9. Statement by the United States Representative (Lodge) to the First Committee of the General Assembly, October 10, 1957, *Documents on Disarmament, 1945-1959*, vol. 2, 901-2.

10. Letters, J. Albert Robbins to Dwight Eisenhower (hereafter DDE), October 9, 1957 (emphasis in original); Paul C. Shererty to DDE, October 6, 1957 (emphasis in original); Thomas R. Mitchell to DDE, undated, 145-F Earth Circling Satellites—Space Travel—Flying Saucers—Outer Space (hereafter 145-F) (1-3), box 1155, General File (hereafter GF), Record of Dwight D. Eisenhower as President, Dwight D. Eisenhower Library (hereafter DDEL). For an analysis of the congressional and media reactions to *Sputnik*, Divine, *Sputnik Challenge*, xiv-xvi; Mieczkowski, *Eisenhower's Sputnik Moment*, chap. 1; McDougall, *Heavens and the Earth*, chap. 6.

11. Donald N. Michael, "The Beginning of the Space Age and American Public Opinion," *Public Opinion Quarterly* 24, no. 4 (Winter 1960): 573-82; Joseph Goldsen, "Public Opinion and Social Effects of Space Activity" (RAND Corporation Research Memorandum RM-2417-NASA, July 20, 1959); Sharon Weinberger, *The Imagineers of War: The Untold Story of DARPA, the Pentagon Agency That Changed the World* (New York, 2017), 32-33.

12. Eugene Exman to DDE, December 18, 1957, Star of Hope Correspondence, 1957, box 32, JMP.

13. Letter, Women's Prayer Crusade for World Order and Peace to DDE, November 22, 1957, 145-F (3), box 1155, GF, Record of Dwight D. Eisenhower as President, DDEL.

14. "Satellite for Peace" recommended in Letter, Henry S. Chasman to DDE, October 11, 1957; "Star of Hope" in several letters, and "Freedom Sphere in Letter, John S. Hayes to DDE, November 14, 1957, 145-F (3), box 1155, GF, Record of Dwight D. Eisenhower as President, DDEL.

15. Letter, Henry S. Chasman to DDE, October 11, 1957, 145-F (3), box 1155, GF, Record of Dwight D. Eisenhower as President, DDEL.

16. Eugene Exman to DDE, December 18, 1957, Star of Hope Correspondence, 1957, box 32, JMP.

17. Star of Hope, "By-laws for the Regulation, Except as Otherwise Provided by Statute or its Articles of Incorporation of Star of Hope, Inc.," n.d., Star of Hope planning/organizational, box 33, JMP; "Star of Hope Proposals," n.d. (spring 1958); "Moral Equivalent to War," in John McConnell to DDE, undated draft (1954); Louise Toness, "Star of Hope: A Coordinated Program to Wage Peace," 145-F (6), box 1156, GF, Records as President, DDEL.

18. John McConnell, "Purpose," n.d., Star of Hope proposals/ideas, box 33, JMP. Consider, as another example of the organization's goals, a June 1958 poem written by Stetler Wright, a Star of Hope member: "Over the weary, aching, world/ Flaming it's friendly light/ Emblem of friendship's noble worth/ Our Star of Hope takes flight./ Burn your message of cheer, good will/ And make the dark clouds blend/ Star of Hope dispel doubt and fear/ United the hearts of all men." See Stetler Wright (Oakland, CA), "Star of Hope," June 6, 1958, Star of Hope Correspondence, 1958, box 32, JMP.

19. Summary of Response to Star of Hope Idea Presented First in an Editorial in the Toe Valley View, October 31, 1957; Bernard Fedler to Arlene Francis, November 29, 1957; Charles and Esther Quedens to Arlene Francis, n.d.; and Mrs. Kenneth Courter to Arlene Francis, n.d.; Sam J. Ervin, Jr. (Raleigh, NC) to DDE, November 25, 1957; W. Kerr Scott (US Senate D-NC) to DDE, November 27, 1957; and Chester Bowles to Peter Hill, January 17, 1958, each in Star of Hope Correspondence, 1957, box 32, JMP; Adlai Stevenson to Peter Hill, February 10, 1958; Eleanor Roosevelt to John McConnell, July 2, 1958; Henry M. Jackson to John McConnell, December 1, 1958; and Peter Hill, Summary of Star of Hope Activities, January 26, 1958, each in Star of Hope Correspondence-1958, box 32, JMP.

20. Letter, John McConnell to Khrushchev, October 1, 1958, Star of Hope Correspondence-195, box 32, JMP; McConnell, *Earth Day*, 285-86.

21. Fred Singer, "A Statement of Conscience," *NYHT*, November 16, 1957; Nate Haseltine, "Singer Urges Christmas Satellite," *WP*, November 14, 1957; S. F. Singer, "A Reply to Sputnik," November 18, 1957, in *The Next Ten Years in Space, 1959–1969: Staff Report of the Select Committee on Astronautics and Space Exploration* (Washington, DC, 1959), 180, 184.

22. Raymond Loewy, "Let's Launch a Star of Good Will," *NYHT*, December 10, 1957; "Raymond Loewy Continues Work on 'Star of Good Will,'" *Desert Sun* (Palm Springs, CA), February 8, 1958; "Loewy Unveils Satellite Designed to Spiritually Link Men of All Nations," *Desert Sun*, February 26, 1958.

23. US Senate, "Extension of Remarks of Hon. Lyndon B. Johnson: United States Satellite's Need of Good Will Stressed," *Congressional Record*, January 29, 1958, 85th Congress, 2nd Session., Vol. 104, part 1, appendix, A759.

24. Or Rosenboim, *The Emergence of Globalism: Visions of World Order in Britain and the United States, 1939–1950* (Princeton, NJ, 2017), 1–5; Perrin Selcer, *The Postwar Origins of the Global Environment* (New York, 2018), 2.

25. John McConnell, "Proposal: Eisenhower and Khrushchev to Trade Places," May 1958, Star of Hope Correspondence-1958, 32, JMP.

26. Mrs. Ruth Finkelstein to Peter Hill, November 29, 1957; Dorothy Auerbach to Arlene Francis, November 29, 1957, Star of Hope Correspondence, 1957, box 32, JMP; Richard Philbrick, "Star of Christ No Satellite, Say Churches," *Chicago Daily Tribune*, December 21, 1957.

27. Lewis M. Haskins to Peter Hill, December 4, 1957, Star of Hope Correspondence, 1957, box 32, JMP.

28. Peter Hill to Edwin Theodore Dahlberg, December 21, 1957, Star of Hope Correspondence, 1957, box 32, JMP.

29. Ivan Whitkov to DDE, August 1, 1955, 145-F (1), box 1155, GF, Record of Dwight D. Eisenhower as President, DDEL.

30. Alfred Amer to DDE, October 14, 1957, 145-F (1), box 1155, GF, Record of Dwight D. Eisenhower as President, DDEL.

31. John McConnell, "Grow Up or Blow Up," *Geneva Diplomat*, October 1, 1958, Star of Hope—news and articles, box 33, JMP.

32. Norman Cousins, "Sense and Satellites," *Saturday Review*, October 19, 1957.

33. Malvina Lindsay, "Go-It-Alone Trend in Space, a Peril," January 16, 1958, *WP*; Malvina Lindsay, "Not Easy to Be a Space Citizen," *WP*, October 19, 1957.

34. Malvina Lindsay, "Missiles or Stars for Outer Space?," *WP*, December 7, 1957.

35. McDougall, *Heavens and the Earth*, 189.

36. Joseph M. Golden and Leon Lipson, "Some Political Implications of the Space Age," RAND P-1435, February 1958.

37. DDE, Annual Message to the Congress on the State of the Union, January 10, 1957, American Presidency Project, http://www.presidency.ucsb.edu/ws/?pid=11029; *American Foreign Policy: Current Documents, 1957*, 1316–1323; Department of State, *Department of State Bulletin*, February 11, 1957, 225, 227; Paul B. Stares, *The Militarization of Space: US Policy, 1945–1984* (Ithaca, NY, 1985), 54.

38. Letter from President Eisenhower to the Soviet Premier (Bulganin), January 12, 1958, *Documents on International Aspects of the Exploration and Use of Outer Space, 1954–1962: Staff Report Prepared for the Committee on Aeronautical and Space Sciences, United States Senate* (Washington, DC, 1963), 52–53 (emphasis added).

39. News Conference Remarks by Secretary of State Dulles Regarding Outer Space, January 16, 1958, *DSB*, February 3, 1958, 166–67.

40. John Rigden, "Eisenhower, Scientists, and Sputnik," *Physics Today* 60, no. 6 (June 1, 2007): 47–52; James R. Killian, Jr., *Sputnik, Scientists, and Eisenhower: A Memoir of the First Special Assistant to the President for Science and Technology* (Cambridge, MA: MIT Press, 1977), chap. 5; George B. Kistiakowsky, *A Scientist in the White House: The Private Diary of President Eisenhower's Special Assistant for Science and Technology* (Cambridge, MA, 1976); Zuoyue Wang, *In Sputnik's Shadow: The President's Science Advisory Committee and Cold War America* (Piscataway, NJ, 2008), 2.

41. Dwight Eisenhower, *Crusade in Europe* (Garden City, NY, 1948), 443; Benjamin Greene, *Eisenhower, Science Advice, and the Nuclear Test Ban Debate* (Stanford, CA, 2007), 10; Ira Chernus, *General Eisenhower: Ideology and Discourse* (East Lansing, MI, 2002), 126; Andrew P. N. Erdmann, "'War No Longer Has Any Logic Whatever': Dwight D. Eisenhower and the Thermonuclear Revolution," in *Cold War Statesmen Confront the Bomb: Nuclear Diplomacy since 1945*, ed. John L. Gaddis (New York, 1999).

42. McDougall, *Heavens and the Earth*, 164–69; NASA, "The Birth of NASA: November 3, 1957–October 1, 1958," Monographs in Aerospace History no. 10, accessed June 9, 2020, https://www.hq.nasa.gov/office/pao/History/monograph10/nasabrth.html.

43. James R. Killian to DDE, "Memorandum on Organizational Alternatives for Space Research and Development," December 30, 1957, accessed June 30, 2017, https://history.nasa.gov/monograph10/doc2.pdf.

44. L. A. Minnich, Jr., "Legislative Leadership Meeting, Supplementary Notes," February 4, 1958, in *Documentary History of the Dwight D. Eisenhower Presidency*, vol 21, *NASA and the US Space Program* (Bethesda, MD, 2013), 28–31.

45. Stares, *Militarization of Space*, 42.

46. Minnich, Jr., "Legislative Leadership Meeting," 28–31.

47. S. Paul Johnson to James R. Killian, "Activities," February 21, 1958, with attached: "Preliminary Observations on the Organization for the Exploration of Outer Space," February 21, 1958, Militarization of Outer Space Collection (MOSC), Digital National Security Archive (DNSA).

48. Robert L. Rosholt, *An Administrative History of NASA, 1958–1963* (Washington, DC, 1966), 9.

49. Memorandum for the President, "Organization for Civil Space Programs," March 5, 1958, with attached: "Summary of Advantages and Disadvantages of Alternative Organizational Arrangements," accessed June 3, 2017, https://history.nasa.gov/monograph10/doc5.pdf; McDougall, *Heavens and the Earth*, 171.

50. Memorandum of Conversation with the President, March 5, 1958, MOSC, DNSA; McDougall, *Heavens and the Earth*, 171; President's Advisory Committee on Government Organization, Executive Office of the President,

Memorandum for the President, "Organization for Civil Space Programs," March 5, 1958, 3.

51. Special Message to the Congress Relative to Space Science and Exploration, April 2, 1958, *Public Papers of the Presidents of the United States: Dwight Eisenhower: 1958*, doc. 64, 269; Memorandum for the Secretary of Defense Chairman, the National Advisory Committee for Aeronautics, April 2, 1958, in US House of Representatives, "Hearings before the Select Committee on Astronautics and Space Exploration," 85th Congress, 2nd Session, April 15–May 12, 1958 (Washington, DC, 1958), 967-68.

52. President's Scientific Advisory Committee (PSAC), "Introduction to Outer Space," March 26, 1958 (Washington, DC, 1958).

53. US Congress, House of Representatives, Subcommittee on National Security and Scientific Developments Affecting Foreign Policy, "Relative to the Establishment of Plans for the Peaceful Uses of Outer Space," 85th Congress, 2nd Session, May 20, 1958.

54. US Congress, House of Representatives, Subcommittee on National Security and Scientific Developments Affecting Foreign Policy, "Relative to the Establishment of Plans for the Peaceful Uses of Outer Space," 85th Congress, 2nd Session, May 20, 1958.

55. US Congress, "Relative to the Establishment of Plans."

56. US Congress, "Relative to the Establishment of Plans."

57. "Main Problems in the Senate Bill Establishing a Federal Space Agency," July 7, 1958, NASA's Origins and the Dawn of the Space Age, accessed September 25, 2018, https://history.nasa.gov/monograph10/doc7.pdf; "The Birth of NASA: November 3, 1957–October 1, 1958," NASA's Origins and the Dawn of the Space Age: Monographs in Aerospace History no.10, accessed September 25, 2018, https://www.hq.nasa.gov/office/pao/History/monograph10/nasabrth.html.

58. "National Aeronautics and Space Act of 1958," Public Law 85-568, 72 Stat., 426, in *Exploring the Unknown: Selected Documents in the History of the US Civil Space Program*,, vol. 1, *Organizing for Exploration* (Washington, DC, 1995), doc. II-17, 335.

59. Cousins, "Sense and Satellites," 26–27.

60. Federation of American Scientists, "Nuclear Test Ban, U.N. Control for Space Research, and U.N. Police Force—First Steps Toward Peace," *BAS* 14, no. 3 (1958): 125; Charles and Norma Herzfeld, "For UN Control of Outer Space," *New Leader*, December 30, 1957, 9–10; Gabriel Almond, "Public Opinion and the Development of Space Technology," in Joseph M. Goldsen, "International Political Implications of Activities in Outer Space: A Report of a Conference, October 22–23, 1959" (Rand Corporation, May 5, 1960).

61. Statement by the United States Representative (Lodge) to the General Assembly, February 14, 1957, *DSB*, March 11, 1957, 423

62. Robert F. Futrell, *Ideas, Concepts, Doctrine*, vol. 1, *Basic Thinking in the United States Air Force* (Maxwell AFB, AL, 1971), 548–49.

63. James M. Gavin, *War and Peace in the Space Age* (New York, 1958), 286–87; James M. Gavin, "Space Strategy and US Defense," *Life*, August 11, 1958, 114; Donald Cox and Michael Stoiko, *Spacepower: What It Means to You* (Philadelphia, PA, 1958), 187–94.

64. Gabriel Almond, "Public Opinion and the Development of Space Technology," in "International Political Implications of Activities in Outer Space"; "Anarchy in Space?" *World Today* 14, no 9. (September 1958): 390-98.

65. Paul Boyer, *By the Bomb's Early Light: American Thought and Culture at the Dawn of the Atomic Age* (Chapel Hill, NC, 1987), chap. 2; Paul Kecskemeti, "Outer Space and World Peace," in *Outer Space and World Politics*, ed. Joseph M. Goldsen (New York, 1963), 40; Lawrence H. Fuchs, "Nations in the Future: Organization for Survival," *Western Political Quarterly* 9, no. 1 (1956): 11-20.

66. Lincoln P. Bloomfield, *Accidental Encounters with History (and Some Lessons Learned)* (Cohasset, MA, 2005), 217; Bryan Marquard, "Lincoln P. Bloomfield, 93; Helped Create Tools to Contain Conflict," *Boston Globe*, November 12, 2013; Lincoln P. Bloomfield, "The United States, the United Nations, and the Creation of Community," *International Organization* 14, no. 4 (1960): 503-8.

67. Lincoln P. Bloomfield, "The Quest for Law and Order," in *Outer Space: Prospects for Man and Society*, rev. ed., ed. Lincoln P. Bloomfield (New York, 1968), 114-15.

68. Joseph M. Goldsen, *Outer Space and World Politics*, 18-19 (emphasis in original).

69. "Soviet Proposal on the Question of Banning the Use of Cosmic Space for Military Purposes, Elimination of Foreign Military Bases on the Territories of Other Countries, and International Cooperation in the Study of Cosmic Space, March 15, 1958," *Documents on Disarmament*, 1945-1959, vol. II, 976-77; "Dulles Rejects Russian Bid to Link Space, US Bases," *The Sun*, March 19, 1958; "Dulles Rejects Russian Plan on Space Control: Secretary in Effect Accuses Soviets of Stealing US Idea and Killing It," *LAT*, March 19, 1958.

70. Memorandum of a Conversation, Department of State, Washington, DC, March 20, 1958, *FRUS*, vol. II, doc. 438, 833.

71. National Security Council Report 5814/1, "Statement of Preliminary US Policy on Outer Space," August 18, 1958, *FRUS*, vol. II, doc. 442, 856.

72. National Security Council Report 5814/1, "Statement of Preliminary US Policy on Outer Space," *FRUS*, vol. II, doc. 442, 845-62.

73. Letter, Lodge to Hammarskjold, September 2, 1958, US Congress, Senate, "Documents on International Aspects of the Exploration and Use of Outer Space, 1954-1962: Staff Report Prepared for the Committee on Aeronautical and Space Sciences (Washington, DC, 1963), 81-82; Telegram from the Department of State to the Mission at the United Nations, *FRUS*, vol. II, doc. 443, 863-864.

74. McDougall, *Heavens and the Earth*, 184-85; "United Nations Establishes Committee on Peaceful Uses of Outer Space," *Department of State Bulletin* 40, January 5, 1959, 24-33; Telegram from the Department of State to the Mission at the United Nations, Washington, DC, November 19, 1958, *FRUS*, vol. II, doc. 457, 875; Telegram from the Mission at the United Nations to the Department of State, November 20, 1958, *FRUS*, vol. II, doc. 450, 872.

75. Address by Secretary of State Dulles to the General Assembly, September 18, 1958, *Department of State Bulletin* 39, no. 1006 (October 6, 1958), 528-29; DDE, Address Before the 15th General Assembly of the United Nations, New York City, September 22, 1960, *APP*, accessed March 8, 2017, http://www.presidency

.ucsb.edu/ws/?pid=11954; Statement by Lodge to the First Committee of the General Assembly, October 10, 1957, *Documents on Disarmament, 1945-1959*, vol. 2, 901-2; Address by Senator Lyndon B. Johnson to the First Committee of the General Assembly, November 17, 1958, *Department of State Bulletin* 39, no. 1016 (December 15, 1958): 977-80.

76. Address by Senator Lyndon B. Johnson to the First Committee of the General Assembly, November 17, 1958, *Department of State Bulletin* 39, no. 1016 (December 15, 1958): 977-80.

77. Thomas J. Hamilton, "Space-Talk Boycott Another Blow to U.N.," *NYT*, May 10, 1959; McDougall, *Heavens and the Earth*, 185.

78. United Nations General Assembly, Resolution 1348 (XIII) Question of the Peaceful Use of Outer Space, 792nd Plenary Meeting, December 13, 1958, 5-6.

79. Circular Airgram from the Department of State to Certain Diplomatic Missions, Washington, DC, April 28, 1959, *FRUS*, vol. II, doc. 458, 884-86.

80. Statement by the United States Representative (Lodge) to the General Assembly, February 14, 1957, *DSB*, March 11, 1957, 423.

81. McDougall, *Heavens and the Earth*, 174; Stares, *Militarization of Space*, 57; John Lewis Gaddis, *The Long Peace: Inquiries Into the History of the Cold War* (New York, 1989), chap. 7; Sean N. Kalic, *US Presidents and the Militarization of Outer Space, 1946-1967* (College Station, TX, 2012), 81-82.

82. USIA, Office of Research and Analysis, "Impact of US and Soviet Space Programs on World Opinion," July 7, 1959, US President's Committee on Information Activities Abroad (Sprague Committee) Reports, 1959-1961.

83. Malvina Lindsay, "Earth in Crisis—A Space View," *WP*, October 4, 1958.

84. Karl G. Harr, Memorandum to the President, "US Position on Outer Space," November 25, 1958, MOSC, DNSA.

85. McConnell, *Earth Day*, 286.

4. Lunartics!

1. US Army, *Project Horizon*, vol. 1, *Summary and Supporting Considerations*, March 20, 1959; US Army, *Project Horizon*, vol. 2, *Technical Considerations and Plans*, March 20, 1959; "Soldiers, Spies and the Moon: Secret US and Soviet Plans from the 1950s and 1960s," National Security Archive (hereafter NSA), http://nsarchive.gwu.edu/NSAEBB/NSAEBB479/; Jeffrey T. Richelson, "Shootin' for the Moon," *Bulletin of Atomic Scientists* 56, no. 5 (September/October 2000): 22-28.

2. US Army, *Project Horizon*, vol. 1, 2.

3. US Army, *Project Horizon*, vol. 1, 3, 61.

4. Memorandum, White House to J. R. Killian. "Preliminary Observations on the Organization for the Exploitation of Outer Space," February 21, 1958, Militarization of Outer Space, 1945-1991 Collection, Digital National Security Archive (hereafter MOSC, DNSA).

5. "Control of the World," *NYT*, January 9, 1958.

6. Simon Ramo, "Choosing Our Space Goals," in *Space Weapons: A Handbook of Military Astronautics*, ed. James H. Straubel et al. (New York, 1959), 164.

7. "Space and National Security," undated (but around December 1960), 4–6 Space/Nuclear, box 45, Thomas D. White Papers hereafter (TWP), Manuscripts Division (MD), Library of Congress (LOC).

8. The Military Mission in Space: A Summary of Views by Key Officials, 7–6 Nuclear Testing and Space 1962, box B137, Curtis E. LeMay Papers (CLP), MD, LOC.

9. Dwayne A. Day, "Take Off and Nuke the Site from Orbit (It's the Only Way to Be Sure)," *Space Review*, June 4, 2007, http://www.thespacereview.com/article/882/1.

10. Stephen B. Johnson, "The History and Historiography of National Security Space," in *Critical Issues in the History of Spaceflight* (Washington, DC, 2006); Dwayne A. Day, *Eye in the Sky: The Story of the Corona Spy Satellites* (Washington, DC, 1999); Paul B. Stares, *The Militarization of Space: US Policy, 1945–1984* (Ithaca, NY, 1985); Jeffery T. Richelson, *America's Space Sentinels: DSP Satellites and National Security* (Lawrence, KS, 2001); William Burrows, *Deep Black: Space Espionage and National Security* (New York, 1988); Robert M. Dienesch, *Eyeing the Red Storm: Eisenhower and the First Attempt to Build a Spy Satellite* (Lincoln, NE, 2016); Sean N. Kalic, *US Presidents and the Militarization of Outer Space, 1946–1967* (College Station, TX, 2012).

11. Johnson, "History and Historiography," 481–572.

12. Steffen Hantke, *Monsters in the Machine: Science Fiction Film and the Militarization of America after World War II* (Jackson, MS, 2016), 4; Howard E. McCurdy, *Space and the American Imagination* (Washington, DC, 1997), chap. 3; Keith M. Booker, *Monsters, Mushroom Clouds, and the Cold War: American Science Fiction and the Roots of Postmodernism, 1946–1964* (Westport, CT, 2001); David Seed, *American Science Fiction and the Cold War: Literature and Film* (Chicago, IL, 1999); Mike Ashley, *Transformations: The Story of the Science-Fiction Magazines 1950 to 1970*, vol. 2 (Liverpool, UK, 2005); David Pringle, "What Is This Thing Called Space Opera?," in *Space and Beyond: The Frontier Theme in Science Fiction*, ed. Gary Westfahl (Westport, CT, 2000).

13. *The Space Ace*, no. 1 (1951), box 34, Lilly Library, Indiana University, Bloomington (hereafter LLIU); *Space Man*, no. 5 (June–August 1963), box OS13, LLIU; "Victory Viceroy," *Out of This World*, vol. 1, no. 11 (January 1959), box OS11, LLIU.

14. John Dower, *War without Mercy: Race and Power in the Pacific War* (New York, 1987).

15. "Easy Victory," *Space War*, vol. 1, no. 19 (March 1963), box OS13, LLIU; *Space Man*, no. 2 (May–July 1962), box OS13, LLIU.

16. McCurdy, *Space and the American Imagination*, chap. 2; Joseph Kaplan and Wernher von Braun, *Across the Space Frontier* (New York, 1952), Wernher von Braun et al., *Conquest of the Moon* (New York, 1953); Chelsey Bonestell, Willy Ley, and Wernher von Braun, *The Exploration of Mars* (New York, 1956).

17. Michael J. Neufeld, *Von Braun: Dreamer of Space, Engineer of War* (Washington, DC, 2007).

18. Speech delivered during the Symposium on Space Medicine at the University of Illinois, Chicago, March 3, 1950, Speeches and Writings File, 1949–50, box 46, Wernher von Braun Papers (WvB), MD, LOC.

19. Speech at ARS Luncheon, May 6, 1953, in Indiana, Speeches and Writings File, 1951-1955, box 46, WvB; "Scientists Says Moon Military Base Useless," *LAT*, March 13, 1958; Speech before the Delta Council in Cleveland, Mississippi, May 19, 1959, Speeches and Writings Files, May-June 1959, box 47, WvB; Space Superiority as a Means for Achieving World Peace, December 1952 in San Diego, California, Speeches and Writings File, 1951-55, box 46, WvB.

20. Robert S. Richardson, "Rocket Blitz from the Moon," *Collier's Weekly*, October 23, 1948, 25.

21. Robert Granville, "Death Ray Weapons," *Space World* 2, no. 5 (May 1962): 48-49; Robert Granville, "The Weapons—Part I," *Space World* 2, no. 2 (February 1962): 34; Stephen Gorove, "On the Threshold of Space: Toward Cosmic Law," *New York Law Forum* 4 (1958): 307-8 (emphasis in original).

22. Marvin Kepler, "The Weapons, Part II," *Space World* 2, no. 3 (February 1962): 37.

23. Donald Cox and Michael Stoiko, "The Need for a United Nations Space Law," in US Senate, *Legal Problems of Space Exploration: A Symposium: Prepared for the Use of the Committee on Aeronautical and Space Sciences* (Washington, DC, 1961), 246; Thomas J. Hamilton, "U.N. Space Agreement Bogged by 'Cold War,'" *NYT*, November 16, 1958.

24. M. N. Golovine, *Conflict in Space: A Pattern of War in n New Dimension* (New York, 1962), 88-96.

25. Golovine, *Conflict in Space*, 96-105.

26. Carroll Kilpatrick, "'Moon's' Military Role Weighed," *WP*, October 10, 1957.

27. James B. Edson, "Astronautics and the Future," *BAS* 14, no. 3 (March 1958): 105.

28. Albert C. Stillson, "Space Control—How . . . and How Much?," *Air Force Magazine* (May 1959), accessed June 15, 2020, http://www.airforcemag.com/MagazineArchive/Pages/1959/May%201959/0559spacecontrol.aspx.

29. Alexander P. deSeversky, "Air Power, Missiles, and National Survival," *Air Power Historian* 5, no. 1 (January 1958): 21; Donald Cox, "Space Power: When US Air Force Becomes US Space Force," *Missiles and Rockets* (June 1957): 65-69.

30. Robert S. Richardson, "Space Fix," *Astounding Science Fiction* 31 no. 1 (April 1943).

31. Nordin Yusof, *Space Warfare: High-Tech War of the Future Generation* (Melaka, Malaysia, 1999), 268-69; Simon P. Worden and Bruce P. Jackson, "Space, Power, and Strategy," *National Interest*, no. 13 (Fall 1988): 43-52; G. Harry Stine, *Confrontation in Space* (New York, 1981), 55-58.

32. Albert C. Stillson, "The Military Control of Outer Space," *Journal of International Affairs* 13, no. 1 (1959), 73-74; Edson, "Astronautics and the Future," 105.

33. Homer A. Boushey, "Some Military Implications of Astronautics," in *The Impact of Air Power: National Security and World Politics*, ed. Eugene M. Emme (Princeton, NJ, 1959), 871-72; Homer Boushey, "Lunar Base Vital Says AF General," *Army-Navy-Air Force Register*, February 8, 1958; Homer Boushey, "The Space Frontier," in *Space Weapons*, 45.

34. "Mars, Moon Bases Foreseen in 20 Years," *AWISP* 68, no. 26, June 30, 1958; "US Base on Moon Vital, Air Force Expert Says," *BG*, January 29, 1958; "A Shot at the Moon," *Time*, March 10, 1958; "A Shot at the Moon," *Time*, March 10, 1958.

35. Louis Kraar, "Westinghouse Seeks to Build Power Station on Moon, Make Electricity from Sunlight," *WSJ*, September 11, 1958; Harold S. Roberts, "Progress Made on Moon Plant," *The Sun*, July 19, 1964.

36. "Model for a Moon-Base Displayed," *LAT*, June 22, 1962.

37. Robert B. Rigg, "Outer Space and National Defense," *Military Review* 39, no. 2 (May 1959): 22.

38. "Space Strategy Hinges on Time Factors, *AWIST* 69, no. 16 (October 20, 1958): 53-55.

39. Rigg, "Outer Space and National Defense," 22; Golovine, *Conflict in Space*, 136.

40. Thomas D. White to Lt. Gen. Ennis C. Whitehead, March 17, 1960, 4-5 Missiles/Space/Nuclear, box 36, TWP.

41. Thomas Schelling, "Military Uses of Outer Space: Bombardment Satellites," in Joseph M. Goldsen, *Outer Space in World Politics* (New York, 1963), 101; Philip Siekman, "The Fantastic Weaponry," in *Reflections on Space*, ed. Oscar H. Rechtschaffen (Air Force Academy, CO, 1964), 265; Edson, "Astronautics and the Future," 105; Golovine, *Conflict in Space*, 110, 115, 136;Donald Cox and Michael Stoiko, *Spacepower: What It Means to You* (Philadelphia, PA, 1958), 149; Klaus Knorr, "The International Implications of Outer Space Activities," in *Outer Space in World Politics*, 134-37.

42. Arthur C. Clarke, "On the Morality of Space," *Saturday Review*, October 5, 1957; Marguerite Higgins, "How Many Pearl Harbors?," in *The Challenge of the Sputniks*, ed. Richard Witkin (New York, 1957), 49-50; Lincoln P. Bloomfield, ed., *Outer Space: Prospects for Man and Society*, rev. ed. (New York, 1968), 3; Joseph M. Goldsen, "Outer Space in World Politics," in *Outer Space and World Politics*, 5; Walter Sullivan, "The International Geophysical Year," *International Conciliation* 111 (1959): 259; Lloyd V. Berkner, "Earth Satellites and Foreign Policy," *Foreign Affairs* 6, no. 2 (January 1958): 221-31.

43. Alfred Thayer Mahan, *The Influence of Sea Power Upon History, 1660–1783*, intro. Louis M. Hacker (New York, 1957).

44. Dandridge Cole, "The Panama Hypothesis," *Astronautics* 6, no. 6 (June 1961), 36-39; Golovine, *Conflict in Space*, 112-13; Cox and Stoiko, *Spacepower*, 102; "Why We Must Beat the Russians to the Moon," *Space World* 2, no. 3 (February 1962), 30.

45. Jacob D. Hamblin, *Arming Mother Nature: The Birth of Catastrophic Environmentalism* (New York, 2013), 142.

46. Stine, *Confrontation in Space*, 58.

47. Peter Ritner, "Reflections on Sea and Space," *Saturday Review*, June 22, 1957, 11-12.

48. The Military Mission in Space: A Summary of Views by Key Officials, 7-6 Nuclear Testing and Space 1962, box B137, CLP.

49. Robert F. Futrell, *Ideas, Concepts, Doctrine*, vol. 1, *Basic Thinking in the United States Air Force* (Maxwell AFB, AL, 1971), 553-54.

50. Stephen Rothstein, "Ideas as Institutions: Explain the Air Force's Struggle with Its Aerospace Concept" (PhD diss., Fletcher School of Law and Diplomacy, 2006), 158.

51. Thomas White, "Space Control and National Security," *Air Force Magazine* 41, no. 4 (April 1958): 83; Thomas D. White, "Air and Space are Indivisible," *Air Force Magazine* 41, no. 3 (March 1958): 40–41; Thomas D. White, "The Inevitable Climb to Space," *Air University Quarterly Review* 10, no. 4 (Winter 1958-59): 3–4; Thomas D. White, "At the Dawn of the Space Age," *Air Power Historian* 5, no. 1 (January 1958), 17; Alexander P. deSeversky, "Air Power, Missiles, and National Survival," *Air Power Historian* 5, no. 1 (January 1958): 22.

52. Rothstein, "Ideas as Institutions," 163.

53. David N. Spires, "The Air Force and Military Space Missions: The Critical Years, 1957–1961," in *The US Air Force in Space, 1945 to the 21st Century*, ed. R. Cargill Hall and Jacob Neufeld (Washington, DC, 1998), 39–40.

54. Emme, *Impact of Air Power*, 844.

55. "A Job for the Military, Too," *NYT*, October 8, 1961.

56. Straubel et al., *Space Weapons*, 2.

57. "Space and National Security," undated but around December 1960, 4–6 Space/Nuclear, box 45, TWP.

58. Statement by General Bernard Schriever Before the House Committee on Armed Services, n.d., box 25, Bernard A. Schriever Papers (BSP), MD, LOC.

59. "A Job for the Military, Too."

60. Bernard Schriever, "Does The Military Have a Role in Space?," *in Space: Its Impact on Man and Society*, ed. Lillian Levy (New York, 1965), 63.

61. Memorandum, Henry Jackson to James S. Lay (executive secretary of the NSC), June 15, 1960, MOSC, DNSA.

62. Eugene Zuckert, "Military Space Program," *Aviation Week and Space Technology*, March 25, 1963, 17.

63. Chester Ward, "National Sovereignty in Space," *Proceedings (American Bar Association. Section of International and Comparative Law)* (August 24, 26, 1958), 42–45; James H. Doolittle, "Impact of the Present World Situation on the Development of Peaceful Uses of Space," in *Peacetime Uses of Outer Space*, ed. Simon Ramo (New York, 1961), 33; Walter McDougall, "Technocracy and Statecraft in the Space Age—Toward the History of a Saltation," *American Historical Review* 87, no. 4 (October 1982): 1010–40.

64. Thomas D. White, "Space Control and National Security," 80–83; Thomas D. White, "Space Control and National Security," in *Space Weapons*, 11–17; Thomas D. White, "Aerospace Power ... Today and Tomorrow," *Air Force Magazine* 41, no. 11 (November 1958): 51–54; Thomas D. White, "The Space Frontier," in *Space Weapons*, 20; Rigg, "Outer Space and National Defense," 22 (emphasis in original).

65. Dan C. Ogle, "The Threshold of Space," *Air University Quarterly Review* 10, no. 2 (Summer 1958): 3.

66. US Congress, *Hearings before the Select Committee on Aeronautics and Space Exploration, House of Representatives, 85th Congress, 2nd Session, April 16, 1958* (Washington, DC, 1958), 99–102.

67. Walter McDougall, *... the Heavens and the Earth: A Political History of the Space Age* (New York, 1985), 339.

68. Nicholas M. Sambaluk, *The Other Space Race: Eisenhower and the Quest for Aerospace Security* (Annapolis, MD, 2015).

69. McDougall, *Heavens and the Earth*, 335–37.

70. "Scientists Says Moon Military Base Useless," *LAT*, March 13, 1958; Seymour Korman, "Raps View of Space as War Asset," *Chicago Daily Tribune*, March 21, 1958; "Military Base on Moon Unsound, Scientist Says," *LAT*, May 15, 1958; "Moon on the Potomac," *WSJ*, August 15, 1958; Hanson W. Baldwin, "Conquest of the Moon," *NYT*, April 14, 1958.

71. "Moon Held Too Valuable for Military Use," *LAT*, August 15, 1958.

72. Arthur C. Clarke, *The Exploration of Outer Space* (1951; repr., Greenwich, CT, 1959), 177.

73. "Moon Discounted as Arms Base," *NYT*, April 29, 1958; "Moon on the Potomac."

74. Peter Ritner, "Reflections on Sea and Space," *Saturday Review*, June 22, 1957, 11–12; Phil Casey, "Lunartics Cry for Moon but US Darkens Hopes," *WP*, November 21, 1958; John Brunner, letter to the editors, *Space World* 3, no. 2 (July 1962): 4.

75. "Planning Implications for National Security of Outer Space in the 1970s, Basic National Security Policy, Planning Task I (1)," January 30, 1964, space, Outer, 11/63-4/64 Volume I [3 of 3], box 37, National Security File (NSF), Subject File, LBJL (emphasis in the original).

76. H. Bruce Franklin, *War Stars: The Superweapon and the American Imagination* (New York, 1988).

5. The Cosmic Bomb

1. L. Reiffel, Armour Research Foundation, Illinois Institute of Technology, *A Study of Lunar Research Flights*, vol. 1 (Kirkland Air Force Base, NM: Air Force Special Weapons Center, June 19, 1959); "Soldiers, Spies and the Moon: Secret US and Soviet Plans from the 1950s and 1960s," National Security Archive (hereafter NSA), accessed May 2, 2017, http://nsarchive.gwu.edu/NSAEBB/NSAEBB479/; Vince Houghton, *Nuking the Moon: And Other Intelligence Schemes and Military Plots Left on the Drawing Board* (New York, 2019), 279–83.

2. Reiffel, *A Study of Lunar Research Flights*; James MacNees, "US Bomb on Moon Seen Sputnik Answer," *The Sun*, November 5, 1957.

3. Sidney Shallett, "New Age Ushered: Day of Atomic Energy Hailed by President, Revealing Weapon," *NYT*, August 7, 1945; Luther Huston, "No Cut in the Army Is Planned as a Result of New Bomb Use," *NYT*, August 8, 1945; "Secret War Nipped Reich Cosmic Bomb," *NYT*, August 10, 1945; Major George Fielding Eliot, "Atomic Bomb Seen Increasingly Necessity for Military Training," *NYHT*, October 26, 1945; James Marshall, "World in Mortal Peril: Promotion of Good Will Called Only Safeguard," *NYHT*, August 15, 1945.

4. Fred Kaplan, *The Wizards of Armageddon* (Stanford, CA, 1991); Daniel Ellsberg, *The Doomsday Machine: Confessions of a Nuclear War Planner* (New York, 2017); Richard Rhodes, *Arsenals of Folly* (New York, 2007).

5. Ed Regis, *Monsters: The Hindenburg Disaster and the Birth of Pathological Technology* (New York, 2015), chap. 9; Scott Kaufman, *Project Ploughshare: The Peaceful Use*

of *Nuclear Explosives in Cold War America* (Ithaca, NY, 2013); Paul R. Josephson, *Red Atom: Russia's Nuclear Power Program from Stalin to Today* (Pittsburg, PA, 2005), chap. 8.

6. H. Bruce Franklin, *War Stars: The Superweapon and the American Imagination* (New York, 1988), 3-4.

7. United Nations, General Assembly Resolution 1884 (XVIII): Stationing Weapons of Mass Destruction in Outer Space, October 17, 1963, *Documents on Disarmament: 1963* (Washington, DC, 1964), 538; Raymond L. Garthoff, "Banning the Bomb in Outer Space." *International Security* 5, no. 3 (Winter 1980): 25-40; Stephen Buono, "'This Grim Game': Kennedy and Arms Control for Outer Space," *Diplomatic History* 43, no. 5 (November 2010): 840-66.

8. Asif Siddiqi, "The Soviet Fractional Orbital Bombardment System," *Quest* 7, no. 4 (2000): 28; Paul N. Stares, "US and Soviet Military Space Programs: A Comparative Assessment," *Daedalus* 114, no. 2 (1985) 127-45.

9. ACDA, "A US-Soviet Arrangement Concerning the Placing in Orbit of Weapons of Mass Destruction," October 1, 1963, Space Activities: Space Bombs in Orbit, 1963: September-November, box 308, NSF, John F. Kennedy Library (JFKL); Peter Hill, "Love in Action Is Our Star of Hope!" December 1957, Star of Hope Correspondence 1957, box 32, JMP.

10. "First a Light Man Had Never Seen . . ." *Life* 53, no. 3 (July 20, 1962), 26-29; "Explosion's Effect on Radio and Radar Studied by Experts," *The Sun* (Baltimore, MD), July 10, 1962; "Report H-Bomb Fired 400 Mi. Up," *Chicago Daily Tribune*, July 10, 1962; John A. Osmundsen, "Blast Makes Visible Fields of Magnetism In Sky Over Samoa," *NYT*, July 10, 1962; "Report H-Bomb Fired 400 Mi. Up," *Chicago Daily Tribune*, July 10, 1962; Daniel G. Dupont, "Nuclear Explosions in Orbit," *Scientific American* 290, no. 6 (June 2004): 100-107; Mark Wolverton, *Burning the Sky: Operation Argus and the Untold Story of the Cold War Nuclear Tests in Outer Space* (New York, 2018), 188-89.

11. Elisheva R. Coleman, "Greek Fire: Nicholas Christofilos and the Astron Project in America's Early Fusion Program," *Journal of Fusion Energy* 30, no. 3 (June 2011): 238-56; Abigail Foerstner, *James Van Allen: The First Eight Billion Miles* (Iowa City, IA, 2007):188; Wolverton, *Burning the Sky*, 22-23.

12. Herbert York, *Making Weapons, Talking Peace: A Physicist's Odyssey from Hiroshima to Geneva* (New York, 1987), 131.

13. James Rodger Fleming, "Iowa Enters the Space Age: James Van Allen, Earth's Radiation Belts, and Experiments to Disrupt Them," *Annals of Iowa* 70, no. 4 (Fall 2011): 310-11.

14. Defense Nuclear Agency, DNA 6039F, *Operation Argus, 1958* (Washington, DC, 1982), 16.

15. US Department of Energy, NV0059492, Briefing to Admiral Arleigh Burke, Chief of Naval Operations, *The Argus Experiment*, July 29, 1958, DOE OpenNet, www.osti.gov/opennet; Frank H. Shelton, *Reflections of a Physicist: Project Argus* (Colorado Springs, CO, 2000), 17, slide 58; Wolverton, *Burning the Sky*, 23; Ann Finkbeiner, *The Jasons: The Secret History of Science's Postwar Elite* (New York, 2006), 56; Zuoyue Wang, *In Sputnik's Shadow: The President's Science Advisory Committee and Cold War America* (Piscataway, NJ, 2008), 109; Lisa M. Mundey, "The Civilization of a Nuclear Weapons Effects Test: Operation ARGUS," *Historical Studies in the Natural Sciences* 42, no. 4 (September 2012): 291.

16. Robert J. Watson, *History of the Office of the Secretary of Defense*, vol. 4, *Into the Missile Age, 1956–1960* (Washington, DC: Office of the Secretary of Defense Historical Office, 1997): 699–700; Donald A. Stricland, "Scientists as Negotiators: The 1958 Geneva Conference of Experts," *Midwest Journal of Political Science* 8, no. 4 (November 1964): 372–84.

17. Lisa Mundey, "Civilianization of a Nuclear Weapon," 296.

18. York, *Making Weapons*, 149.

19. Mundey, "Civilianization of a Nuclear Weapon," 296; George H. Ludwig, *Opening Space Research: Dreams, Technology, and Scientific Discovery* (Washington, DC, 2011), 369.

20. Defense Nuclear Agency, DNA 6039F, 18.

21. Wolverton, *Burning the Sky*, 87, 97–98; James A. Van Allen, Carl E. McIlwain, and George H. Ludwig, "Satellite Observations of Electrons Artificially Injected into the Geomagnetic Field," *Proceedings of the National Academy of Sciences of the United States of America* 45, no. 8 (August 15, 1959): 1152–71; Christian Brahmnstedt, ed., *Defense's Nuclear Agency, 1947–1997* (Washington, DC, 2002): 145–46.

22. James R. Killian, Memorandum for the President, "Preliminary Results of the ARGUS Experiment," November 3, 1958, Declassified Document Reference System (DDRS); Memorandum for the White House, October 17, 1958, DDRS.

23. Wolverton, *Burning the Sky*, 110.

24. N. C. Christofilos, "The Argus Experiment," *Proceedings of the National Academy of Sciences of the United States of America* 45, no. 8 (August 15, 1959): 1144–52.

25. Walter Sullivan, *Assault on the Unknown: The International Geophysical Year* (New York, 1961), 143; Mark Wolverton, "How the World Learned about the Pentagon's Sky-High Nuclear Testing," *The Atlantic*, November 24, 2018.

26. Wolverton, *Burning the Sky*, 93, 112.

27. Hanson W. Baldwin, "US Atom Blasts 300 Miles Up Mar Radar, Snag Missile Plan," *NYT*, March 19, 1959; George B. Kistiakowsky, *A Scientist in the White House: The Private Diary of President Eisenhower's Special Assistant for Science and Technology* (Cambridge, MA, 1976), 72; York, *Making Weapons*, 149–50.

28. Nuclear Explosions in Space: Hearing before the Committee on Science and Astronautics, US House of Representatives, 86th Congress, 1st Session, April 10, 1959 (Washington, DC, 1959).

29. Baldwin, "US Atom Blasts 300 Miles Up."

30. Fleming, "Iowa Enters the Space Age," 324.

31. Katherine B. Faulkner, "To the Editor," *NYT*, March 30, 1959; "Of Bombs and Men," *BG*, March 20, 1959.

32. US Senate, The Nature of Radioactive Fallout and Its Effects on Man, Hearings before the Special Subcommittee on Radiation of the Joint Committee on Atomic Energy, Congress of the United States, 85th Congress, 1st Session, May 27, 28, 29, and June 3, 1957 (Washington, DC, 1957); US Congress, Biological and Environmental Effects of Nuclear War, Hearings before the Special Subcommittee on Radiation of the Joint Committee on Atomic Energy, 86th Congress, 1st Session, June 22, 23, 24, 25, and 26, 1959 (Washington, DC, 1959), 56–60; Barton Sumner, "Space Taint Could Twist Weather," *BG*, January 21, 1962; Samuel

Glasstone, ed., *The Effects of Nuclear Weapons* (Washington, DC, 1957), 446-49; Edward Gamarekian, "Expert Says One Bomb Can End Satellite Life around World," *WP*, April 30, 1959; Brahmnstedt, *Defense's Nuclear Agency*, 135-37.

33. Harold K. Jacobson and Eric Stine, *Diplomats, Scientists, and Politicians: The United States and the Nuclear Test Ban Negotiations* (Ann Arbor, 1966), 89-90.

34. Martha Smith-Norris, "The Eisenhower Administration and the Nuclear Test Ban Talks, 1958-1960," *Diplomatic History* 27, no. 4 (September 2004): 530; Robert Divine, *Blowing on the Wind: The Nuclear Test Ban Debate, 1954–1960* (New York, 1978), chap. 10; Thomas F. Soapes, "A Cold Warrior Seeks Peace: Eisenhower's Strategy for Nuclear Disarmament," *Diplomatic History* 4, no. 1 (Winter 1980): 67-68; Brahmnstedt, *Defense's Nuclear Agency*, 147.

35. Andrei Sakharov, *Memoirs*, trans. Richard Lourie (New York, 1992), 215-27; Sergei Khrushchev, *Nikita Khrushchev and the Creation of a Superpower* (Philadelphia, PA, 2001), 514-15.

36. Anatoly Zak, "The 'K' Project: Soviet Nuclear Tests in Space," *Nonproliferation Review* 13, no. 1 (March 2006): 143-50; Pavel Podvig, ed., *Russian Strategic Nuclear Forces* (Cambridge, MA, 2001), 449-52.

37. Zak, "The 'K' Project," 143.

38. V. N. Greesai, et. al., "Response of Long Lines to Nuclear High-Altitude Electromagnetic Pulse (HEMP)," *IEEE Transactions on Electromagnetic Compatibility* 40, no. 4 (November 1998): 348-54; Fleming, "Iowa Enters the Space Age," 319.

39. "Report H-Bomb Fired 400 Mi. Up," *CT*, July 10, 1962; Robert C. Toth, "The Big Blast," *NYHT*, July 10, 1962; "Explosion's Effect on Radio and Radar Studied by Experts," *The Sun* (Baltimore, MD), July 10, 1962; Dupont, "Nuclear Explosions in Orbit,"100.

40. John Lear, "The Facts about the 1962 Space Bomb," *SR/Research*, August 6, 1963, 47; Walter Sullivan, "Radiation Belt Made by H-Bomb Produces Hiss Audible on Radio," *NYT*, October 14, 1962; Robert J. Toth, "In Outer Space—Hell to Pay," *NYHT*, August 29, 1962.

41. David S. Portree, "Starfish and Apollo (1962)," *Wired*, March 21, 2012, https://www.wired.com/2012/03/starfishandapollo-1962/.

42. Wolverton, *Burning the Sky*, 174-75.

43. Richard Hollingham, "The Cold War Nuke That Fried Satellites," *BBC.com*, September 11, 2015, https://www.bbc.com/future/article/20150910-the-nuke-that-fried-satellites-with-terrifying-results.

44. Phyllis Schlafly and Chester Ward, *The Gravediggers* (Alton, IL, 1964); Phyllis Schlafly and Chester Ward, *The Betrayers* (Alton, IL, 1968); Phyllis Schlafly and Chester Ward, *Kissinger on the Couch* (Westport, CT, 1975); Phyllis Schlafly and Chester Ward, *Ambush at Vladivostok* (Alton, IL, 1976).

45. Phyllis Schlafly and Chester Ward, *Strike from Space: A Megadeath Mystery* (Alton, IL, 1965), 50 (emphasis in original).

46. "The Polyansky Report on Khrushchev's Mistakes in Foreign Policy, October 1964," October 1964, *The National Security Archive*, https://digitalarchive.wilsoncenter.org/document/115108

47. Jeff Sutton, *Bombs in Orbit* (New York, 1959); Jeff Sutton, *H-Bomb over America* (New York, 1967).

48. Everett C. Dolman, *Astropolitik: Classical Geopolitics in the Space Age* (London: Frank Cass, 2002), 8.

49. "Scientists Say Moon Military Base Useless," *LAT*, March 13, 1958.

50. Report on Political and Informational Aspects of Satellite Reconnaissance Policy, July 5, 1962, NSAM, 156, box 336, NSF, JFKL.

51. "Sun Gun," *Time* 46, no. 2 (July 9, 1945), 58-60; "The German Space Mirror," *Life* 19, no. 4 (July 23, 1945), 78-80; Gladwin Hill, "Nazis' Scientists Planned Sun 'Gun' 5,100 Miles Up," *NYT*, June 29, 1945; John O'Reilly, "Weapons the Nazis Almost Used: A Nightmare of War Horrors," *NYHT*, June 29, 1945.

52. Louis N. Ridenour, "Pilot Lights of the Apocalypse," *Fortune* 33 (January 1946): 116-17, 219.

53. Douglas Aircraft Company, *Preliminary Design of an Experimental World-Circling Spaceship*, Report No. SM-11827 (Santa Monica, CA, 1946), 10.

54. Rand Corporation, "Conference on Methods for Studying the Psychological Effects of Unconventional Weapons," Santa Monica, California, January 26-28, 1949, MOSC, DSNA; Sean N. Kalic, *US Presidents and the Militarization of Outer Space, 1946–1967* (College Station, TX, 2012), 18-24.

55. National Security Council Report, NSC 5520, May 20, 1955, *FRUS*, 1955-1957, United Nations and General International Matters, vol. 11, ed. Lisle A. Rose (Washington, DC, 1998), doc. 340, 725.

56. PSAC, "Introduction to Outer Space," March 26, 1958, *Exploring the Unknown: Selected Documents in the History of the US Civil Space Program*, vol. 1, *Organizing for Exploration*, ed. John M. Logsdon (Washington, DC, 1995), 332; James R. Killian, *Sputnik, Scientists and Eisenhower: A Memoir of the First Special Assistant to the President for Science and Technology* (Cambridge, MA, 1977), 297.

57. Cargill Hall, "Origins of US Space Policy: Eisenhower, Open Skies, and Freedom of Space," in *Exploring the Unknown*, 229, fn. 72; Kistiakowsky, *Scientist in the White House*, 229-46; "Satellite Unsuited to Be Bomb Carrier," *BG*, March 27, 1958; "Satellite Is Seen A Flop As Carrier of H-Bomb," *The Sun*, October 9, 1957; "Satellites Have No Military Future," *LAT*, March 23, 1958.

58. Paul B. Stares, *The Militarization of Space: US Policy, 1945–1984* (Ithaca, NY, 1985), 49, 59.

59. Rand Corporation, *Space Weapons, Earth Wars* (Santa Monica, CA, 2002), 10-11.

60. Robert W. Buchheim and the Rand Corporation, *Space Handbook* (New York, 1958), 255-57; Edward Gamarekian, "Bomb Role Held Possible For Satellites: RAND Study Also Says A-Weapons Usable in Space," *WP*, January 9, 1959.

61. Brent D. Ziarnick, "Tough Tommy's Space Force: General Thomas S. Power and the Air Force Space Program" (master's thesis, Air University, 2016), 71; Dwayne A. Day, "Nuking the Site From Orbit: When the Air Force Wanted a Base on the Moon," *Space Review*, November 4, 2019, https://www.thespacereview.com/article/3826/1; Dwayne A. Day, "Take Off and Nuke the Site From Orbit (It's the Only Way to Be Sure . . .)," *Space Review*, June 4, 2007, https://www.thespacereview.com/article/882/1.

62. "The USAF Reports to Congress: A Quarterly Review Staff Report," *Air University Quarterly Review* 10, no. 1 (Spring 1958): 49.

63. James M. Gavin, *War and Peace in the Space Age* (New York, 1958), 223–24; Thomas C. Schelling, "The Military Uses of Outer Space," 97–113; M. N. Golovine, *Conflict in Space: A Pattern of War in n New Dimension* (New York, 1962), 93; Miles Marvin, "Satellites with Bombs Possible," *LAT*, May 17, 1959.

64. Golovine, *Conflict in Space*, 93; Frank Bristow, "Satellite Bomb, Good or Evil?," *LAT*, October 2, 1960.

65. Schelling, "Military Uses of Outer Space," 97–113.

66. Paul V. Bartlett and Relf A. Fenley, "The Case for a Manned Space Weapon System," *Air University Quarterly Review* 10, no. 4 (Winter 1958–59): 45; Marvin, "Satellites with Bombs Possible."

67. Astronautics and Space Exploration: Hearings Before the Select Committee on Astronautics and Space Exploration, 85th Congress, 2nd Session, April 15–May 12, 1958, 1180–81; "The USAF Reports to Congress: A Quarterly Review Staff Report," 49.

68. Max Frankel, "Khrushchev's Weapon An 'Orbital H-Bomb,' US Scientists Suggest: Soviet Boss Decides to Cut Army," *Atlanta Constitution*, January 15, 1960; "Orbital H-Bomb Seen as New Soviet Weapon," *LAT*, January 15, 1960.

69. Telegram from the Executive Secretary of the National Security Council (Smith) to the President's Special Assistant for National Security Affairs (Bundy)," September 16, 1964, *FRUS*, 1964–1968, vol. 10, *National Security Policy*, ed. David S. Patterson (Washington, DC, 2001), doc. 51.

70. Siddiqi, "Soviet Fractional Orbiting Bombardment System," 23; Siddiqi, *Sputnik and the Soviet Space Challenge* (Gainesville, FL, 2003), 321–22.

71. Director of Central Intelligence, National Intelligence Estimate (NIE) 11-1-62, *The Soviet Space Program*, December 5, 1962, CIA Electronic Reading Room, doc. 0000283833, 3–4, 21–22; Memorandum for Director of Intelligence McCone, September 25, 1963, *FRUS*, 1961–1963, vol. 5, *Soviet Union* (Washington, DC, 1998), doc. 357.

72. Siddiqi, "Soviet Fractional Bombardment System," 24; Air Power Australia, Technical Report APA-TR-2010-0101, "The Soviet Fractional Orbital Bombardment System Program," January 2010, http://www.ausairpower.net/APA-Sov-FOBS-Program.html; Braxton Eisel, "The FOBS of War," *Air Force Magazine* (June 2005): 72–75; Steven J. Zaloga, *The Kremlin's Nuclear Sword: The Rise and Fall of Russia's Strategic Nuclear Forces, 1945–2000* (Washington, DC, 2002), 112–14; Nicholas Johnson, *Soviet Military Strategy in Space* (Coulsdon, UK, 1987), 131–48.

73. Siddiqi, *Sputnik and the Soviet Space Challenge*, 253–54; "Reds Retrieve Space Capsule with Live Animals Aboard," *WP*, August 21, 1960.

74. Miles Marvin, "Can Russ Orbit Satellite Bomb? Retired General Speculates upon Feat . . ." *LAT*, August 25, 1960.

75. Stares, *Militarization of Space*, 74–75; Herbert L. Sawyer, "The Soviet Space Controversy, 1961 to 1963" (PhD diss., Fletcher School of Law and Diplomacy, 1969), 34.

76. Mark Eastwood, "Anti-Nuclear Activism and Electoral Politics in the 1963 Test Ban Treaty," *Diplomatic History* 44, no. 1 (January 2020): 133.

77. Sarah Bridger, *Scientists at War: The Ethics of Cold War Weapons Research* (Cambridge, MA, 2015), 31–32.

78. Stares, *Militarization of Space*, 73.

79. John F. Kennedy, Address in New York City Before the General Assembly of the United Nations, September 25, 1961, *APP*, accessed February 1, 2018, http://www.presidency.ucsb.edu/ws/?pid=8352; US Department of State, *Freedom from War: The United States Program for General and Complete Disarmament in a Peaceful World* (Washington, DC, 1961), 7, 15; Stares, *Militarization of Space*, 82.

80. Kendrick Oliver, *Kennedy, Macmillan and the Nuclear Test-Ban Debate, 1961–63* (London, 1998); Ronald J. Terchek, *The Making of the Test Ban Treaty* (New York, 1970); Glenn Seaborg, *Kennedy, Khrushchev and the Test Ban* (Berkeley, CA, 1983).

81. James Clay Moltz, *The Politics of Space Security: Strategic Restraint and the Pursuit of National Interests*, 3rd ed. (Stanford, CA, 2019), 130–33; Meeting on the Dominic Nuclear Test Series, September 5, 1962, in Timothy Naftali and Philip Zelikow, *John F. Kennedy: The Great Crises*, vol. 2, *September–October 21, 1962*, 81–100.

82. "Russians Accuse US on Test Plan," *NYT*, June 5, 1962; "Space Atom Dangerous, Russia Says," *Atlanta Constitution*, June 4, 1962.

83. "Charges US Tests Peril Astronauts," *CT*, September 11, 1962.

84. Ernest B. Furgurson, "Titov Assails US Nuclear Plan as Space 'Sabotage,'" *The Sun* (Baltimore, MD), June 5, 1962.

85. Ernest B. Furgurson, "Gagarin Seeks US Support In Opposing High A-Blasts," *The Sun* (Baltimore, MD), July 12, 1962; Gagarin Statement, Moscow TASS, June 5, 1962, Foreign Broadcast Information Service (FBIS).

86. Georgi Pokrovsky, "Crime in Space," *New Times* no. 25, June 20, 1962, 9–11.

87. Wolverton, *Burning the Sky*, 176

88. Alekseyev on US "Adventurism," Moscow, June 4, 1962, FBIS; High Altitude Tests Endanger World, Cairo Domestic Service, June 19, 1962, FBIS; Matyash on Indian Reaction, Moscow TASS, July 15, 1962, FBIS; US High-Altitude Tests Stir Protests, Moscow Voice and Press Transmissions, June 8, 1962, FBIS.

89. "Thant Raps US Nuclear Space Blasts: Sees Planned Tests As Sign of Psychosis," *NYT*, June 6, 1962.

90. "High A-Test Halt Is Asked by Scientists," *CDT*, June 18, 1962; "Britons Deplore Atom Test in Sky," *NYT*, May 17, 1963; Robert Toth, "Will New A-Test in Space Backfire? Yes! Say British—Bunk! Says US," *BG*, May 8, 1962.

91. Furgurson, "Gagarin Seeks US Support"; "Explosion's Effect on Radio and Radar Studied by Experts," *The Sun*, July 10, 1962; Lear, "The Facts about the 1962 Space Bomb," 46; Arthur Veysey, "Macmillan Defends US Nuclear Tests in Space," *CDT*, May 9, 1962.

92. Lisa Ruth Rand, "Orbital Decay: Space Junk and the Environmental History of Earth's Planetary Borderlands (PhD diss., University of Pennsylvania, 2016), 126; "Soviet Protests Space Needles as War Move," *WP*, May 18, 1963.

93. Toth, "Will New A-Test in Space Backfire?"

94. Veysey, "Macmillan Defends US Nuclear Tests."

95. Memo, McGeorge Bundy to Dean Rusk, n.d. (spring 1962), NSAM 156, Re: Negotiations on disarmament and peaceful uses of outer space (NSAM 156), box 336, NSF, JFKL; Editorial Note, *FRUS*, 1961–1963, vol. 7, doc. 226.

96. Memorandum for the President, "Recommended US Position on a Separate Ban on Weapons of Mass Destruction in Outer Space," July 12, 1962, NSAM 156, box 336, NSF, JFKL.

97. Garthoff, "Banning the Bomb in Space," 29.

98. NSAM 183, Space Program of the United States, August 27, 1962, box 338, NSF, JFKL.

99. Excerpt from Remarks by Deputy Secretary of Defense Gilpatric, South Bend, Indiana, September 5, 1962, NSAM 192, Re: "A Separate Arms Control Measure for Outer Space" (NSAM 192), box 339, NSF, JFKL; Editorial Note, *FRUS*, 1961–63, vol. 7, doc. 565, 563–64.

100. Address at Rice University in Houston on the Nation's Space Effort, September 12, 1962, *APP*, accessed January 22, 2018, http://www.presidency.ucsb.edu/ws/index.php?pid=8862; John Logsdon, *John F. Kennedy and the Race to the Moon* (Washington, DC, 2010) 149–51.

101. Garthoff, "Banning the Bomb in Space," 31–32.

102. Speech by Mr. Gromyko, September 19, 1963, UN doc. A/PV.1208.

103. Address before the 18th General Assembly of the United Nations, September 20, 1963, AAP, https://www.presidency.ucsb.edu/documents/address-before-the-18th-general-assembly-the-united-nations.

104. Statement by the United States Representative (Stevenson) to the First Committee of the General Assembly: Stationing Weapons of Mass Destruction in Outer Space, October 16, 1963, *Documents on Disarmament: 1963*, 536.

105. Arthur C. Clarke, "On the Morality of Space," *The Saturday Review*, October 5, 1957; Lyndon B. Johnson address before meeting of CBS affiliates, January 14, 1958, "Space Notebook [removed from binder]," box 359, CMTE on Aeronautical and Space Sciences, United States Senate, 1949–61, Papers of LBJ, LBJL; US House of Representatives, "Outer Space—the Road to Peace; Observations on Scientific Meetings and International Cooperation," 86th Congress, 2nd Session (Washington, DC, 1960); "Space: Zone of Peace," *NYT*, October 5, 1963.

106. Leonard Meeker to U. Alexis Johnson, Longer-Range Prospects in the Law of Outer Space, April 25, 1962, "Celestial Bodies Treaty Negotiating History—4, box 71, Office Files of Joseph Califano, Presidential Papers, 1963–1969, LBJL.

107. Walter McDougall, . . . *the Heavens and the Earth: A Political History of the Space Age* (New York, 1985), 335; Stares, *Militarization of Space*, 90; Kalic, *US Presidents*, 82.

108. John Lewis Gaddis, *The Long Peace: Inquiries Into the History of the Cold War* (New York, 1989), chap. 7.

109. ACDA, "A US-Soviet Arrangement Concerning the Placing in Orbit of Weapons of Mass Destruction," October 1, 1963, Space Activities: Space Bombs in Orbit, 1963: September–November, box 308, NSF, JFKL; McDougall, *Heavens and the Earth*, 342.

6. A Celestial Magna Carta

1. Thomas Graham, Jr. and Damien J. LaVera, eds., *Cornerstones of Security: Arms Control Treaties in the Nuclear Era* (Seattle, WA, 2003), 35-40.

2. John W. Finney, "Senate Approves Treaty to Limit Space Arms, 88-0," *NYT*, April 26, 1967.

3. Hal Brands, "Progress Unseen: US Arms Control Policy and the Origins of Détente, 1963-1968," *Diplomatic History* 30, no. 2 (April 2000): 255, 262; Helen Caldicott and Craig Eisenwrath, *War in Heaven: The Arms Race in Outer Space* (New York, 2007), 17.

4. Walter McDougall, ... *the Heavens and the Earth: A Political History of the Space Age* (New York, 1985), 416; Everett C. Dolman, *Astropolitik: Classical Geopolitics in the Space Age* (London, 2002), 8; Robin Ranger, *Arms and Politics, 1958–1978: Arms Control in a Changing Political Context* (New York, 1982), 8.

5. Lyndon Johnson, Remarks at the Signing of the Treaty on Outer Space, January 27, 1967, *APP*, accessed July 2, 2018, http://www.presidency.ucsb.edu/ws/?pid=28205.

6. United Nations Committee on the Peaceful Uses of Outer Space (COPUOS), Legal Subcommittee (LS), A/AC.105/C.2/SR.57, *Summary Record of the 57th Meeting*, July 12, 1966; Andrew G. Haley, *Space Law and Government* (New York, 1963), 394; COPUOS, LS, A/AC.105/C.2/SR.62, LS Summary Records, July 19, 1966.

7. Haley, *Space Law and Government*, 394.

8. COPUOS, LS, A/AC.105/C.2/SR.62, Summary Record of the 62nd Meeting, July 19, 1966.

9. Emile Laude, "Questions Pratiques," *Revue Juridique Internationale de Locomotion Aérienne*, vol. 1 (Paris, 1910): 16-18. Partial translation in Stephen E. Doyle, *The Origins of Space Law and the International Institute of Space Law of the International Astronautical Federation* (San Diego, CA, 2002), 1.

10. V. A. Zarzar, "Mezhdunarodnoye Publichnoye Vozdushnoye Pravo" (Public International Air Law), in *Problems of Air Law*, A Symposium of Works by the Air Law Sections of the USSR and RSFSR Unions of Societies for Assisting Defense and Aviation and Chemical Construction, Moscow, December 1926, vol. 1, 90-103.

11. Vladimir Mandl, *Das Weltraum-Recht: Ein Problem der Raumfahrt* (Leipzig, 1932).

12. Vladimir Kopal, "Vladimir Mandl: Founding Writer on Space Law," *Smithsonian Annals of Flight* no. 8-10 (1974): 89-90.

13. Philip C. Jessup, "Jurisdiction," *Naval War College Review* 8, no. 5 (January 1956): 28.

14. American Bar Association, Report of the Committee on Law of Outer Space, 14.B Outer Space 14.B.11 International Space Law 1959, part 2 of 2, box 343, RG 59 General Records of the Department of State, Office of the Secretary Special Assistant to Secretary of State for Atomic Energy and Outer Space, General Records Relating to Atomic Energy Matters, 1948-1962, USNA.

15. Ann L. Beresford, "Astronautics in World Affairs," *World Affairs* 124, no. 1 (Spring, 1961): 5.

16. P. W. Quigg, "Open Skies and Open Space," *Foreign Affairs* 37, no. 1 (October 1958): 95-106; Andrew G. Haley, "Law Must Precede Man into Space," *Missiles and Rockets* 2, no. 11 (November 1957): 67-70; Stuart H. Loory, "Lawyer Says Space Law Is Needed Now: Fears War If Rocket 'Incidents' Continue," *NYHT*, December 8, 1960; Matthew J. Corrigan, "Outer Space Lawyers: Eagles or Turtles?," *American Bar Association Journal* 51, no. 9 (September 1965): 858; D. G. Brennan, "Why Outer Space Control?," *BAS* 15, no. 5 (May 1959): 198-202.

17. US House of Representatives, *Hearings before the Committee on Science and Astronautics*, "International Control of Outer Space," March 5, 6, and 11, 1959, 86th Congress, 1st Session (Washington, DC, 1959), 67.

18. Chester Ward, "Projecting the Law of the Sea into the Law of Space," in *Space Law: A Symposium*, 163; McDougall, *Heavens and the Earth*, 188.

19. Chester Ward, "National Sovereignty in Space," *Proceedings of the American Bar Association. Section of International and Comparative Law* (August 24 and 26, 1958): 42-45 (emphasis in original).

20. House of Representatives, "International Control of Outer Space," 96.

21. McDougall, *Heavens and the Earth*, 188.

22. Andrew G. Haley, "The Rule of Law in the Space Age," *Foreign Policy Bulletin*, September 1, 1958, 189.

23. Andrew G. Haley, "Space Law and Metalaw—A Synoptic View," in *Space Law: A Symposium*.

24. Haley, "Space Law and Metalaw," 158.

25. Elizabeth Mirel, "Golden Rule Invalid in Space," *Science News-Letter* 85, no. 7 (February 15, 1964): 106-7.

26. Haley, *Space Law and Government*, 395; "Space Ethics," *Time* 68, no. 14 (October 1, 1956): 53; Harry Gabbett, "Lawyer Blazes Cosmic Trail," *WP*, January 9, 1957.

27. Mirel, "Golden Rule," 106-7.

28. Haley, "Space Law and Metalaw," 150-62.

29. Haley, "Rule of Law in the Space Age," 189.

30. Haley, "Space Law and Metalaw," 150-62.

31. Andrew G. Haley, "Space Law and Metalaw—Jurisdiction Defined," *Journal of Air Law and Commerce* 24, no. 3 (1957): 286-303; Mirel, "Golden Rule," 106-7.

32. Haley, *Space Law and Government*, 394.

33. Haley, "Space Law and Metalaw," 150-62.

34. Haley, "Rule of Law in the Space Age," 189.

35. Haley, "Rule of Law in the Space Age," 190.

36. American Bar Foundation, *Report to the National Aeronautics and Space Administration on the Law of Outer Space* (Chicago, IL, 1961), 5.

37. William A. Hyman, *Magna Carta of Space* (Amherst, MA, 1966), 304A.

38. Hyman, *Magna Carta of Space*, 18.

39. "William Hyman, Lawyer, Is Dead: Early Advocate of Rules for Outer Space Was 72," *NYT*, July 11, 1966.

40. Ryan A. Musto, "The Eisenhower Administration and the Origins of Regional Denuclearization: An International History (PhD diss., George Washington University, 2019).

41. Paul G. Dembling and Daniel M. Arons, "The United Nations Celestial Bodies Convention," *Journal of Air Law and Commerce* 32 (January 1966): 424-25.

42. Albert Gore, "To Rule Space: Law or Might?," *NYT*, November 10, 1963.

43. Arthur J. Goldberg Address to Committee I of the UN General Assembly on December 18, 1965, *DSB*, January 31, 1966, 163-67.

44. Position Paper for US Participation in Legal Subcommittee of UN Committee on the Peaceful Uses of Outer Space, July 15, 1960, "Celestial Bodies Negotiating History—4," box 71, Office Files of Joseph Califano, Presidential Papers, 1963-1969 (hereafter FJC), Lyndon B. Johnson Presidential Library (LBJL).

45. "United States Calls for Treaty on Exploration of the Moon," May 7, 1966, *DSB* LIV, no. 1384 (June 1966): 900; Letter, Arthur J. Goldberg to Kurt Waldheim, May 9, 1966, *DSB* LIV, no. 1406 (June 6, 1966): 900-901; ACDA, *Documents on Disarmament, 1966* (Washington, DC, 1966), 304, 347-55.

46. Eilene Galloway, "United States and USS.R. Draft Treaties on the Moon and Other Celestial Bodies—A Comparative Analysis," Space Treaty Proposals by the United States and USS.R., Staff Report Prepared for the Use of the Committee on Aeronautical and Space Sciences, US Senate, July 1966 (Washington, DC, 1966); Department of State, "Transcript of Background Press and Radio New Briefing, January 27, 1967," Background on Outer Space Treaty Amb Golberg/Sisco/Meeker, 1/27/67, RG 59, Background and Press Briefings 1961-1975, box 6, USNA.

47. Galloway, "United States and USS.R. Draft Treaties," 16-26; Goldberg to Rusk, July 22, 1966, "Celestial Bodies Treaty Negotiating History—2," box 71, FJC, LBJL; Roman Gaither to Leonard Meeker, "Comparison of Soviet and United States Drafts, Outer Space Agreement, June 17, 1966," "Celestial Bodies Treaty Negotiating History—3," box 71, FJC, LBJL; Department of State, "Transcript of Background Press and Radio New Briefing, January 27, 1967."

48. Legal Sub-Committee on the Peaceful Uses of Outer Space, July 12, 1966 to August 24, 1966, S-0858-0001-02-00001, United Nations Archives (UNA); Department of State, "Transcript of Background Press and Radio New Briefing, January 27, 1967."

49. "Toward Lunar Coexistence," *WP*, August 28, 1966; "Background Papers for Forthcoming Outer Space Talks in New York, August 29, 1966," Celestial Bodies Treaty Negotiating History—1, box 71, FJC, LBJL; Memorandum from the Executive Secretary of the Department of State (Read) to the President's Special Assistant (Rostow), August 10, 1966, *FRUS*, 1964-1968, vol. 11, doc. 145.

50. Memorandum of a Conversation, Secretary's Delegation to the Twenty-First Session of the United Nations General Assembly, September-October 1966, September 22, 1966, *FRUS* vol. 11, doc. 151, 365-68; Memo, Joseph Sisco to Dean Rusk, *FRUS*, vol. 11, doc. 154, 383; "Text of Goldberg's Address on Vietnam, Africa and Space," *NYT*, September 23, 1966.

51. Memorandum for Walt Rostow, "Negotiation of an Outer Space Treaty," August 10, 1966, MOSC, DSNA.

52. Telegram, Goldberg to DoD, August 4, 1966, "Celestial Bodies Negotiating History (hereafter CBNH)—2," box 72, FJC, LBJL.

53. Department of State, "Transcript of Background Press and Radio New Briefing, January 27, 1967."

54. Rusk to Goldberg, July 29, 1966, CBNH–2, box 71, FJC, LBJL.

55. Memorandum of a Conversation, Secretary's Delegation to the Twenty-First Session of the United Nations General Assembly, September–October 1966, September 22, 1966, *FRUS*, vol. 11, doc. 151, 365–68; Memo, Joseph Sisco to Dean Rusk, *FRUS*, vol. 11, doc. 154, 383; "Text of Goldberg's Address on Vietnam, Africa and Space."

56. Administrative History of the Department of State during the Johnson Administration, vol. 1, chap. 10: The United Nations, section E: Outer Space Treaty, White House, n.d.; Background Papers for Outer Space Talks in New York; Proposal That Nuclear Weapons Not Be Placed in Orbit, August 29, 1966; Summary of Ambassador Arthur Goldberg's Talks with Soviet Foreign Minister Andrei Gromyko Regarding: US-Soviet Outer Space and Celestial Bodies Treaty, US Bombing of North Vietnam, Nuclear Nonproliferation, Chinese Membership in the UN. Department of State, October 4, 1966. All documents available in the Declassified Document Reference System (DDRS).

57. Position Paper for US Participation in Legal Subcommittee of UN Committee on the Peaceful Uses of Outer Space, n.d., CBNH–4, box 71, FJC, LBJL.

58. Paul Kennedy, *The Parliament of Man: The Past, Present, and Future of the United Nations* (New York, 2006), 125; Mark Mazower, *No Enchanted Palace: The End of Empire and the Ideological Origins of the United Nations* (Princeton, NJ, 2009), chap. 4.

59. Dolman, *Astropolitik*, 100.

60. Dembling and Arons, "United Nations Celestial Bodies Convention," 539.

61. US Senate, "Treaty on Outer Space: Hearings before the Committee on Foreign Relations," 90th Congress, 1st Session, March 7–April 12, 1967 (hereafter "Space Treaty Hearings"), 36.

62. Background Papers for Forthcoming Outer Space Talks in New York, August 29, 1966, CBNH–1, FJC, box 71, LBJL.

63. Telegram, Goldberg to DoD, August 4, 1966, CBNH–2, box 72, FJC, LBJL.

64. Mark Orlove, "Space Out: The Third World Looks for a Way into Outer Space," *Connecticut Journal of International Law* 4, no. 3 (Spring/Summer 1989): 597–634.

65. Excerpts from an Address Delivered on Thursday, 8 December 1966, by President William V. S. Tubman, President of the Republic of Liberia, Relating to the Subject of Outer Space, PO 146/1 (1) Legal Principles on the exploration and use of outer space, S-0442-0116-07, UNA (emphasis added).

66. Excerpts from an Address Delivered on Thursday, 8 December 1966, by President William V. S. Tubman.

67. "Keeping Space Bomb-Free," *BG*, August 8, 1966.

68. "US, Russia Agree on Space Use Pact," *WSJ*, December 9, 1966, 2.

69. Lyndon Johnson, Remarks at the Signing of the Treaty on Outer Space, January 27, 1967, *APP*; Remarks of His Excellency, Anatoly F. Dobrynin, Ambassador of the USS.R. to the United States, at the Ceremony on the Outer

Space Treaty, October 10, 1967, Outer Space Treaty, NSF, box 36, LBJL; "Keeping Space Bomb-Free"; "Space Treaty?," *WP*, July 24, 1966; "A Good Treaty," *WP*, December 10, 1966.

70. "Peace on the Moon, Good Will to Orion," *CT*, December 12, 1966.

71. "Space: No Arms There, but Lots on Earth," *NYT*, December 11, 1966.

72. Joshua Lederberg, "On Cosmic Law: 'An Earthly Start,'" *WP*, December 17, 1966.

73. Cameron K. Wehringer, "The Treaty on Outer Space," *American Bar Association Journal* 54, no. 6 (June 1968): 586–88.

74. M. N. Golovine, *Conflict in Space: A Pattern of War in n New Dimension* (New York, 1962), 110, 115, 136; Donald Cox and Michael Stoiko, *Spacepower: What It Means to You* (Philadelphia, PA, 1958), 149; Philip Siekman, "The Fantastic Weaponry," in *Reflections on Space*, 265, ed. Oscar H. Rechtschaffen (Air Force Academy, CO, 1964); Thomas Schelling, "Military Uses of Outer Space: Bombardment Satellites," in Joseph M. Goldsen, *Outer Space in World Politics* (New York, 1963), 101.

75. "Peace on the Moon, Good Will to Orion."

76. "Peace on the Moon, Good Will to Orion"; Art Buchwald, "A Hawk in Space," *BG*, March 21, 1967.

77. Murrey Murder, "Senators Become Wary on Outer Space Treaty," *WP*, March 8, 967; Memorandum for the Assistant Secretary, ISA, February 2, 1967, DDRS.

78. Finney, "Senate Approves Treaty."

79. "Space Treaty Hearings," 83 (emphasis added).

80. Finney, "Senate Approves Treaty"; "Space Treaty Welcome If . . .," *LAT*, December 12, 1966.

81. "Space Treaty Hearings," 81.

82. Finney, "Senate Approves Treaty," 1.

83. Asif Siddiqi, "The Soviet Fractional Orbiting Bombardment System," *Quest* 7, no. 4 (2000): 25.

84. Statement by Secretary of Defense Robert S. McNamara at Pentagon, November 3, 1967, Bombs in Orbit-General (Ballistic Missiles in Orbit, FOBS, MOBS, etc. [hereafter Bombs in Orbit]), box 11, NSF, Files of Charles E. Johnson, LBJL; "Shadow of the FOBS: Bombs in Orbit," *Science News* 92, no. 21 (November 18, 1967): 487–88; Charles Corddry, "US Reports Russ Orbit Bomb Work," *The Sun*, November 4, 1967.

85. Memorandum for Mr. Rostow (from Spurgeon Keeny), November 8, 1967, Bombs in Orbit-General, Bombs in Orbit, box 11, NSF, FCJ, LBJL.

86. Statement on FOBS, November 3, 1967 (Spurgeon Keeny prepared for [NSC staffer] Dick Moose in the Senate), Bombs in Orbit, box 11, NSF, FCJ, LBJL.

87. Tom Wicker, "In the Nation: The Meaning of the Orbital Bomb," *NYT*, November 5, 1967.

88. Stewart Alsop, "M.I.R.V. and F.O.B.S. and D.E.A.T.H.," *Saturday Evening Post* 241, no. 7 (April 6, 1968), 16.

89. Majorie Hunter, "Dual Motive Seen in A-Bomb Report," *NYT*, November 5, 1967.

90. "Russia's Orbital Bomb," *NYT*, November 5, 1967.

91. Siddiqi, "Soviet Fractional Orbital Bombardment System," 28.

92. Dolman, *Astropolitik*, 8.

93. "Peace on the Moon, Good Will to Orion," 20.

94. Sean Harry Strijdom, "The Use of Outer Space for Military Purposes: Article IV of the Outer Space Treaty" (PhD diss., University of Pretoria, 2013), 37.

95. Lisa Ruth Rand, "Orbital Decay: Space Junk and the Environmental History of Earth's Planetary Borderlands" (PhD diss., University of Pennsylvania, 2016).

96. United Nations Institute for Disarmament Research, *Celebrating the Space Age: 50 Years of Space Technology 40 Years of the Outer Space Treaty: Conference Report 2–3 April 2007* (Geneva, 2007); Annette Froehlich, ed., *A Fresh View on the Outer Space Treaty* (Cham, Switzerland, 2018); Duncan Blake and Steven Freeland, "As the World Embraces Space, the 50-Year-Old Outer Space Treaty Needs Adaptation," *The Conversation*, July 9, 2017, accessed July 2, 2018, https://theconversation.com/as-the-world-embraces-space-the-50-year-old-outer-space-treaty-needs-adaptation-79833.

97. McDougall, *Heavens and the Earth*, 419.

98. James Vedda, "The Outer Space Treaty: Assessing Its Relevance at the 50-Year Mark," (policy paper, Center for Space Policy and Strategy, June 2017).

99. "Senate Treaty Hearings," 90.

100. Ethan Siegel, "The Tragedy of Apollo 1 and the Lessons that Brought Us to the Moon," *Forbes*, January 27, 2016, https://www.forbes.com/sites/startswithabang/2016/01/27/the-tragedy-of-apollo-1-and-the-lessons-that-brought-us-to-the-moon/#33871c9f330f.

7. Stairway to Heaven?

1. Andrew L. Jenks, *The Cosmonaut Who Couldn't Stop Smiling: The Life and Legend of Yuri Gagarin* (DeKalb, IL, 2012), 241–51.

2. William E. Farrell, "The World's Cheers for American Technology Are Mixed with Pleas for Peace," *NYT*, July 21, 1969.

3. "'In Peace for All Mankind,'" *LAT*, August 15, 1969.

4. Sunny Tsiao, *"Read You Loud and Clear!" The Story of NASA's Spaceflight and Tracking and Data Network* (Washington, DC, 2008).

5. John Krige, Angelina Long Callahan, and Ashok Maharaj *NASA in the World: Fifty Years of International Collaboration in Space* (New York, 2013), 3–4; Walter Sullivan, "Moon Landing May Be Biggest Day for Science," *Atlanta Constitution*, July 18, 1969; "Physicist to Speed Solar Wind Project," *NYT*, August 17, 1969; Richard D. Lyons, "Three Apollo Astronauts Poised to Set Out Today on Moon-Landing Mission," *NYT*, July 16, 1969; John Noble Wilford, "Lunar Dust Yields Solar Wind Gases," *NYT*, August 6, 1969; Victor McElheney, "Soviets May Use Moon Laser Reflector First," *BG*, July 20, 1969.

6. Remarks at Andrews Air Force Base on Returning From the Global Tour, August 3, 1969, *APP*, accessed September 1, 2018, http://www.presidency.ucsb.edu/ws/index.php?pid=2178; "Nixon, Returning, Hails Friendship He Found on Trip," *NYT*, August 4, 1969; Rita G. Koman, "Man on the Moon: The US

Space Program as a Cold War Maneuver," *OAH Magazine of History* 8, no. 2 (Winter 1994): 42–50; Karsten Werth, "A Surrogate for War—The US Space Program in the 1960s," *Amerikastudien/American Studies* 49, no. 4 (2004): 563–87.

7. Roger D. Launius, *Apollo's Legacy: Perspectives on the Moon Landings* (Washington, DC, 2019), 196; Asif Siddiqi, "Competing Technologies, National(ist) Narratives, and Universal Claims: Toward a Global History of Space Exploration," *Technology and Culture* 51, no. 2 (April 2010): 425–43.

8. George Dugan, "Mental Power as Force of Peace Tied to Man's Triumph in Space," *NYT*, July 21, 1969.

9. Daniel Immerwahr, "Twilight of Empire," *Modern American History* 1, no. 1 (March 2018): 132.

10. Richard S. Lewis, *The Voyages of Apollo: The Exploration of the Moon* (New York, 1974), 3–5.

11. Inaugural Address of Richard M. Nixon, January 20, 1969, *APP*, https://www.presidency.ucsb.edu/documents/inaugural-address-1.

12. "Vietnam: One Week's Dead," *Life* 66, no. 25, June 27, 1969, 20–32; Dale Carter, *The Final Frontier: The Rise and Fall of the American Rocket State* (New York, 1988), 232; "The Massacre at My Lai," *Life* 67, no. 23 (December 5, 1969): 36–45.

13. Robert Donovan, "Afterglow Lights Problems: Moon Voyage Turns Men's Thoughts Inward," *LAT*, December 29, 1968.

14. Launius, *Apollo's Legacy*, 107.

15. Inaugural Address of Richard Nixon.

16. Gil Scott Heron, "Whitey on the Moon," on *Small Talk at 125th and Lenox* (Ace Records, 1970); Jenna M. Loyd, "'Whitey on the Moon': Space, Race, and the Crisis of Black Mobility," in *Mobile Desires: The Politics and Erotics of Mobility Justice*, ed. Liz Montegary and Melissa Autumn White (London, 2015), 41–52.

17. Michael Allen, *Live From the Moon: Film, Television and the Space Race* (London, 2009), 143, 154; Neil A. Armstrong, "The Moon Had Been Awaiting Us a Long Time," *Life* 67, no. 8 (August 22, 1969): 26; "Don Oberdorfer, "Nixon Telephones the Moon: He Asserts Space Feat Will Spur Global Peace," *WP*, July 21, 1969; Andrew Chaikin, *A Man on the Moon: The Voyages of the Apollo Astronauts* (New York, 2007), 213.

18. Alexander C. T. Geppert, "Moontime: Global Synchronicity in the Age of Space," Rethinking Apollo: Technopolitics, Globality, and the Space Age, 133rd Meeting of the American Historical Association, Chicago, IL, January 3, 2019.

19. Robert Poole, *Earthrise: How Man First Saw the Earth* (New Haven, CT, 2008); Benjamin Lazier, "Earthrise; or, The Globalization of the World Picture," *American Historical Review* 16, no. 3 (2011): 602–30; Kelly Oliver, *Earth and World: Philosophy after the Apollo Mission* (New York, 2015); Denis Cosgrove, "Contested Global Visions: One-World, Whole-Earth, and the Apollo Space Photographs," *Annals of the Association of American Geographers* 84, no. 2 (1994): 270–94; Joshua Yates, "Mapping the Good World: The New Cosmopolitans and Our Changing World Picture," *Hedgehog Review* 11, no. 3 (Fall 2009): 7–27; Sheila Jasanoff, "Image and Imagination: The Formation of Global Environmental Consciousness," in *Changing the Atmosphere: Expert Knowledge and Environmental Governance*, ed. Clark A. Miller and Paul N. Edwards (Cambridge, MA, 1996); Robin Kelsey, "Reverse Shot: Earthrise and Blue Marble in the American Imagination," in

New Geographies 4: Scales of the Earth (Cambridge, MA, 2011). For a complete list works linking space exploration to the emergence political and environmental globality, see Neil H. Maher, *Apollo in the Age of Aquarius* (Cambridge, MA, 2017), 241, n. 4.

20. Chaikin, *Man on the Moon*, 212.
21. Poole, *Earthrise*, 133–34.
22. "Riders on the Earth," *BG*, December 28, 1968, 6.
23. "Small World," *NYT*, December 24, 1968, 22.
24. Poole, *Earthrise*, 6.
25. Walter Orr Roberts, "After the Moon, the Earth!," *Science* 167, no. 3914 (January 2, 1970): 13; Ron Garan, *The Orbital Perspective: Lessons in Seeing the Big Picture from a Journey of 71 Million Miles* (Oakland, CA, 2015); Frank White, *The Overview Effect: Space Exploration and Human Evolution* (New York, 1987). "Footprints on the Moon," *BG*, July 21, 1969.
26. Teasel Muir-Harmony, "Project Apollo, Cold War Diplomacy and the American Framing of Global Interdependence (PhD diss., MIT, 2009), 200; see also Teasel Muir-Harmony, *Operation Moonglow: A Political History of Project Apollo* (New York, 2020).
27. Michael Collins, *Carrying the Fire: An Astronaut's Journey* (New York, 2009), 470; White, *Overview Effect*, 37, 187.
28. Remarks to Apollo 11 Astronauts aboard the *USS Hornet* Following Completion of Their Lunar Mission, July 24, 1969, *APP*, accessed September 1, 2018, http://www.presidency.ucsb.edu/ws/?pid=2138.
29. Statement by the Secretary-General Welcoming the Apollo 11 Astronauts, August 13, 1969, Operational Files of the Secretary-General: U Thant: Speeches, Messages, Statements, and Addresses, S-0885-0001, UNA; "U Thant Hails Apollo, Urges World Harmony," *Chicago Tribune*, July 22, 1969.
30. NASA, "50 Years Ago: Apollo 11 Astronauts Return from Around the World Goodwill Tour," November 5 2019, https://www.nasa.gov/feature/50-years-ago-apollo-11-astronauts-return-from-around-the-world-goodwill-tour.
31. Teasel Muir-Harmony, "Selling Space Capsules, Moon Rocks, and America: Spaceflight in US Public Diplomacy," in *Reasserting America in the 1970s: US Public Diplomacy and the Rebuilding of America's Image Abroad*, ed. Hallvard Notaker, Giles Scott-Smith, and David J. Snyder (Manchester, UK, 2016), 131.
32. Oral History Transcript, Geneva B. Barnes, interviewed by Glen Swanson, Washington, DC, March 26, 1999, Johnson Space Center Oral History Project.
33. Norman Mailer, *Of a Fire on the Moon* (New York, 1970), 398; Roger D. Launius, "Heroes in a Vacuum: The Apollo Astronaut as Cultural Icon," *Florida Historical Quarterly* 87, no. 2 (Fall 2008): 174–209.
34. Muir-Harmony, "Selling Space Capsules," 128–30; Trevor Rockwell, "Space Propaganda 'For All Mankind': Soviet and American Responses to the Cold War 1957–1977 (PhD diss., University of Alberta, 2012).
35. Muir-Harmony, "Project Apollo," 197.
36. Eric Benson, "One Small Step for Mankind: Was Apollo 11 a Beginning or an End?," *Texas Monthly* 47, no. 7 (July 2019): 20; Amitai Etzioni, *Moondoggle: Domestic and International Implications of the Space Race* (New York, 1964).

37. John M. Logsdon, *After Apollo? Richard Nixon and the American Space Program* (London, 2015), 12.

38. Launius, *Apollo's Legacy*.

39. Maher, *Apollo in the Age of Aquarius*, 11-14, 48-53; "Letters to the Times: Readers View Apollo's Achievements in Light of Earthly Needs, Benefits," *LAT*, July 24, 1969.

40. James Jeffrey, "The Dark Side of the Moon Mission," *The Progressive*, July 18, 2019, https://progressive.org/dispatches/dark-side-of-moon-mission-jeffrey-190718/.

41. "Reaching into Space: Eminent Thinkers Mull Import of Moon Voyage for Mankind's Future," *WSJ*, July 18, 1969.

42. David Streitfeld, "Footprints in the Cosmic Dust," *WP*, July 20, 1989.

43. Dean Rusk, "The Unfinished Business of Peace," *American Scientist* 57, no. 4 (Winter 1969): 545, 553-63.

44. Flora Lewis, "Real Wisdom from Moon Trip May Be of Ourselves," *LAT*, June 22, 1969; "Man on the Moon: Mixed Emotions," *Science News* 96, no. 4 (July 26, 1969); Max Lerner, "Ironies Abound in Space Effort," *Washington Evening Star*, July 23, 1969.

45. Alexis C. Madrigal, "Moondoggle: The Forgotten Opposition to the Apollo Program," *The Atlantic*, September 12, 2012, https://www.theatlantic.com/technology/archive/2012/09/moondoggle-the-forgotten-opposition-to-the-apollo-program/262254/.

46. George Lichtheim, "Beneath the Moon," *Commentary* 47, no. 3 (March 1, 1969): 75.

47. G. L. Mehta, "Stray Thoughts on the Moon," *Times of India*, August 10, 1969.

48. GDR Papers, "Scientists Comment on Apollo," July 24 Press Opinion, Foreign Broadcast Information Service (FBIS).

49. Hanoi in Vietnamese to South Vietnam 0330 GMT, July 30, 1969, FBIS; Hanoi VNA International Service in English 0608 GMT, July 26, 1969, B, FBIS; Edward K. Wu, "Hanoi Lauds Apollo Feat: But Rejects Nixon Words on Peace Hopes," *The Sun*, July 27, 1969, 4.

50. W. Fred Boone, *NASA Office of Defense Affairs: The First Five Years* (Washington, DC, 1970), 249-51; Maher, *Apollo in the Age of Aquarius*, 60; Thomas O'Toole, "NASA's Role in War Grows," *WP*, 4 December 1967; Thomas O'Toole, "NASA Cut $282 Million More," *WP*, August 19, 1967, A4.

51. Boone, *NASA Office of Defense Affairs*, 249-51; Andrew Butrica, *Beyond the Ionosphere: Fifty Years of Satellite Communication* (Washington, DC, 1997), 65-69; Maher, *Apollo in the Age of Aquarius*, 60-65; O'Toole, "NASA's Role in War Grows."

52. Maher, *Apollo in the Age of Aquarius*, 64; Seymour J. Deitchman, "The 'Electronic Battlefield' in the Vietnam War," *Journal of Military History* 72, no. 3 (July 2008): 869-87.

53. O'Toole, "NASA's Role in War Grows."

54. Rudy Abramson, "Apollo Paradox: Peace Symbol Rooted in War," *LAT*, July 15, 1969; "Intellectuals Divided over Apollo," *BG*, July 27, 1969.

55. Mailer, *Of a Fire on the Moon*, 63-81.

56. Matthew Tribbe, *No Requiem for the Space Age: The Apollo Moon Landings and American Culture* (New York, 2014), 104; Oriani Fallaci, *If the Sun Dies* (New York, 1966).
57. Tribbe, *No Requiem*, 114–15.
58. Saul Bellow, *Mr. Sammler's Planet* (New York, 1969), 6; Tribbe, *No Requiem*, 113.
59. William D. Atwill, *Fire and Power: The American Space Program as Postmodern Narrative* (Athens, GA, 1994), chap. 6.
60. "What Is Thomas Pynchon's Gravity's Rainbow About?," *Martin Paul Eve* (blog), March 7, 2011, https://eve.gd/2011/03/07/what-is-thomas-pynchons-gravitys-rainbow-about/.
61. Immerwahr, "Twilight of Empire"; Pierre J. Huss, "Let's Claim the Moon!," *Mechanix Illustrated* (February 1957), 70–72.
62. "Flag-Planting Gives No Claim to Moon," *WP*, July 18, 1969.
63. Eugene Brooks, "Legal Aspects of the Moon Landings," *International Lawyer* 4, no. 3 (April 1970): 416–17.
64. Rusk, "Unfinished Business"; Immerwahr, "Twilight of Empire."
65. Anne M. Platoff, "Where No Flag Has Gone Before: Political and Technical Aspects of Placing a Flag on the Moon," NASA Contractor Report 188251, August 1993, accessed May 29, 2020, https://www.hq.nasa.gov/alsj/alsj-usflag.html.
66. Courtney G. Brooks, James M. Grimwood, and Loyd S. Swenson, Jr., *Chariots for Apollo: A History of Manned Lunar Spacecraft* (Washington, DC, 1979), chap. 13.
67. Muir-Harmony, "Selling Space Capsules," 132.
68. "House Votes US Flag Use on Moon," *Chicago Tribune*, June 11, 1969.
69. "Michel Insists US Flag Be 1st on Moon," *Chicago Tribune*, March 26, 1969; US Senate, Congressional Record, June 12, 1969, 91st Congress, vol. 115, part 12, 15521.
70. Anne M. Platoff, "Flags in Space: NASA Symbols and Flags in the US Manned Space Program," *Flag Bulletin: International Journal of Vexillology* 46, no. 5–6 (2007): 169–70; US Senate, Congressional Record, June 16, 1969, 91st Congress, vol, 115, part 12, 15859.
71. "What Flag(s) on the Moon?," *NYT*, June 23, 1969.
72. Lincoln Harrice, "Letter to the Editor: Flags on the Moon," *NYT*, June 6, 1969; Stephen Borsody, "Letters to the Editor of the Times: Flags on Moon," *NYT*, July 6, 1969; Darius Jhabvala, "Envoys Say UN Flag Should Fly on Moon," *BG*, July 16, 1969; Glenn W. Coate, "US Flag on Moon," *Chicago Tribune*, June 27, 1969; Philip Howitt, "Appropriate Moon Flag," *LAT*, June 12, 1969; Max Frankel, "Moon Wide Open to Human Strife: Rivalry Expected to Develop Despite Space Treaty," *NYT*, July 21, 1969.
73. Charles Reagan Wilson, "American Heavens: Apollo and American Civil Religion," *Journal of Church and State* 26, no. 2 (Spring 1984): 214; Jeffrey St. John, "Confusion in America: Space Effort: No Apologies Necessary," *LAT*, April 9, 1972; Buzz Aldrin and Malcolm McConnell, *Men from Earth* (New York, 1989), 274; Allen, *Live from the Moon*, 57.
74. Arthur C. Clarke, "Will Advent of Man Awaken a Sleeping Moon?," *NYT*, July 17, 1969.

75. Alexander MacDonald, *The Long Space Age: The Economic Origins of Space Exploration from Colonial America to the Cold War* (New Haven, CT, 2017).

76. Robert Jastrow, "The Moon Is a Rosetta Stone," *NYT*, November 9, 1969.

77. Megan Black, *The Global Interior: Mineral Frontiers and American Power* (Cambridge, MA, 2018), chap. 6.

78. C. L. Sulzberger, "Foreign Affairs: From the Moon to the Earth," *NYT*, July 20, 1969; J. E. S. Fawcett, "The Politics of the Moon," *World Today* 25, no. 8 (August 1969): 357–62.

79. *Report of the Second Annual Meeting of the Working Group on Extraterrestrial Resources* (Washington, DC, 1963), v; *Proceedings of the Working Group on Extraterrestrial Resources, 4th Annual Meeting* (Washington, DC, 1966), 10.

80. Black, *The Global Interior*, chap. 6; Thomas Atchison and Clifford W. Schultz (Bureau of Mines), "Bureau of Mines Research," *Proceedings of the Sixth Annual Working Group on Extraterrestrial Resources* (Washington, DC, 1969).

81. "Mineral Mining on Moon Studied: US Seeking to Develop Resources for Astronauts," *NYT*, July 4, 1965; Robert A. Wright, "Can We Mine the Moon?," *NYT*, July 27, 1969; Joseph L. Myler, "US Seeks Means to Tap Mineral Riches in Space," *WP*, July 4, 1965.

82. R. L. Schmidt, "Developing a Lunar Drill: A 1969 Status Report," *Proceedings of the Seventh Annual Working Group on Extraterrestrial Resources* (Washington, DC, 1970); "Ways Sought to Tap Resources on Moon," *WP*, November 20, 1969; Walter Sullivan, "What Earthly Use in the Moon? What Earthly Use Is the Moon?," *NYT*, August 28, 1966.

83. "Ways Sought to Tap Resources on Moon"; Kathleen Teltsch, "Nations Beginning to Consider Who Owns What on the Moon," *NYT*, June 29, 1969; Black, *Global Interior*, chap. 6.

84. A/8391, Request for the Inclusion of an Item in the Provisional Agenda of the Twenty-Sixth Session, Preparation of a Treaty Concerning the Moon, May 27, 1971; Letter, Andrei Gromyko to U Thant, Preparation of a Treaty Concerning the Moon," May 27, 1971.

85. COPUOS, Legal Subcommittee, A/AC.105/C.2/L.91, Draft Treaty Relating to the Moon, May 6, 1974; "US, at U.N., Urges Broader Moon Pact," *NYT*, April 11, 1972.

86. Stanley B. Rosenfield and Delbert D. Smith, "The Moon Treaty: The United States Should Not Become a Party," *Proceedings of the Annual Meeting (American Society of International Law)* 74 (April 17–19, 1980): 162–70.

87. Ed Zuckerman, "Homesteading in Space," *BG*, July 15, 1979. Patrick McCray, *The Visioneers: How a Group of Elite Scientists Pursued Space Colonies, Nanotechnologies, and a Limitless Future* (Princeton, NJ, 2012), chap. 3; Michael A. G. Michaud, *Reaching for the High Frontier: The American Pro-Space Movement, 1972–84* (Westport, CT, 1986), chap. 4; Gerard K. O'Neill, *The High Frontier: Human Colonies in Space* (New York, 1976).

88. William J. Broad, "Earthlings at Odds Over Moon Treaty," *Science, New Series* 206, no. 4421, November 23, 1979, 915.

89. Fabio Tronchetti, *The Exploitation of Natural Resources of the Moon and Other Celestial Bodies* (Leiden, 2009), 57.

90. S. Neil Hosenball and Pierre Hartman, "The Dilemma of Outer Space Law," *American Bar Association Journal* 60, no. 3 (March 1974): 302.

91. "Toward an 'Open' Moon," *NYT*, June 14, 1971.

92. Statement of Leigh Ratiner, Counsel to the L-5 Society, Accompanied by Gerald Driggers, President, L-5 Society, July 31, 1980, in US Senate, "The Moon Treaty: Hearings before the Subcommittee on Science, Technology, and Space of the Committee on Commerce, Science, and Transportation United States Senate (hereafter Moon Treaty Hearings), 96th Congress, 2nd Session, July 29 and 31, 1980 (Washington, DC, 1980), 105–7; "Way Out," *WP*, October 31, 1979.

93. Letter, Alexander Haig to Robert B. Owen, June 4, 1980, "Moon Treaty Hearings," 219–20.

94. Charles Sheffield [President of American Astronautical Society], "AAS against Lunar Agreement," *L-5 News* 5, no. 1 (January 1980).

95. Keith Henson, "Bulletin from the Moon Treaty Front," *L-5 News* 5, no. 1 (January 1980).

96. Helen Dewar, "Would-Be Space Colonists Lead Fight against Moon Treaty," *WP*, October 30, 1979.

97. Letter, Jacob Javits and Frank Church to Secretary of State Cyrus Vance, *L-5 News* 4, no. 12 (December 1979): 5.

98. Arel Lucas, "Ratiner Attacks Treaty," *L-5 News* 4, no. 12 (December 19979): 7. The five opposed were Church, Javits, S. I. Hayakawa (R-CA), Dick Stone (D-FL), and Richard Luger (R-IN).

99. "UN Moon Treaty Falling to US Opposition Groups," *L-5 News* (March 1982): see http://space.nss.org/l5-newsun-moon-treaty-falling-to-us-opposition-groups/.

100. Paul Kramer, "How Not to Write the History of US Empire." *Diplomatic History* 42, no. 5 (November 2018): 919.

101. Frankel, "Moon Wide Open to Human Strife."

102. William E. Farrell, "The World's Cheers for American Technology Are Mixed with Pleas for Peace," *NYT*, July 21, 1969; Frankel, "Moon Wide Open to Human Strife."

Conclusion

1. "Moon 'Overlord' Sends Passports to Astronauts," *WP*, December 27, 1968.

2. "Mangan Dies at 73; Founded Nation of Celestial Space," *Sun*, July 16, 1970.

3. Virgiliu Pop, *Unreal Estate: The Men Who Sold the Moon* (Cardiff, UK, 2006), 33.

4. Dean Stump, "Celestia: Nation of Celestial Space," https://nationofcelestialspace.com/history/.

5. National Space Act of 1958, 51 U.S.C. Chap. 201 (1958).

6. William C. Selover, "Nobody Owns Moon, Legal Experts Say," *CSM*, July 19, 1969.

7. Robin Dickey, *The Rise and Fall of Space Sanctuary* (El Segundo, CA, 2020), 9; Aaron Bateman, *Weapons in Space: Technology, Politics, and the Rise and Fall of the Strategic Defense Initiative* (Cambridge, MA, 2024), 27, 32–33.

8. Useful introductions can be found in James Clay Moltz, *The Politics of Space Security: Strategic Restraint and the Pursuit of National Interests*, 3rd ed. (Stanford, CA, 2019), chaps. 5–7; Joan Johnson-Freese, *Heavenly Ambitions: America's Quest to Dominate Space* (Philadelphia, PA, 2009), chap. 3; Aaron Bateman, *Weapons in Space: Technology, Politics, and the Rise and Fall of the Strategic Defense Initiative* (Cambridge, MA, 2024), chap. 1; and later chapters of Paul B. Stares, *The Militarization of Outer Space: US Policy, 1945–1984* (Ithaca, NY, 1985).

9. Joan Hoff, "The Presidency, Congress, and the Deceleration of the US Space Program in the 1970s," in *Spaceflight and the Myth of Presidential Leadership* (Chicago, IL, 1997).

10. Herbert Friedman, "Adrift in Space," *Proceedings of the American Philosophical Society* 140, no. 1 (March 1996): 10–21; Roger D. Launius, *Apollo's Legacy: Perspectives on the Moon Landings* (Washington, DC, 2019).

11. Roger D. Launius, "NASA and the Decision to Build the Space Shuttle, 1969-72," *The Historian* 57, no. 1 (Autumn 1994): 17–34.

12. Launius, *Apollo's Legacy*, 198.

13. Dwayne A. Day, "Doomed from the Start: The Manned Orbiting Laboratory and the Search for a Military Role for Astronauts," *Space Review*, June 17, 2019; Stares, *Militarization of Outer Space*, appendix I, table I, 258.

14. Paul Stares, *Space and National Security* (Washington, DC, 1987), 86; Nicholas L. Johnson, *The Soviet Year in Space, 1983* (Colorado Springs, CO, 1983), 39; Stares, *Militarization of Outer Space*, 143, 210.

15. Stares, *Militarization of Outer Space*, 203–4.

16. Stares, *Space and National Security*, 99–113.

17. Stares, *Space and National Security*, 96–99; UNIDIR, *Disarmament: Problems Related to Outer Space* (New York, 1987), 180–81; John Pike, "Anti-Satellite Weapons and Arms Control," *Arms Control Today* 13, no. 11 (December 1983): 1, 4–7.

18. Stares, *Militarization of Outer Space*, 219.

19. U.S. Senate, "Arms Control and the Militarization of Space," *Hearings before the Subcommittee on Arms Control, Oceans, International Operations and Environment of the Committee on Foreign Relations*, Senate, 97th Congress, September 20, 1982 (Washington, DC, 1982), 9.

20. Ronald Regan, Remarks at Edwards Air Force Base, California, on Completion of the Fourth Mission of the Space Shuttle Columbia, July 4, 1982, *APP*, https://www.presidency.ucsb.edu/documents/remarks-edwards-airforce-base-california-completion-the-fourth-mission-the-space-shuttle.

21. Ronald Reagan, Address to the Nation on Defense and National Security, March 23, 1983, *APP*, https://www.presidency.ucsb.edu/documents/address-the-nation-defense-and-national-security.

22. Thomas Karas, *The New High Ground: Systems and Weapons of Space Age War* (New York, 1983).

23. Frances Fitzgerald, *Way Out There in the Blue: Reagan, Star Wars, and the End of the Cold War* (New York, 2000), 481–84.

24. Konstantin Lantratov, trans. Asif Siddiqi, "The 'Star Wars' Which Never Happened," *Quest Magazine* 14, no. 1 (2007); Dwayne A. Day and Robert G. Kennedy III, "Soviet Star Wars," *Air and Space Magazine*, January 2010, https://www.airspacemag.com/space/soviet-star-wars-8758185/; Peter J. Westwick, "'Space-Strike Weapons' and the Soviet Response to SDI," *Diplomatic History* 32, no. 5 (2008): 955–79.

25. Peter Anson Bt. and Dennis Cummings, "The First Space War: The Contribution of Satellites to the Gulf War," *RUSI Journal* 136, no. 4 (Winter 1991): 41–53.

26. Larry Greenemeier, "GPS and the World's First 'Space War,'" *Scientific American*, 8 February 2016; Marcia S. Smith, CRS Report for Congress, *Military and Civilian Satellites in Support of Allied Forces in the Persian Gulf* (Washington, DC, 1991).

27. Greenemeier, "World's First 'Space War.'"

28. William J. Perry, "Desert Storm and Deterrence," *Foreign Affairs* 70, no. 4 (Fall 1991): 66–82.

29. For a partial list, see Bleddyn Bowen, *War in Space: Strategy, Spacepower, Geopolitics* (Edinburgh, UK, 2020), 13, n. 5.

30. Steven J. Burger, "Not Ready for the First Space War: What about the Second?," *Naval War College Review* 48, no. 1 (Winter 1995): 73–83; Colin S. Gray, "The Influence of Space Power upon History," *Comparative Strategy* (October–December 1996); Judson J. Jussel, *Space Power Theory: A Rising Star* (Maxwell AFB, AL, 1998), 10–12.

31. David E. Lupton, *On Space Warfare* (Maxwell AFB, AL, 1998), 29, 34.

32. Niall Ferguson, "Cold War II," *Boston Globe*, March 11, 2019; Michael Hirsh, "Global Cold War," *Foreign Policy*, June 27, 2022; Robert Kaplan, "A New Cold War Has Begun," *Foreign Policy*, January 7, 2019.

33. Demetri Sevastopulo, "China Tests New Space Capability with Hypersonic Missile," *Financial Times*, October 16, 2021, https://www.ft.com/content/ba0a3cde-719b-4040-93cb-a486e1f843fb.

34. Andrew Jones, "'We're in a Space Race,' NASA Chief Says US 'Better Watch Out' for China's Moon Goals," *Space.com*, January 5, 2023, https://www.space.com/nasa-bill-nelson-china-space-race-moon.

35. Tim Marshall, "China's Bid to Win the New Space Race," *Wired*, April 12, 2023, https://www.wired.co.uk/article/china-space-race.

36. DoD, *National Security Space Strategy: Unclassified Summary* (Washington, DC, 2011), i.

37. David Montgomery, "Trump's Excellent Space Force Adventure," *Washington Post Magazine*, December 3, 2019, https://www.washingtonpost.com/magazine/2019/12/03/trumps-proposal-space-force-was-widely-mocked-could-it-be-stroke-stable-genius-that-makes-america-safe-again/.

38. United States Space Force, *Spacepower: Doctrine for Space Forces* (June 2020), 16, 26.

39. CNN Special Report, Season 37, Episode 11, "War in Space: The Next Battlefield," aired November 29, 2016, 9 p.m. ET, CNN.

40. Everett C. Dolman, *Astropolitik: Classical Geopolitics in the Space Age* (London, 2002), 4–5.

41. Bleddyn E. Bowen, *Original Sin: Power, Technology and War in Outer Space* (New York, 2022); Daniel Deudney, *Dark Skies: Space Expansionism, Planetary Geopolitics, and the Ends of Humanity* (New York, 2020), xix.

42. Quoted in Barton Beebe, "Law's Empire and the Final Frontier: Legalizing the Future in the Early *Corpus Juris Spatialis*," *Yale Law Journal* 108, no. 7 (May 1999):1754–55.

Index

Page references followed by an f indicate a figure.

Abernathy, Ralph, 232
Abramson, Rudy, 237
Adler Planetarium, Chicago, 2f, 253
Advanced Research Projects Agency (ARPA), 98, 102, 142, 158, 170, 172
Aelita (A. Tolstoi), 40
Aerojet Engineering Corporation, 76, 79
Aerojet-General Corporation, 135
Air Force Magazine, 133, 135, 141
Air Force Space Command, 131
Air University, 141
Air University Quarterly, 172
Aldiss, Brian, 150
Aldrin, Edwin "Buzz," 223–24, 227, 230, 233, 238, 241
Almond, Gabriel, 105
Alsop, Stewart, 218
Alter, Dinsmore, 149
AMC, 72
American Association for the Advancement of Science (AAAS), 161
American Astronautical Society, 247
American Interplanetary Society (AIS), 43, 45, 52, 54, 57
American Machine and Foundry, Inc. (AMF), 146
"American Moon" (song), 241
American Physical Society, 161
American Rocket Society (ARS), 52, 62
Anders, Bill, 13–14, 227–28
Anderson, Rudolf, 185
Anglo-Peruvian Company, 21
A.N.N.A. satellite, 116
Antarctic Treaty, 110, 204, 206, 221, 250
Anticipations (Wells), 26
Apollo 1, 222
Apollo 8, 4, 13–15, 227–28, 232, 252
Apollo 11, 5, 147, 223–26, 229–32, 234, 238, 240–42, 249–51, 253

Apollo 13, 243
Apollo Service and Lunar Excursion Modules, 74–75
Apollo-Soyuz Test Project, 219
Arid Zone Program, 78
Ariel I, 166
Arlene Francis Show, 88, 91
Armed Forces Special Weapons Project (AFSWP), 159, 163
Arms Control and Disarmament Agency (ACDA), 155, 183, 186, 257
arms race
 FOBS and, 218
 LTBT and, 177
 NASA avoidance of, 85
 Outer Space Treaty (OST), 193, 213
 sanctuary politics and, 4, 157
 Soviet Union and, 95, 186, 199, 256, 259–60
 to space, 5–6, 113, 126, 146, 150, 155, 164, 175, 178, 180, 184, 221, 258, 264
 UN Resolution 1884 and, 184–86
Armstrong, Neil, 223–24, 227, 230, 233, 238
Army Air Forces (AAF), 72, 97, 141
Arnold, Henry "Hap," 76, 141
Aron, Raymond, 48
"Artificial Modification of the Earth's Radiation Belt" (Singer), 161
Association of Inventors, 40
Astounding Science Fiction, 133
Astronautics (magazine), 138
Atomic Energy Commission (AEC), 97, 145, 158–59, 166
Atoms for Peace Conference (1958), 89
Auf Zwei Planeten. *See* Two Planets
Aviation Incorporated, 135
Aviation Week and Space Technology, 128, 145, 148

INDEX

Baldwin, Hanson, 62, 160–61
Ballistic Missile Early Warning System (BMEWS), 173, 217f
Baltimore Sun, 54, 88
Ban on Bombs in Space (BBS), 191
Bartlett, Paul V., 172
Battle of Dorking, The (Chesney), 34
Becker, Loftus E., 200
Bellow, Saul, 237–38
Bell Telephone, 72
Bennett, Wallace F., 240
Bethe, Hans, 96
Biden, Joe, 261
Binder, Otto, 130
Black, Megan, 243
Bloomfield, Lincoln P., 9, 106–7
Blue Marble, 14–15
Boeing, 72, 141, 257
Bogdanov, Aleksandr, 14, 18–19, 35, 38
Bohr, Niels, 53
Bonestell, Chelsey, Jr., 130
Borman, Frank, 13–14, 227–28
Boston Globe, 228
Boston Post, 30
Boushey, Homer A., 134–35, 140, 142, 144–45, 191, 242, 261
Bowles, Chester, 88
Boyle, Hal, 62
Bradley, Omar, 155
Brands, Hal, 193
Bridges, Styles, 102
Brilliant Pebbles, 258–59
Bristow, Frank, 171
British Interplanetary Society (BIS), 54–55, 57–58, 60–61, 63, 68–69, 71, 73
British Projectile Development Establishment, 77
Brundage, Percival, 99
Bryce, James, 45–46
Buchheim, Robert, 170
Buell, Raymond Leslie, 46
Bulganin, Nikolai, 95, 103, 108, 185
Bulletin (AIS), 43
Bureau of Budget (BoB), 99–100
Bush, George H. W., 259

Caldicott, Helen, 193
Calley, William, 234
Carne, Judy, 241
Carpenter, Scott, 178
Carr, E. H., 47–48

Carrying the Fire (Collins), 229
Carson, Rachel, 179
Carter, Jimmy and his administration, 256–57, 260
Catholic Association for International Peace, 104
CBS Evening News, 232
Celestia (Nation of Celestial Space), 1–3, 56, 73–74, 252, 268n1
Central Intelligence Agency (CIA), 114, 159, 173, 178, 181, 183, 217
Chaffee, Roger, 222
Chelomei, Vladimir N., 174
Chesney, George Tomkyns, 34
Chicago Herald, 88
Chicago Tribune, 213, 219
China, 7, 240, 248, 260–62
"choke point" or Panama hypothesis, 137–39, 261
Christian Science Monitor, 88, 199f
Christofilos, Nicholas, 157–62
Church, Frank, 247–48
Clarke, Arthur C.
 background of, 58
 BIS address, 60–61, 63
 ground-controlled approach (GCA) radar, 58
 interplanetary discourse and, 15, 57
 Interplanetary Project and, 175
 Lasser and, 64
 military strength and space, 65
 on moon bases, 149
 "morality of space," 184, 253, 264–65
 nursery of human life, 80–81
 photograph of, 59f
 predictions of war, 62
 space exploration and peace, 16–17, 58–61, 73–74, 128, 249
 Stapledon and, 68
 symbolism of space, 265
 United States and, 73
Clarke, Arthur C., publications
 2001: A Space Odyssey, 228
 "Challenge of the Spaceship, The," 63–64
 Childhood's End, 58, 65–66, 73
 Exploration of Space, The, 51, 64, 128
 Islands in the Sky, 58
 Preludes to Space, 58, 66–67
 Sands of Mars, The, 58
Cleator, Philip Ellaby, 58

INDEX 319

Cleaver, Arthur "Val," 55f, 56–57, 73, 175
Cold War
  fears of future war, 5–6, 62, 73, 91–93
  "first" Cold War, 7
  military space spending, 7, 115, 147–48
  military technologies and, 125
  nuclear testing and, 163
  Outer Space Treaty (OST) and, 196, 221
  Second Cold War, 7, 261–64, 271n25
  space exploration and peace, 41–42, 51, 71, 85
  See also nuclear weapons; space race
Cole, Dandridge, 133, 138
Collier's, 62, 128–29
Collins, Michael, 229–30, 233
Committee on the Peaceful Uses of Outer Space (COPUOS), 112–13, 123, 178, 191, 193, 205–7, 209–10, 219–20, 244
"Common Sense of World Peace, The" (Wells), 30
Conference on Discontinuance of Nuclear Weapons Tests, 162–63
Conflict in Space (Golovine), 131, 135–36
Conquest of Outer Space, The (Ley), 128
Conquest of Space, The (Lasser), 44–45
Convair, 72
cooperative space projects, 191
Cornils, Ingo, 32
Corréa, Henrique Alvim, 23f
cosmic philosophies, 39–40, 51, 54
Cosmopolitan (magazine), 30
Cosmos (Malina), 79
Cousins, Norman, 92–93, 100, 104
Cox, Donald, 105, 131
Cultural and Scientific Congress for World Peace (1949), 73

Darwinism, 21, 46, 60, 69
Das Weltraum-Recht (The Law of Outer Space) (Mandl), 197–98
Day, Dwayne, 125
Dean, Patrick, 192f
Defense Nuclear Agency (DNA), 159
defensive systems
  antiballistic missile (ABM), 164, 193, 219, 256
  ballistic missile defense (BMD), 155, 257, 259
  FOBS and, 174
  NAVSTAR GPS, 260
  nuclear deterrence, 134, 163, 261

  Projects SMART and SLOMAR, 132
  research and development (R&D), 215, 257
  satellite interceptor (SAINT), 123, 132, 147
  satellite protection for area defense (SPAD), 132
defensive systems, Air Force studies, 171
Democratic party, 225–26
Department of Defense (DoD), 88, 97, 99, 102–3, 115–16, 255, 260–61, 263
  See also military and space
deSeversky, Alexander P., 133
Dickinson, G. Lowes, 45
Dickinson, William L., 218
Die Rakete (The Rocket), 41, 51
Die Rakete zu den Planetanraumen (The Rocket into Interplanetary Space) (Oberth), 36
Die Starfield Company (Ley), 42
Dillon, Clarence, 113
Director of Defense Research and Engineering (DDR&E), 142
disarmament and demilitarization of outer space, 4, 105, 108
Dobrynin, Anatoly, 183, 192f, 196, 212, 216
Dolman, Everett C., 194, 219, 263
Dominic Nuclear tests, 166, 177–79
Doolittle, James, 146
Dooner, Pierton, 35
Dornberger, Walter, 30, 50, 52, 130, 137
Dostoevsky, Fyodor, 37
Douglas Aircraft Company, 72, 169
Dr. Strangelove (Kubrick), 125, 145
Dryden, Hugh, 242
du Bois, W. E. B., 46–47
DuBridge, Lee A., 96, 148–49
Dulles, John Foster, 95, 108, 110, 115

Earthrise, 13–15, 228f, 234, 252
Earth vs. The Flying Saucers (movie), 127
Edison's Conquest of Mars (Serviss), 30–31
Edson, James B., 132, 134, 137
Ed Sullivan Show, 241
Effects of Nuclear Weapons (Glasstone), 162
Eighteen Nation Committee on Disarmament (ENCD), 181, 212
Einstein, Albert, 78
Eisendrath, Craig, 193

Eisenhower, Dwight D. and his administration
  ARPA created, 142, 158
  Bulganin and, 95, 103, 108
  demilitarization of outer space, 107, 264
  Killian and, 98–99
  McConnell and, 89, 91
  NASA and, 97, 102
  nuclear testing moratorium, 155, 159, 162–63, 185
  nuclear weapons and, 96, 153
  "Open Skies" proposal, 185
  Project Horizon and, 122
  satellite technology and, 114, 170
  Soviet Union and, 85
  space based weaponry and, 5
  on space exploration, 110–11
  space race and, 92
  Sputnik and, 94
  weaponization of space rejected, 140
  See also Committee on the Peaceful Uses of Outer Space
Electric, General, 72, 98, 133, 170
Electric Spaceship, The (Oberth), 73
Ervin, Sam, 88
Etzioni, Amitai, 233
European Anarchy, The (Dickinson), 45
Evening Journal (New York), 30
Evins, Joe, 240–41
Experiment in Autobiography (Wells), 23
Exploration of Space, The (Clarke), 51
Explorer I, 138, 151, 198
Explosives Act (1875), 55
"Extra-Terrestrial Relays" (Clarke), 58

Fallaci, Oriana, 237
Farnsworth, R. L., 62–63, 66
Federal Bureau of Investigation (FBI), 79–80
Fedorenko, Nikolai, 206, 208
Fedorov, Nikolai Federovich, 37–38
Fenley, Relf A., 172
Ferguson, James, 235
fiction, post Great War, 34–35
"Final War, The" (Spohr), 42
Finch, Edward R., 4
First Men in the Moon, The (Wells), 14, 24, 25f, 26–27
Fisher, Adrian, 183
Forbidden Planet (movie), 127

Ford, Gerald and his administration, 166, 254, 256
Foreign Affairs, 261
Forman, Ed, 74
Fortune (magazine), 168
Four-Power Disarmament Conference, 95
Frankel, Max, 241
Franklin, H. Bruce, 154
Frau im Mond (Lang), 41
Friendship 7, 231
From the Earth to the Moon (Verne), 21, 36
Fulbright, J. W., 210
Furnas, C. C., 149

G-20 Summit (2022), 261
Gagarin, Yuri, 174, 178, 204, 223
Gallup polls, 87, 233
Garan, Ron, 229
Garthoff, Raymond L., 181
Gavin, James M., 104
Geiss, Johannes, 224
Geppert, Alexander, 227
Germany, 40–41, 52–53, 197, 237–38
Gernsback, Hugo, 42–43
Gilpatric, Roswell, 182
Glasstone, Samuel, 162
Glenn, John, 178, 231
Global Positioning System (GPS), 260–61, 263
Global Protection Against Limited Strikes (GPALS), 259
Global South, 5, 46, 196, 209–12, 220, 230, 236, 247–49
Goddard, Robert, 30, 36, 40, 44, 52, 74, 76
Goldberg, Arthur, 192f, 205–6, 212, 215
Goldsen, Joseph M., 107
Goldwater, Barry, 8, 247
Goldwater conservatism, 167
Golovine, M. N., 131, 134–35, 137, 171
Goodyear, 72
Gorbachev, Mikhail, 259
Gore, Albert, 205
Gorove, Stephen, 130
Graham, Billy, 88
Grant, Madison, 46
Granville, Robert, 130
Gravity's Rainbow (Pynchon), 238
Gray, Colin, 261
Great War, 26–27, 47–48, 58
Green, Harold, 181
Grissom, Virgil "Gus," 222

INDEX 321

Gromyko, Andrei, 183–84, 206–8
Grotius, Hugo, 202
Guggenheim Aeronautical Laboratory (GALCIT), 74, 76

Haig, Alexander, 247–48
Hailsham, Quintin Hogg, 166
Haley, Andrew G., 195, 201–3
Hammarskjold, Dag, 105, 109
Harbou, Thea von, 41
Harr, Karl G., 116
Harris polls, 255
Heavens and the Earth, The (McDougall), 6, 270n18
Heinemann, William, 20
Henson, Keith, 245
Hersey, John, 61
Herter, Christian, 108, 110, 113
Herzfeld, Norma and Charles, 104
Hill, Peter, 89, 92
Hiroshima (Hersey), 61
Ho Chi Minh, 234
Hotline Agreement, 183, 194
House Armed Services Committee, 143
House Committee on Science and Aeronautics, 185
House Select Committee on Astronautics and Space Exploration, 170
Hoyle, Fred, 179
Hsue-Shen Tsien, 76
Hughes Aircraft Company, 72
Huxley, Thomas Henry, 21–22, 26
Hyman, William A., 203–4
Hynek, Allen, 87

Icaromenippus (Lucian), 24
India, 7, 21, 112, 209–10, 260
Influence of Sea Power upon History, The (Mahan), 137–38
"Influence of Space Power upon History, The" (Gray), 261
Institute for Aeronautical Sciences, 98–99
Inter-American Bar Association, 203
intercontinental ballistic missile (ICBM)
  building better missiles, 104
  countermeasures and deterrence, 186
  detection and destruction of, 259
  disarmament and, 108, 115
  electronic components in, 160
  FOBS and, 217
  manned satellite cost comparison, 172

new insights for, 54
nuclear-armed bombardment satellites (NABS), 171
plasma shields and, 158
Positive Control Bombardment System, 131
R-7 Semyorka, 151
Random Barrage System (RBS), 132
Russian R-7 missile, 82
science discovery of, 65
Soviet Union and, 83, 94, 165
Soviet Union and FOBS, 173–74
Space Defense program and, 256
submarine-launched ballistic missile (SLBM), 129, 216, 218
targets for, 148
tracking satellites and, 136
X-20 Dyna-soar and, 141
International Geophysical Year (IGY), 85, 92, 109, 112–13, 158–60
International Interplanetary Commission, 45
international law, 204
  See also Antarctic Treaty
International Organization (Bloomfield), 106
International Relations (Buell), 46
international relations (IR)
  Apollo 11 view of earth and, 229
  Huxley's ideas on, 26
  international peace, 51, 56
  "Interplanetary School," 17–18, 47–48, 51, 72–73, 249
  Machiavellianism, 45–46
  outer space as an escape, 19–20
  pessimism for peace, 7
  sanctuary paradigm and, 9
  science exploration and peace, 65
  space exploration and peace, 47, 55–56, 65, 201
  space exploration impact, 44–45
  "strategic restraint," 253
  United States space policy, 108
  World War I and, 45, 47
International Space Station, 263
Interplanetary Project, 57, 175, 251
In the Days of the Comet (Wells), 27, 28f, 29–30, 36, 81
In the Fourth Year (Wells), 27
Invaders from Mars (movie), 127
Invasion of New York, The (Palmer), 35
Invisible Man, The (Wells), 20

INDEX

Island of Dr. Moreau, The (Wells), 20, 22, 24
It! The Terror from Outer Space (movie), 127
Izvestiya, 40, 161

Jackson, Henry "Scoop," 88, 144
JAG Journal, 200
Javits, Jacob, 247–48
Jessup, Philip C., 198
jet-assisted takeoff (JATO), 76–77
Jet Propulsion Laboratory (JPL), 75f, 76–78, 98, 141, 159, 186, 242
Johnson, John A., 200
Johnson, Lyndon B. and his administration, 123, 150, 192f, 206, 212, 218, 253
   See also Vietnam war
Johnson, Lyndon B. (senator), 4, 98, 111, 112f, 140, 184
Johnson, Roy, 98, 170, 172, 175
Johnson, S. Paul, 99, 102, 123
Johnson, Stephen, 125
Johnson, U. Alexis, 240
Joint Chiefs of Staff (JCS), 215
Journal of the British Interplanetary Society, 72

Kant, Immanuel, 201
Kash, Donald, 7
Kaysen, Carl, 177
Keating, Kenneth, 100–101
Kecskemeti, Paul, 106
Keeny, Spurgeon, 218
Kennedy, John F. and his administration, 5, 9, 89, 123, 135, 140, 147, 175–77, 180–85, 241, 250, 264
Kennedy, Paul, 209
Khariton, Yuri, 164
Khrushchev, Nikita
   flag planting on moon and, 241
   FOBS and, 173–74
   joint lunar mission, 186
   Kennedy and, 176
   McConnell proposal and, 91
   nuclear testing moratorium, 155, 159, 163
   space weaponry and, 8, 85, 94, 164, 191
   Star of Hope and, 89
   test ban resolutions, 176
Khrushchev, Sergei, 164
Killian, James R., 85, 96, 98–99, 157–58, 160
King, Martin Luther, Jr., 225
Kissinger, Henry, 166
Kistiakowsky, George, 98

Knorr, Klauss, 137
Komarov, Vladimir, 223
Korolev, Sergei, 30, 173–74
Krasovskiy, V. I., 161
Kubrick, Stanley, 145, 228

L-5 Society, 246–47
Lang, Fritz, 41
Lasser, David, 14–15, 43, 44f, 47, 50, 52, 57, 64, 80, 195
Lasswitz, Kurd, 31, 32f, 33–35, 39, 42, 45, 65, 67
Last Days of the Republic (Dooner), 35
Laude, Emil, 197
Launius, Roger, 232
Lawrence Livermore National Laboratory, 157–58, 258
Leary, Timothy, 247
Lederberg, Joshua, 214
LeMay, Curtis, 140, 170
Lewis, C. S., 63–64, 68
Lewis, Flora, 233
Lewis, Richard, 225
Ley, Willy, 41–42, 52, 73, 80, 128, 130–31
Liberia, 211
Lichtheim, George, 233
Life (magazine), 1, 128, 156, 168, 226, 255
Limited Test Ban Treaty (LTBT) (1963), 147, 155, 175–77, 180–84, 186, 191, 194, 204, 216
Lindsay, Malvina, 93–94, 100, 115
Lipp, James, 169
Lipp, Ralph E., 173
Lodge, Henry Cabot, 86, 95, 104, 109–11, 113
Loewy, Raymond, 90–91, 106
London, Jack, 35
Look (magazine), 228
Los Angeles Times, 171, 224, 233
Lovecraft, H. P., 68
Lovell, Bernard, 162, 179
Lovell, James, 13–14, 227
Lucas, George, 150
Lucky Dragon incident, 162
Luna 10, 222
Luna I, 198
Luna II, 239
Luna IX, 205
Lupton, David, 261

Mackenzie, Jeanne and Norman, 22
Mackinder, Halford, 45, 124, 130

INDEX 323

MacLeish, Archibald, 13, 232
Macmillan, Harold, 159, 166
Mahan, Alfred Thayer, 124, 137–39, 261
Mailer, Norman, 231, 237
Major Inapak the Space Ace (movie), 127
"Make Our Satellite a Symbol of Hope!" (McConnell), 83–84
Malina, Frank Joseph, 74–75, 75f, 76–80
Mandl, Vladimir, 197–98
Mangan, James T., 1, 2f, 3, 5–6, 9, 51, 56, 73–74, 252–53, 265
Manhattan Project, 53, 122, 173
Man into Space (Oberth), 41
manned spaceflight, 74, 111, 142, 210, 242, 254
Mars Project (von Braun), 50
Martin Company, 72, 135, 248
Martin-Marietta, 243
Massachusetts Institute of Technology (MIT), 43, 96, 106, 146
McCarthy, Eugene, 214–15, 220
McConnell, John, 83–84, 88–92, 94, 100, 104, 106–7, 116, 221
McCormack, John, 100–102
McCray, W. Patrick, 79
McDonnell Aircraft, 72
McDougall, Walter, 6–7, 9, 194
McElroy, Neil, 98, 158
McNamara, Robert, 147, 166, 217, 217f, 218–19
Mead, Margaret, 144
Meinel, Carolyn, 245
Menter, Martin, 238
Michael, Donald N., 87
Michel, Robert, 240–41
military and space
  Lagrange points, 138–39
  lunar military base, 129–30, 134–37
  Operation Desert Storm effect, 261
  Outer Space Treaty (OST), 196
  programs cancelled, 147
  Project Horizon moon military base, 121, 122f, 123, 125–26, 128, 140
  research and development (R&D), 72, 98, 123, 140, 149, 151
Moltz, James Clay, 253
Mongolia, 210
Moon Car, The (Oberth), 73
moon exploration
  Apollo Lunar Surface Drill (ALSD), 243
  claim to territory, 250
  Earthrise, 228f

lunar mining interest, 242–44, 249–50
Luna X, 205
moon landing, 227–29
national flag raising, 238–42
politics, military and, 237, 249
program cost and domestic needs, 232–34
Surveyor 1, 205
symbolic act of, 248–49
Tranquility Base, 224, 227, 250, 252
Moon Treaty (1979), 196, 244–48
Morgenthau, Hans, 48
Morozov, Platon, 178
Mr. Sammler's Planet (Bellow), 237–38
Mumford, Lewis, 61
Munro, Leslie, 108–9
Murrow, Edward R., 62
mutually assured destruction (MAD), 257

National Academy of Science Committee on Air Corp Research, 75–76
National Academy of Sciences (NAS), 97, 158
National Advisory Committee for Aeronautics (NACA), 97, 99–102, 225
National Aeronautics and Space Act of 1958, 102–3, 169, 172, 236
National Aeronautics and Space Administration (NASA)
  Air-Delivered Seismic Intrusion Detector (ADSID), 236
  Apollo 11 "Giantstep" tour, 230–31
  Apollo Lunar Surface Drill (ALSD) and, 243
  Apollo Paradox, 237
  Apollo program budget, 232
  Apollo spacecraft, 135, 224
  Applications Technology Satellite (ATS), 235
  budget cutbacks, 255
  Chinese competition, 262
  civilian or military, 5, 8, 97–101, 114, 141, 191
  Committee on Symbolic Activities for the First Lunar Landing, 239
  Department of Defense and, 123
  Earthrise photo and, 228
  in fiction, 127
  FOBS and, 218
  imperialism charge, 238, 242
  legislation for, 102
  "Limited Warfare Committee," 235

INDEX

lunar flag raising, 225–26, 241
Lunar Module (LM), 223, 239–40
lunar resources and, 244
A.N.N.A. satellite, 116
peaceful uses of outer space and, 85, 103, 114, 227
planning after moon landing, 254
Project Argus, 160
religion and, 91
satellite overflight question, 200
Starfish radiation issue, 166
technology usefulness, 249
Vietnam war and, 234–37
WAC Corporal Rocket, 75f
Working Group on Extraterrestrial Resources (WGER), 242–43
See also National Advisory Committee for Aeronautics
National Intelligence Estimate (NIE), 62, 173
National Science Foundation (NSF), 97
National Security Council (NSC), 94, 108–9, 116, 123, 129, 169, 216, 218
National Space Act (1958), 108, 169, 172
Nature (magazine), 128
Naval Research Laboratory (NRL), 98, 141
Navigation Signal Timing and Ranging (NAVSTAR), 260
Nelson, Bill, 262
Nervo, Padilla, 184
Neufeld, Michael, 40
Newsweek, 161
New Times (newspaper), 178
New World of Islam, The (Stoddard), 46
New Worlds for Old (Wells), 26
New Yorker, 50, 61
New York Herald Tribune, 88, 90
New York Times
  Apollo 8 and, 228
  Argus and, 160–62
  flag planting on moon and, 241
  FOBS and, 217f
  future space conflict, 131, 144
  moon landing, 13
  moon mineral resources, 242
  Moon Treaty (1979) and, 246
  Outer Space Treaty (OST), 213, 218
  space as zone of peace, 185
  Vietnam war, 234
Next War, The (Wallace), 35
Nicholas II (tsar), 18

Nixon, Richard, 98, 224–27, 229–30, 232, 234, 238, 249, 255
Non-Proliferation of Nuclear Weapons Treaty(NPT), 193–94, 212
North Atlantic Treaty Organization (NATO), 7, 138, 174, 204, 263
Northrop Space Laboratories, 243
Northrup, 72
Nuclear Science and Engineering Corporation, 145
nuclear weapons
  arms race, 164, 199
  Astron, 157
  Bikini Atoll nuclear test, 162
  Fractional Orbital Bombardment System (FOBS), 7–8, 155, 173–75, 196, 216–19, 221, 255
  IRBMs, 160, 186, 256
  moratorium on, 155, 159, 163–64, 167–68, 185–86
  nuclear tests in space, 156–57
  Project Argus, 159–62, 164, 166, 177
  Soviet threat, 166–67, 183
  Starfish Prime, nuclear test, 156–57, 165–66, 177–80
  UN prohibition in outer space, 205
  See also Limited Test Ban Treaty

Oberth, Hermann, 15, 30, 36, 40–44, 73
Ogle, Dan C., 145
O'Neill, Gerard K., 245
On the Beach (Shute), 162
On the Origin of the Species (Darwin), 21
"On the Possibility of Establishing a Plasma Shield" (Christofilos), 158
Opel, Fritz von, 41
Operation Desert Storm, 260–61
Oppenheimer, J. Robert, 96
Orwell, George, 30, 51
Outer Space (Bloomfield), 107
Outer Space Treaty (OST), 200
  approval and skepticism, 193–94, 209, 212, 213f, 214
  compromise of, 197
  developing world and, 210
  draft issues, 206–8, 220
  importance of, 196, 219–22
  inspection provision, 207, 214–16
  military viewpoint of, 215–16, 255
  moon claims and, 238, 241
  moon landing and, 239

moon resources and, 250
Moon Treaty (1979) and, 246
ratification of, 140, 193, 214–16
space sanctuary and, 194–95, 215, 253
space war and, 123, 125
State Department and, 244
tracking facilities and, 207–9
Treaty on Principles Governing the Activities of States in the Exploration and Use of Outer Space, 192
UNGA resolutions in, 204
United Nations and, 8, 86, 193, 212
violations of, 218
White House ceremony, 192f
Outline of the History of the Rocket (Ley), 42

Paine, Thomas O., 232, 239–40
Parsons, John "Jack," 30, 74
Passing of the Great Race, The (Grant), 46
Paul VI (pope), 204, 224
Pearson's Magazine, 20
Pendray, G. Edward, 43
Perel'man, Yakov Isidorovich, 39
Perry, William J., 260
Philosophy of the Common Task, The (Filosofiia obshchego dela) (Fedorov), 37
"Pilot Lights of the Apocalypse" (Ridenour), 168
Pinto, M. C. W., 246
Pius XII (pope), 82
Pokrovsky, Georgi, 178
political realism, 18, 47
Post-Cuba Negotiations with the USSR, 183–84
Power, Thomas S., 136, 242
Powers, Francis Gary, 185
Preliminary Design of an Experimental World-Circling Spaceship, 169
President's Scientific Advisory Committee (PSAC), 96, 98, 100, 146, 158, 169
Project Apollo, 237, 249, 255
Project Argus, 159–62, 164, 166, 177
Project Excalibur, 258
Protazanov, Yakov, 40
Putin, Vladimir, 262
Putt, Donald, 140, 142, 146, 170, 174
Pynchon, Thomas, Jr., 9, 238

Quarles, Donald, 170

Rabi, I. I., 96
race relations, 46–47
radiation pollution, 157, 159–60, 162, 166, 177
Radio Moscow, 174
RAF Quarterly, 59
Ramo, Simon, 149
RAND (Research and Development Corporation), 94, 107, 169–70, 186, 242
Rapacki, Adam, 204
Ratiner, Leigh, 246
Reagan, Ronald and his administration, 125, 194, 221, 247–48, 256–60
red-fuming nitric acid (RFNA), 75, 77
Red Star (Bogdanov), 14, 18–19
Republican party (GOP), 98
Revolt against Civilization (Stoddard), 46
Rhodes, Cecil, 271n30
Richardson, Robert S., 129–30, 133
Ridenour, Louis, 168–69
Rigg, Robert B., 135
Rising Tide of Color, The (Stoddard), 46
Ritland, Osmond J., 131
Ritner, Peter, 139, 149–50
Roberts, Walter Orr, 229
Rockefeller, Nelson, 99
"Rocket and the Next War, The" (Lasser), 43–44
Rocket Research Project, 74
Roddenberry, Gene, 150, 202
Roosevelt, Eleanor, 88
Roosevelt, Theodore, 138, 238
Rostow, Eugene, 257
Rostow, Walt, 183
Roudeboush, Richard, 240
Rumsfeld, Donald, 254
Rusk, Dean, 177, 181, 192f, 207, 212, 215, 233, 239
Russia, 18–19, 37–39, 262–63
  See also Soviet Union
Rynin, Nikolai, 39

Sagan, Carl, 153, 224–25
Sakharov, Andrei, 164
SALT. *See* Strategic Arms Limitation Talks/Treaty
sanctuary doctrine, 86, 144, 147
Saturday Evening Post, 218
Saturday Review, 61, 92, 139, 150
Saturn V, 226, 232, 239
Saund, Dalip Singh, 101
Schelling, Thomas, 137, 171–72
Schirra, Willy, 177

INDEX

Schlafly, Phyllis, 166-67
Schlesinger, Arthur, Jr., 253
Schmidt, R. L., 243
Schmitt, Carl, 48
Schofield, Martin B., 104
Schriever, Bernard, 140, 142-44, 146-47, 191, 261
science fiction and science writing
  casual mechanism of, 16
  early space novels, 34-36
  early space travel science, 36, 38-41
  future war stories comparison, 34-35
  "gravity well" and, 133-34, 139
  interplanetary fiction, 14-15, 19, 41
  military and space, 123-24, 150-51
  moral arguments of, 15-16, 36, 41, 80-81
  peaceful uses of outer space and, 48, 85
  space war depictions, 127-28, 130
  See also Bogdanov, Aleksandr; Clarke, Arthur C.; Lasser, David; Stapledon, William Olaf; Wells, Herbert George
Scott, W. Kerr, 88
Scott-Heron, Gil, 226
Seaborg, Glenn T., 89
Senate Foreign Relations Committee, 215, 221, 246, 248
"Sense and Satellites" (Cousins), 104
Serviss, Garrett, 31
Seventh International Congress of Astronautics, 82
Shafer, Raymond, 226
Shapley, Willis, 239
Sheffield, Charles, 247
Shepherd, Alan, 231
Shiel, M. P., 35
Shklovskiy, I. S., 161
Shute, Nevil, 162
Silent Spring (Carson), 179
Singer, S. Fred, 90-91, 106, 161
Slaughterhouse Five (Vonnegut), 237
Society for Space Travel (Verein fur Raumschiffahrt) (VfR), 40-42, 50-51, 54
Society for the Study of Interplanetary Communications (OIMS), 40, 54
Sonnengewehr (Sun Gun), 168
Southern Christian Leadership Conference, 232
Soviet Union
  air law conference (1926), 197
  antisatellite (ASAT) testing, 255-56
  Cuban Missile Crisis, 156, 165, 167, 183, 185
  Eisenhower and, 107-8
  Experimental Design Bureau (OKB), 174
  FOBS and, 173-74, 217-19
  Globalnaya Raketa-1 (Global Rocket 1, or GR-1), 173
  Laika, dog orbited, 198
  military and space, 124, 129, 142-43
  Ministry of Medium Machine Building, 164
  nuclear testing and moratorium, 163-64
  Outer Space Treaty (OST) and, 206-7, 212, 214, 218-19
  political collapse of, 259-60
  R&D laser studies, 259
  "Soviet World Domination under Preparation," 143f
  space travel and, 52, 122, 151
  System K, 164-65
  "Tsar Bomba" blast, 163-64
  UN proposal, 111-12
  view of NASA, 103
  Vostok program, 174
space boosters, 128, 131, 139, 242, 245
space diplomacy, 230-32
space exploration
  Tubman moratorium proposal, 211
  UN resolutions, 8
space for peace
  Apollo and, 226
  arms control and, 171
  arms race and, 177
  Cold War and, 185
  Mangan and, 6, 74, 253
  military and space, 137, 144
  post-Sputnik advocates, 85, 94-96
  satellite technology and, 210
  Space Act and, 103
  United Nations and, 107, 109
  U. S. and Soviets and, 110, 113-16
Space Frontier (von Braun), 128
space law, 86, 105, 109, 130, 193-98, 200-204
Space Law and Government (Haley), 201
Space Man, 128
Space Opera (Aldiss), 150
Spacepower (Cox, Stoiko), 105
space race (1957-1969), 9, 84, 86-87, 90, 92, 114-15, 198

space sanctuary
  Apollo 11 and, 5, 226
  or dark side of space, 9
  Earthrise photo and, 14
  etymology of, 4, 185
  failure of, 261
  idea of, 6
  military and political assessment of, 155
  nuclear tests in space, 157
  Reagan administration and, 257
  space policy and, 18, 250-51, 253-54
space warfare
  American Artemis Program, 262
  civilian casualties and, 136-37
  Cold War and, 151
  depictions of, 124, 127-31
  Desert Storm and, 261
  failure of, 146-50
  First Gulf War as, 260
  military and civilian disagreement, 183
  military study of, 125-26, 153
  options of, 263-64
  orbital bombardment, 129-30
  Outer Space Treaty (OST), 192
  programs cancelled, 147
  security and insecurity, 152
  UN proposal, 111
  USAF and, 140
  US Space Force, 263
  See also military and space
Space warfare, 128
Space World, 128, 130-31, 149
Spohr, Carl, 42
Sputnik
  China and, 7
  fears of future war after, 8
  globalists and, 91-92, 94, 96, 100, 109, 116
  ideas prior to, 1, 4
  military policy after, 124
  post-Sputnik advocates, 5, 90
  reaction to, 86-87, 92, 151
  U.S. space exploration and, 6, 83, 122
Spykman, Nicholas, 124, 130
spy satellites, 114, 125, 147-48, 183-85, 214
Stapledon, William Olaf, 57, 67-71, 69f, 70-73, 80-82, 195, 253
Stapledon, William Olaf, publications
  Last and First Men, 68-73
  Last Men in London, 68
  Latter-Day Psalms, 67

  Modern Theory of Ethics, A, 68
  Odd John, 68
  Star Maker, 68, 73
Stares, Paul, 175
Star of Goodwill, 90
Star of Hope, 84, 88, 89f, 90-92, 104, 107, 116, 284n18
Star Trek (Roddenberry), 150, 202
Star Wars (Lucas), 150
Stevenson, Adlai, 88, 184
Stillson, Albert C., 133-34
Stine, G. Harry, 133, 139
Stoddard, T. Lothrop, 46
Stoiko, Michael, 105, 131
Stolley, Dick, 156
Strand Magazine, The, 24
Strategic Air Command (SAC), 136, 144
Strategic Arms Limitation Talks/Treaty, 193, 255-56
Strategic Defense Initiative Organization (SDIO), 258-59
Strike from Space (Schlafly), 166-67
"Study of Lunar Research Flights, A," 153, 175
Stump, Dean, 252
Subcommittee on National Security and Scientific Developments hearing, 100-102
Suicide Squad, 75, 77
Sullivan, Walter, 160-61
Sulzberger, C. L., 242

Teller, Edward, 163, 258
Tereshkova, Valentina, 186
Thant, U, 179, 229-30
Thurmond, Strom, 216
Time (magazine), 61, 71, 128, 158
Time Machine, The (Wells), 20, 22
Times of India, 233, 253
Titov, Gherman, 174, 178
Today Show, 88
Toe Valley Review, 83, 92
Tolkien, J. R. R., 63
Tolstoi, Aleksey, 40
Tolstoy, Leo, 37
Toynbee, Arnold J., 233
True History (Lucian), 24
Truman, Harry S., 96
Trump, Donald, 7, 263
Tsiolkovsky, Konstantin, 15, 17, 30, 36, 38-40, 47, 50, 80, 221

INDEX

Tsukov, Yuri K., 165
Tubman, William V. S., 211
Turner, Frederick Jackson, 45, 63
Twenty Years' Crisis, The (Carr), 47
Twining, Nathan, 140
Two Planets (*Auf zwei Planeten*) (Lasswitz), 31–36, 39, 42, 253
2001: A Space Odyssey (film) (Kubrick), 228

Union of Soviet Socialist Republics. *See* Soviet Union
United Arab Republic (UAR), 112, 210
United Kingdom, Explosives Act (1875), 55
United Nations (UN)
  arms control and, 86
  ban on weapons in space, 105, 155, 175
  Declaration of Legal Principles, 187
  Disarmament Committees, 176
  founding of, 51
  Moon Treaty (1979) and, 245
  nuclear weapons and, 59–60
  role in outer space, 104–7, 109–13, 115–16, 131, 144
  UN Declaration of Legal Principles, 221
  wealth disparity issues, 209
United Nations Convention on the Law of the Sea (UNCLOS), 246, 250
United Nations Educational, Scientific and Cultural Organization (UNESCO), 78–79
United Nations General Assembly (UNGA)
  cultural and economic exchange programs, 105
  disarmament and demilitarization of outer space, 181
  Kennedy address, 175–76
  Lyndon Johnson and, 111, 112f, 123
  moon treaty and, 205, 248
  negative law resolutions and outer space, 204–5
  "nuclear free zone" in Central and Eastern Europe, 204
  Outer Space Treaty (OST), 193, 206, 212, 215, 219
  Space committee, 110
  space exploration resolutions, 8
  "Stationing Weapons of Mass Destruction in Outer Space," 184–85
  test ban resolutions, 109, 176, 184, 186, 191
  world peace and, 85

United States, 53–54, 71–72, 108–9, 116, 145–47, 155, 260
United States Bureau of Mines (USBM), 242–43
United States Information Agency (USIA), 115, 231
United States Space Force (USSF), 263
United Technologies Corporation, 247
Universal Declaration of Human Rights, 51
"Unparalleled Invasion, The" (London), 35
U. S. Air Force (USAF)
  Air-Delivered Seismic Intrusion Detector (ADSID), 236
  Air Force Office of Scientific Research, 135
  ASAT countermeasures and, 143, 256
  Ballistic Missiles Division, 125, 170
  China and, 262
  Directorate of Advanced Technology, 140, 145
  Dyna-Soar program, 7–8, 123, 141, 147
  Kennedy's moon landing pledge, 135
  lunar military base idea, 134
  Manned Orbiting Laboratory (MOL), 7–8, 123, 147, 216, 255
  manned satellite bombers, 172
  military and space, 124, 126, 140–41
  military space spending, 148, 151, 214
  miniature homing vehicle (MHV), 256
  nuclear explosion on moon, 153–54
  nuclear weapons and, 175
  Project Orion, 145
  research and development (R&D) at, 145
  satellite technology and, 170–71
  Soviet competition and, 142–44
  Space Command (1982), 257
  Systems Requirements (SR) studies, 125
  X-series aircraft, 141
U. S. Army
  Army Ballistic Missile Agency (ABMA), 98, 141
  Army Ordnance Department, 72, 77
  Counter Intelligence Corps (CIC), 78
  Project Horizon and, 121, 122f, 135, 147
  rocketry and, 53
  space weaponry and, 175
  von Braun and, 50
  See also Army Air Forces; military and space
US Disarmament Administration, 175

INDEX 329

US National Defense Research
  Committee, 76
U. S. Navy Bureau of Aeronautics, 72
U. S. Navy Bureau of Ordnance, 72
US Rocket Society, 62
U. S. State Department, 200
  Bloomfield and, 106
  Committee of Principals, 175
  Giantstep tour and, 230
  lunar interests, 226
  Malina's passport, 79
  manned lunar landing and, 205
  National Intelligence Estimate (NIE) and, 62
  National Security Action Memorandum (NSAM), 180
  nuclear weapons ban and, 183
  Outer Space Treaty (OST), 208, 215, 217, 244
  public diplomacy and, 231
  sanctuary doctrine and, 144
  Soviet proposal for demilitarization, 108–10
  space law and, 247

Valier, Max, 40–41
Van Allen, James, 158–61, 264
Van Allen belts, 138, 158, 161, 165–66, 179, 198
Vance, Cyrus, 216, 247
Verne, Jules, 21, 36, 38, 58, 74
Vietnam war, 8, 167, 193, 208, 212, 226, 234–36, 238, 249, 255
Vitalis, Robert, 46
Vitoria, Francisco de, 201–2
von Braun, Wernher, 30, 33, 50, 52, 73, 79–80, 88, 128–29, 148, 237
von Kármán, Theodore von, 74, 77, 141
Vonnegut, Kurt, 232, 237
Vorwarts (Social Democratic Party), 42
Vought company, 257

Wald, George, 237
Waldheim, Kurt, 206
Wallace, King, 35
Wall Street Journal, 212
Wang, Zuoyue, 96
War and Peace in the Space Age (Gavin), 104
War and the Future (Wells), 26
Ward, Chester, 145, 166–67, 200
"War in Space" (CNN documentary), 263

War in the Air, The (Wells), 26
War of the Worlds, The (Wells), 20–22, 23f, 24, 26, 30–31, 35, 130, 249, 253
War of the Worlds, The (Wells) (movie), 127
War That Will End War, The (Wells), 27
Washington Post, 88, 93, 212, 236, 243
weapons
  antisatellite (ASAT), 7–8, 97, 136, 147, 155, 166
  cis-lunar space and, 124, 133, 138–39
  delayed impact space missiles (DISMs), 131
  military technologies and space, 132–33
  multiple independently targeted reentry vehicles (MIRV), 155
  nuclear-armed bombardment satellites (NABS), 123, 131–32, 136, 154–55, 167, 170, 172–75, 184
  nuclear weapons, 53–54, 59–60, 72, 78, 138, 154–55
  orbital space weapons, 131
  satellite technology and, 168–70
  technology report to LBJ, 150
  V-2 rocket, 49–54, 57–58, 72, 77, 81
  See also defensive systems; intercontinental ballistic missile; nuclear weapons
weapons of mass destruction (WMD), 111, 181, 184, 205
Webb, James, 235–36
Wells, Herbert George (H. G.)
  Clarke and, 67
  dystopia of, 249
  fascination with outer space, 23–24
  influence of, 30, 35, 50, 58
  multiple aspects of, 20
  opening of the space age, 15
  political worldview of, 26–27, 29–30, 45, 264
  socialism of, 80, 221
  United States and, 73
  violence of outer space, 21–22
  War That Will End War, The, 27
  World War I and, 47
Wells, Herbert George (H. G.) publications
  In the Days of the Comet, 27–29
  First Men in the Moon, The, 14, 27
  In the Fourth Year, 27
  War of the Worlds, The, 20–22, 23f, 24, 26, 30–31, 35, 130, 249, 253
  World Set Free, The, 27

INDEX

Western Development Division, 140
Westinghouse Company, 135, 243
Wheeler, Earle "Bus," 215–16, 221
White, Ed, 222
White, Frank, 14, 229
White, Thomas D., 136, 140–41, 172, 191, 261
Wiesner, Jerome, 175
Williams, William Appleman, 248
Winkler, Johannes, 40
Wireless World, 58
without altitude control (WAC), 75, 77–78, 80
Women's Prayer Crusade for World Order and Peace, 87
Wonder Stories, 42–43

Works Progress Administration (WPA), 43
world government, 26, 56, 93, 106
World Set Free, The (Wells), 27
World War II, German ballistic missiles, 50–51
Wright, Stetler, 284n18
Wylie, Philip, 62

Yangel, Mikhail K., 174
Yellow Danger, The (Shiel), 35
Ye Peijian, 262
York, Herbert, 98, 158–59

Zarzar, V. A., 197
Zorin, Valerian, 110, 178
Zuckert, Eugene, 125, 140, 210

www.ingramcontent.com/pod-product-compliance
Lightning Source LLC
Chambersburg PA
CBHW030520230426
43665CB00010B/696